P9-BYR-138

THE SCIENCE OF COMPUTING

About the cover:

The design is composed of a family of 30 curves known as *limaçons*. Each curve is described mathematically, in plane polar coordinates, by the formula $r = b - a \cos t$, as t varies continuously from 0 to 2π radians. The family is generated by increasing b from 0 to 1 while decreasing a from 1 to 0, in 30 equal steps. The large circle is obtained when b is 1 and a is 0, while the small circle is obtained when a is 1 and b is 0. The curve which just touches the small circle is a *cardioid*, for which a and b each have the value $\frac{1}{2}$. An irregular texture is introduced by the incremental plotter that was used to draw the curves.

THE SCIENCE OF COMPUTING

Loren P. Meissner

Computer Science Department
and Lawrence Laboratory
University of California, Berkeley

Wadsworth Publishing Company, Inc.
Belmont, California

Designer: Steve Renick

Editor: Mary Arbogast

Technical Illustrator: John Foster

ISBN 0-534-00302-8
L. C. Cat. Card No. 73-81780
Printed in the United States of America

1 2 3 4 5 6 7 8 9 10—78 77 76 75 74

Acknowledgments

Harcourt Brace Jovanovich, Inc. for permission to reprint an excerpt from Carl Sandburg, *Abraham Lincoln: The Prairie Years and the War Years, One Volume Edition,* 1966.

Lowie Museum of Anthropology, University of California, Berkeley for permission to reprint a portion of a photo of the Wepemnofret stela.

Macmillan London and Basingstoke for permission to reprint an excerpt from W. W. Rouse Ball, *Mathematical Recreations and Essays*, 1939.

Methuen and Company, Ltd. for permission to reprint an excerpt from H. W. Turnbull, *The Great Mathematicians*, 1929.

Time, The Weekly Newsmagazine, for permission to reprint an excerpt from "Pits and Pebbles," *Time*, 14 June 1963. Copyright Time, Inc.

Preface

This book is intended for the student who is especially interested in the newer, non-numerical kinds of computing, and who wants to actually try his hand at some computations beyond the most trivial level of difficulty. It emphasizes "computing" rather than "computers": the characteristics of computing *machines* certainly affect what we compute and how we compute, but in this book the computing machine is kept in the background.

The two central topics are *program structures* and *information structures.* I believe that a student can best learn to use a computer as a tool to solve increasingly complex problems if he understands the building blocks available for constructing computational procedures. For example, the simplest information structure (though not necessarily the most fundamental) is the *cell* whose contents is a single number; and the simplest program structure is the *assignment instruction.* A considerable portion of the first two chapters is therefore devoted to developing the student's appreciation of these two concepts. The essential points are (1) that no number can be obtained from a cell until something has been stored there and (2) that a number once stored in a cell can be obtained repeatedly from the cell. These two facts may seem obvious to an expert but can be a major stumbling block to a student. (I find it more confusing than helpful to introduce the "micro structure" of the cell at this level, before the student can appreciate the simplest ways of getting numbers into and out of the cell as a whole.) An assignment instruction is a portion of a *static* program describing what should happen *dynamically* to the cells when the program is *executed.* Thus, even at the earliest level the student's attention is drawn to the relation between the dynamic computation and the static program. He is led to explore this relationship using program structures and information structures at the appropriate level.

As soon as possible, the student using this textbook should be encouraged to put these structural concepts into practice by actually trying things out on a real computer. At the end of nearly every section, I have repeated the programming examples (illustrated in flowchart form in the body of the text) and translated them into parallel versions in each of three programming languages: Basic, Fortran, and PL/1.* The rudiments of each language are included in this book,

*The names of the three languages are ordinarily written as BASIC, FORTRAN, and PL/I, respectively. I have used the simpler style in this book because I find it less distracting.

but just deeply enough to illustrate the concepts being described. For all students it will be necessary to supplement this material with instructions for preparing a program in computer-readable form and for entering it into a specific computer. Many instructors who want their students to attain further competence in a particular programming language will add a supplemental book dealing with the specific language (such as one of the books listed in Appendix B).

For organizing and sequencing the individual topics, I have adopted a guiding philosophy expounded to me by E. M. McCormick: "Do not start to discuss a topic until you are prepared to finish it." This has resulted in the almost total absence of forward references to topics in later sections, so the discussion up to almost any point is essentially self-contained. This may imply that the user of the text has more difficulty if he wishes to interchange the order of presentation, for example, of two of the chapters. However, Chapter 4 does not greatly depend on Chapters 2 and 3; this chapter (with the possible exception of Section 4.4) might be studied earlier. Also, some of Chapter 7 (especially Section 7.1, except for "Global Lists and Strings" and the sorting example) could follow Section 3.7.

Otherwise, it should not be difficult to construct courses for various purposes on the basis of subsets of the material in this text, if the order of presentation is maintained. One possibility might be a short course based entirely on the first three chapters. A course that emphasizes *programming* might omit sections 2.10, 3.1, 3.2, 3.8, 4.1, 4.2, and 6.3. A course in "theory of computing" for nonmajors could dwell mainly on Chapter 1, the first seven sections of Chapter 2, and sections 3.2 to 3.4, 4.3, 5.1, 5.3, and 7.1.

In the belief that mastery of vocabulary is of paramount importance to a new student of any subject, I have included vocabulary lists at the ends of most sections, and have recapitulated these lists in a glossary (Appendix A).

In accordance with the modern emphasis on non-numerical applications, the mathematical prerequisites assumed for the student using this textbook are nearly trivial. I have used preliminary versions of this material with a mixed group of students from such diverse disciplines as physical and biological sciences, business administration, history, architecture, and comparative literature, as well as mathematics and computer science itself. As a result, I am inclined to agree with those optimists who view computing as a unifying force, tending to counteract the often-lamented split of the academic world into "two cultures": scientific and humanistic.

Special Acknowledgments

Both the author and publisher wish to express their appreciation to the following individuals who actively participated in the development of THE SCIENCE OF COMPUTING.

John D. Stevens	*Department of Computer Science* *Iowa State University*
R. Waldo Roth	*Computing Center* *Taylor University*
John Tartar	*Department of Computer Science* *University of Alberta*
David Falconer	*Department of Computer Science* *Florida Technological University*
Robert M. Aiken	*University Computing Center* *University of Tennessee at Knoxville*
Larry K. Flanigan	*Department of Computer and Communication Sciences* *University of Michigan*
Gerald L. Engel	*Computer Science Department* *Pennsylvania State University*
Thomas Nartker	*Department of Computer Science* *New Mexico Institute of Technology*
Lawrence E. Rosen	*Stanford Center for Information Processing*
William Dempster	*Santa Fe, New Mexico*
Philip Levy	*Oakland, California*

The student reviewers selected from the approximately 750 students who used the preliminary edition at University of California at Berkeley:

William Schmidt

Glenda Walen

Patrick T. Miller

Richard Harrison

Andrew Averill

Harriet W. Zais

Willard D. Normandy

Aino Tossavainen

Marguerite Lee

Dan Wilks

Table of Contents

to Peggy

1 *Introduction*

1.1 *What Do the Many Uses of Computers Have in Common?*

A national magazine (White, 1970) recently featured the "computer revolution" and described many of the ways computers are used today. Some of the computer applications mentioned in that article were:

Department store credit accounting
Banking
Telephone dialing assistance
Inventory control
Crime prevention
Analysis of construction specifications
Payroll calculation
Industrial process control
Space flight
Projection of fashion trends
Theater and airline ticket reservations
Insurance
Securities brokerage accounting
Assistance in education
Modeling and simulation
Library cataloging and retrieval
Monitoring of experiments
Monitoring status of hospital patients
Matching job applicants
Urban traffic control
Warfare
Artificial intelligence
Astrology
Art and music

Although these varied applications differ immensely in subject matter, they all require the storage of *information* that is combined and manipulated in a variety of ways. The commercial applications (accounting, banking, insurance, inventory, payroll) use mostly numerical information, including amounts of money, number of items in stock, and hours worked. These applications also use names, addresses, and other alphabetical, numerical, or symbolic information as identification. Scientific and mathematical applications deal with numbers and with graphical forms of information. Applications in the humanities (Sedelow, 1970) depend less on numbers and make greater use of alphabetical, symbolic, and pictorial information. A major share of the success of the "computer revolution" can be attributed to the development of unified approaches to the processing of all kinds of information, despite the wide variety of sources from which it is obtained and the wide variety of uses for which it is intended.

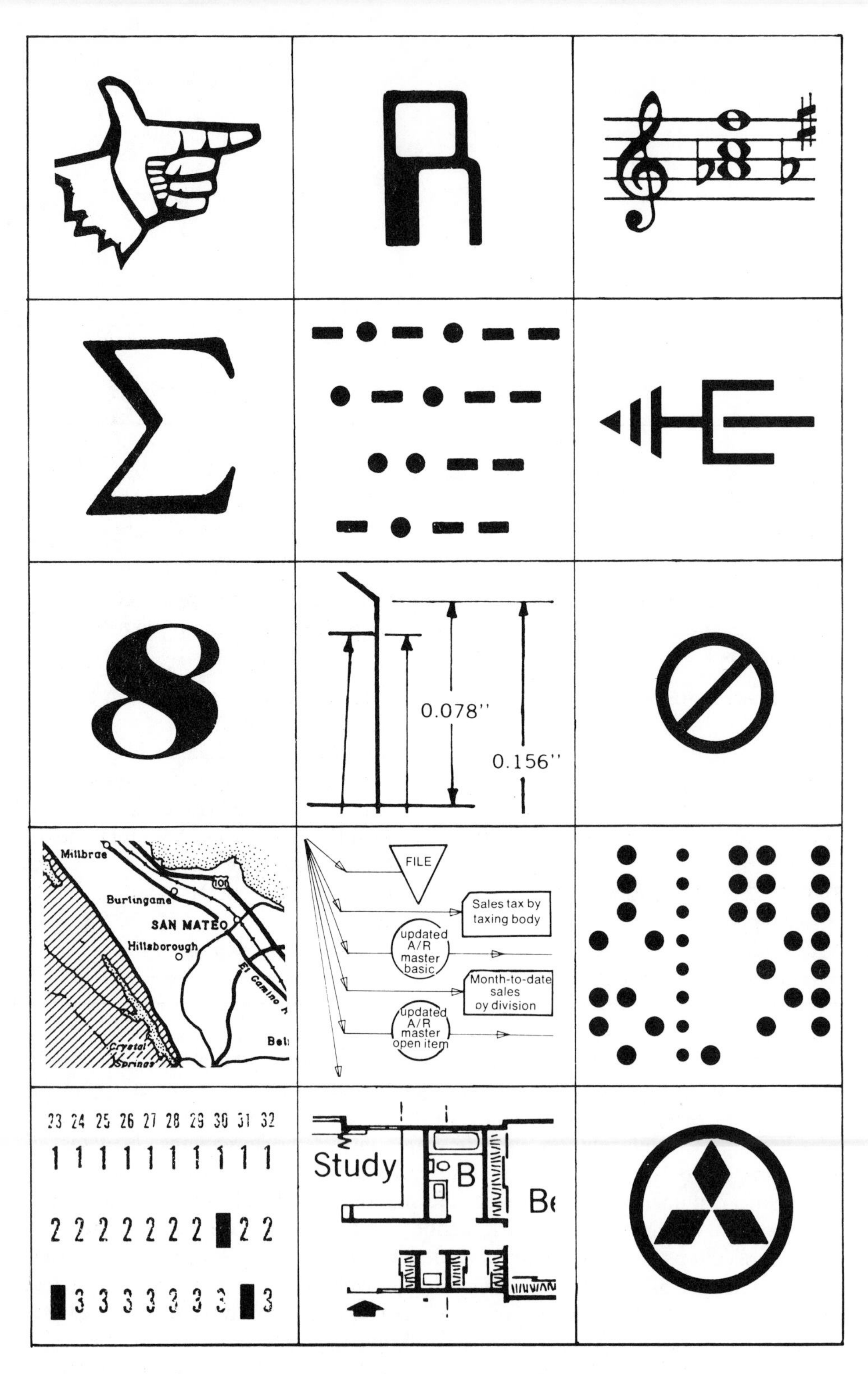

1.1a. Some kinds of alphabetical, symbolic, numerical, and pictorial information.

The underlying purpose of all this storage and manipulation of information is to communicate ideas—to produce a certain image in the brain of a person or to inspire a person (or perhaps an animal or even a machine) to act in a certain way. A banker decides to approve a loan application on the basis of a credit report produced by a computer. The police department analyzes crime statistics and assigns more patrolmen to a high-crime area. Up-to-date information on cancellations allows a theater manager to resell seats that might otherwise have remained empty. An automatic alarm calls the nurse to the bedside of a critically ill patient.

The person who originated the information may also be one of the users—sometimes we store information for our own later use. On the other hand, when a piece of information is placed into storage, nobody knows exactly when or by whom it will be used. For instance, it is possible that information stored for a valid reason by a credit reporting agency may be used in less valid ways later on. Usually, a large number of would-be users can obtain access to stored information, just as books in a library are accessible; definite action and some additional costs are entailed in limiting access to a more restricted group of users.

The science of computing, or *informatics* (Arsac, 1970), which deals with the storage and transformation of information, is not greatly concerned with the "meaning" of information or with the purposes to which it is put. As a branch of physical science, the science of computing is essentially neutral, just as the science of dynamics does not consider why anyone should want to use a lever, or whether any socially desirable goal is reached in a certain case by rolling a wheel up an inclined plane. Still, the social implications of the "computer revolution" cannot be ignored. Because of the revolutionary nature of these information transforming machines and processes, it is difficult for the general public to remain aware of the rapid development of new devices and methods in this field. Knowing about the ways computers affect people's lives thus becomes part of the computer scientist's job. He must be ready to provide intelligible answers to questions from concerned citizens, and even to take the initiative in telling people about new developments whose potential social effects are not being recognized.

Words to Remember

information	symbolic information	alphabetical information
numerical information	pictorial information	

Read a chapter from one of the books in the following bibliography, and write a paragraph giving your reaction. What has been the most important effect of technology on people's lives? What must be done to improve the social effects of technology, and of computers in particular?

*Bibliography: The Social Implications of Computers and Technology**

Conant, J. B., *Modern Science and Modern Man.* New York: Columbia University Press, 1952.
Diebold, J., *Automation: The Advent of the Automatic Factory.* New York: Van Nostrand, 1952.
Vonnegut, K., *Player Piano.* New York: Holt, Reinhart and Winston, 1952.
Wiener, N., *The Human Use of Human Beings.* Boston: Houghton Mifflin, 1954.
Buckingham, W. S., *Automation.* New York: Harper & Row, 1961.
Taube, M., *Computers and Common Sense.* New York: Columbia University Press, 1961.
Charter, S. P. R., *Man on Earth.* Sausalito, Calif.: Angel Island Publications, 1962.
Philipson, M. H., ed., *Automation: Implications for the Future.* New York: Random House, 1962.
Bagrit, Sir L., *The Age of Automation* (BBC Reith Lectures 1964). London: Weidenfeld, 1964.
Ellul, Jacques, *The Technological Society.* New York: Knopf, 1964.
Marx, L., *The Machine in the Garden.* New York: Oxford University Press, 1964.
Mason, E. E., and W. G. Bulgren, *Computer Applications in Medicine.* Springfield, Ill.: Charles C. Thomas, 1964.
Packard, V., *The Naked Society.* New York: David McKay, 1964.
Wiener, N., *God and Golem, Inc.* Cambridge, Mass.: M.I.T. Press, 1964.
Boguslaw, R., *The New Utopians: A Study of System Design and Social Change.* Englewood Cliffs, N.J.: Prentice-Hall, 1965.
Burck, G., *The Computer Age and its Potential for Management.* New York: Harper & Row, 1965.
Mumford, L., *The Myth of the Machine.* New York: Harcourt Brace Jovanovich, 1966.
Schon, D. A., *Technology and Change.* New York: Dell, 1967.
Westin, A., *Privacy and Freedom.* New York: Atheneum, 1967.
Bernstein, G. B., *A Fifteen Year Forecast of Information Technology.* Washington, D.C.: Naval Supply Systems Command, 1969.
Diebold, J., *Man and the Computer.* New York: Frederick A. Praeger, 1969.
Oettinger, A. G., *Run Computer Run.* Cambridge, Mass.: Harvard University Press, 1969.
Ferkiss, V. C., *Technological Man: The Myth and the Reality.* New York: New American Library, 1970.
Martin, J., and A. R. D. Norman, *The Computerized Society.* Englewood Cliffs, N.J.: Prentice-Hall, 1970.
Pylyshyn, Z. W., ed., *Perspectives on the Computer Revolution.* Englewood Cliffs, N.J.: Prentice-Hall, 1970.
Sackman, H., and N. Nie., eds., *The Information Utility and Social Choice.* Montvale, N.J.: AFIPS Press, 1970.
Tavis, I., ed., *The Computer Impact.* Englewood Cliffs, N.J.: Prentice-Hall, 1970.
Toffler, A., *Future Shock.* New York: Random House, 1970.
Whisenand, P., and T. Tamaru, *Automated Police Information Systems.* New York: Wiley, 1970.
Bemer, R. W., ed., *Computers and Crisis.* New York: Association for Computing Machinery, 1971.
Greenberger, M., *Computers, Communications, and the Public Interest.* Baltimore, Md.: Johns Hopkins Press, 1971.
Baer, Robert M., *The Digital Villain.* Reading, Mass.: Addison-Wesley, 1972.
Rothman, S., and C. Mosmann, *Computers and Society.* Chicago: Science Research Associates, 1972.

* Arranged chronologically.

1.2 *How Can Information Be Stored and Transmitted?*

We have said that the purpose of information is to communicate ideas. In this study, however, we shall not be concerned with ideas themselves but only with the physical means used to communicate ideas. Any change in the physical universe that can be detected and used to re-create an idea at a different time or place we call a *mark*. Let us consider some physical properties of a mark, in *time* and in *space*.

A mark may be temporary, as when a person communicates his idea by speaking or flashing a light or waving his arms. If nobody hears or sees the signal, or if a person recognizes it but then more or less ignores it, the physical effect does not extend beyond a small region of space and dies out after a very short period of time. If there is no wide-reaching or long-lasting result, such a mark is of little interest.

One important way for a person to physically extend the scope of a mark is to *transmit* it to some other place. The branch of science that studies the transmission of marks by electromagnetic signals such as telephone, telegraph, or "wireless" (AM or FM radio or television), is usually called "communication theory" or "information theory." The computer-related information sciences, on the other hand, study the *storage* of marks. We may contrast these two approaches by saying that transmission consists almost entirely of changing the location of marks in space, and very little with changes in time. On the other hand, storage allows us to use a mark at a different time, with little change in its position in space.

Of course, the space and time properties of a mark, like the properties of any physical phenomenon, cannot really be separated. It takes some space to store a mark and some time to transmit it. When a man carries a book from one place to another, he is transmitting stored marks. And when a messenger makes a note (even mentally) so that he will not forget his message, he is storing it during transmission.

Any reasonably permanent mark can be used to store information—for instance, a notched stick, a pile of stones, or a finger mark in the sand. The most familiar stored marks, however, are those made on paper (or a similar material) with ink (or a similar substance such as pencil lead). Marks on paper are quite permanent, are very portable, and are easily made accessible to would-be users. Paper and its predecessors, including clay tablets, papyrus rolls, and parchments (animal skins), have been used to store permanent, portable records for at least five thousand years. In the past few decades paper has been challenged as a recording medium by photographic film (including microfilm) and by magnetic tape. But people still prefer to read records from paper because they do not have to use a special viewer or playback device. The newer storage media are used mainly in special situations (such as for less active documents in large libraries) where compactness is more important than decreased accessibility.

Information Storage for Computers Machines (including computers) can also read ink marks from paper, but they have some difficulty in recognizing such marks because of the wide variety of type fonts and symbol shapes that can occur. Records stored on microfilm impose the same difficulty.

1.2a. *Ancient and modern marks.*

Ink marks can be inscribed (on bank checks, for example) in a form that is specially designed to be recognized by a machine but can also be read by people. Holes punched in paper tape or in cards can be interpreted less easily by people but more easily by computers. Records on magnetic tape are completely inaccessible to people but can be read easily and rapidly by computers.

Just as a person must find a library book before he can read it, so a computer must be able to *locate* a desired item of information before reading it. The reader mechanism must be provided with a means for physically moving the paper, film, or magnetic tape to the point where an item to be read is situated. The computer sends signals to the reader, causing the storage medium to move to the proper position. Readers for magnetic discs and drums (which rotate continuously) have controls to insure that information transfer occurs only when the desired mark is in position to be read.

All such readers are expensive. More importantly, however, they are slow in comparison to the internal operating speed of the computer. The computer can digest and generate information at electronic speeds—much faster than the information can be transmitted to or from a moving storage device such as a tape. If a computer had no other storage devices, its operating speed would be severely limited.

To solve this problem, bulk electronic storage devices have been developed. These solid-state devices (such as magnetic cores) are very fast and are reasonably inexpensive and easy to control. Modern computers incorporate millions of these components, all directly accessible through electronic switches. Electromagnetic marks can be stored in these devices and retrieved on demand within a few millionths of a second, because no moving parts are involved. Only in exceptional cases (where more permanent storage is needed, for instance, or where human intervention is required) are the slower devices used.

As we have seen, the marks used by a computer do not have to represent numbers. The meaning depends only on the way they are used. Nevertheless, for historical reasons we use derivatives of the word "compute," despite its numerical connotations, to describe the manipulation of information. Similarly, we call the manipulating machine a *computer*. The sequence of transformations performed on stored marks, to generate whatever new information is appropriate to the application, is called a *computation* or a *computing process*.

Words to Remember

mark	transmit	store
access	computer	computation

References

Arsac, J. J., "Informatics and Computer Education," *First IFIP World Conference on Computer Education*, Vol. 1. New York: Science Associates International, 1970. P. 69.

Sedelow, S. Y., "The Computer in the Humanities and Fine Arts," *Computing Surveys* 2: 89–110 (Jun 1970).

White, P. T., "Behold the Computer Revolution." *National Geographic Magazine* 138: 593–633 (Nov 1970).

2 *Let's Start Programming*

2.1 *How Can We Tell a Computer What to Compute?*

A computation is a sequence of operations on marks. Let us watch as someone punches holes in a lot of cards and feeds them into a reader attached to a computer. The holes in the cards are marks. The marks are sensed in the reader and transmitted (as electromagnetic marks) to the computer, where they are stored temporarily and then manipulated and transformed into new patterns. These new patterns are again stored inside the computer or are sent to a printer, making new marks for someone to read.

Within half an hour or so, the printer of a modern computing center produces reports of several different kinds—although, superficially, all the cards fed into the reader look similar. What makes the computer perform a payroll computation one minute and an inventory control computation the next? The answer is that each different computation is *prescribed* (specified, designated, or controlled) by a different *program.*

In this chapter, we begin the study of programming. The main question is how to create a program to prescribe a desired computation. To learn this skill, it is also necessary to know how to tell, from a given program, what computation would result if it were executed. More generally, the subject of programming is the study of how a program and the corresponding *computation* are related. Briefly stated, a program is the prescription of a computation; a computation is the execution of a program. A computation is a dynamic process executed by the computer and prescribed by a program. A program is a group of words and symbols written by the (human) programmer, which prescribes a particular computation to be performed by the computer.

The dynamic process of a computation involves the manipulation of marks by the computer. The marks are a concrete representation, in electromagnetic form, of specific information. The computer generates new marks by interpreting in a prescribed way the marks that are stored inside it. A bustle of activity and constant change goes on inside the computer as the computation progresses. The program that prescribes this computation process, on the other hand, consists of words and other symbols, and may be written down on paper in a static form. (In the study of programming, concepts often occur in pairs; an idea relating to the dynamic physical world often has a counterpart idea in the static symbolic world. When a new idea is introduced in this chapter, it is important to understand which of these two worlds it belongs to.)

A computation is a sequence of simple steps or *operations*; each operation is prescribed by an *instruction* or program step. The prescription for an operation is an instruction; the execution of an instruction is an operation. Instructions, being parts of the program, are static, while the operations they designate are parts of the computation and so are dynamic.

Every program must be written according to the rules of some *programming language*. These rules define the way symbols must be arranged to form instructions, the way instructions are combined to make a program, and how the instructions of the program will be interpreted when the program is executed. There is no better way to learn to use a language than by writing programs in that language and testing them on a computer. It is assumed that students using this book will be learning either Basic, Fortran, or PL/1[1] as a programming language. The choice among these three languages will depend mostly on the availability of computing facilities.

Basic is the simplest language of these three; it is used mainly at smaller computing sites—ones that use small computers, or ones that consist entirely of typewriter-like consoles connected to a larger computer at a remote site. Fortran, the oldest general purpose language that is still in wide use, has developed over the years into a very efficient "production" language for computations in engineering, statistics, and science. PL/1, which came into use about ten years later than Fortran, includes information handling features that make it usable in all fields of application, from commercial computations to research in the humanities to scientific work.[2] While each of these three programming languages is well suited for use in a certain area of computing, the strength of each language becomes a weakness when we try to use it outside its own area. The simplicity of Basic makes it difficult to use for complicated calculations. The maturity of Fortran means that it does not include features that have become important more recently (in particular, the manipulation of alphabetical text). The richness of PL/1 makes it a sophisticated language, full of pitfalls for the beginner.

Flowcharts The languages we have been discussing (and which we shall refer to collectively as "conventional programming languages") are intended for interpretation by computers. That is, a person is supposed to prepare the instructions so that they can be entered directly into a computer. Therefore, many of the restrictions in the rules of these conventional languages are imposed by the requirements of the computer entry mechanism.

---

[1] Pronounced "P L one."

[2] A number of other languages are also widely available. Algol-60 is the major language in Europe, although in the United States it is used mostly in the special field of numerical analysis. (This language has had great influence on the design of other new languages, notably PL/1.) Cobol is a business-oriented language, and because most of the computing done in the world is commercial, Cobol is a very widely used programming language. Machine-oriented "assembly" languages are used by experienced programmers when they need to supplement a user-oriented language in order to take advantage of an internal feature of a particular computing machine. There are also many special-purpose languages (see Rosen, 1967; Sammet, 1969) with narrower fields of application—for example, Snobol, which is easy to use for the manipulation of alphabetical text.

On the other hand, it is possible to write programs in a *flowchart* language, which may be thought of as a programming language designed for interpretation by people rather than computers. Flowcharts graphically represent the time and space relations among the steps of a calculation, and most persons are able to perceive these relations more readily by looking at a flowchart than by reading the equivalent program written in a conventional language.

In the body of this text, therefore, we introduce and discuss programming concepts using flowchart language. This helps us to avoid tying the discussion to any specific conventional language. At the end of each section, we translate most of the flowcharts into each of the other three languages.

Words to Remember

program	instruction	Fortran
prescribe	programming language	PL/1
execute	Basic	flowchart
operation		

2.2 *Time and Space Relations in Computing*

Time Sequence of Computation Steps When we have two or more actions to perform, the *time sequence* in which we perform them may or may not be important. If we are making two deposits at a bank, we can make either deposit first; but if we are making a deposit and a withdrawal, it is often important that we make the deposit first. The instruction "Put on your shoes and stockings" should certainly not be executed in the sequence that the words might imply. When a cook bakes a cake, it may be that the dry ingredients can be combined in any sequence, but that the mixing of liquids with the dry ingredients must take place later.

To specify the time sequence for carrying out a group of instructions, we can use several different techniques. We can use words that imply time relationships: *then, next, afterward, later*. For instance, "Put on your shoes *after* you put on your stockings." We can number the steps: "(1) Put on your stockings. (2) Put on your shoes." Or (as we most frequently do in this book) we can make a flowchart, using arrows to specify the time sequence:

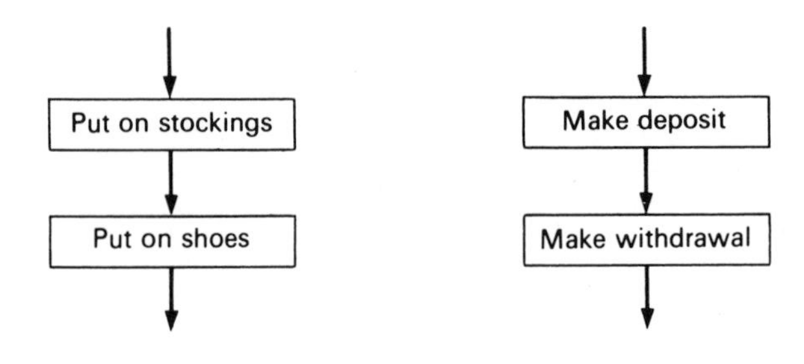

Usually we arrange the boxes of a flowchart vertically, with the arrows leading from each box to the one just below it; however, the arrows guarantee that there would be no confusion if we used a horizontal arrangement such as

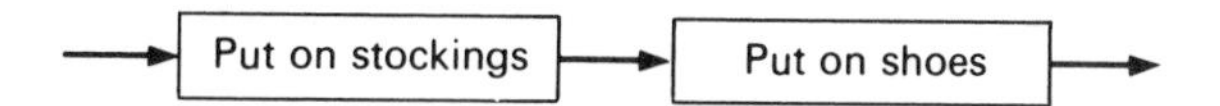

or even an arrangement with the arrows pointing upward or diagonally in some manner. In most cases, we must also indicate the first step to be performed, and often we must say explicitly when the sequence of steps has been completed. Thus a "complete" flowchart for the process "Put on your shoes and stockings" might be:

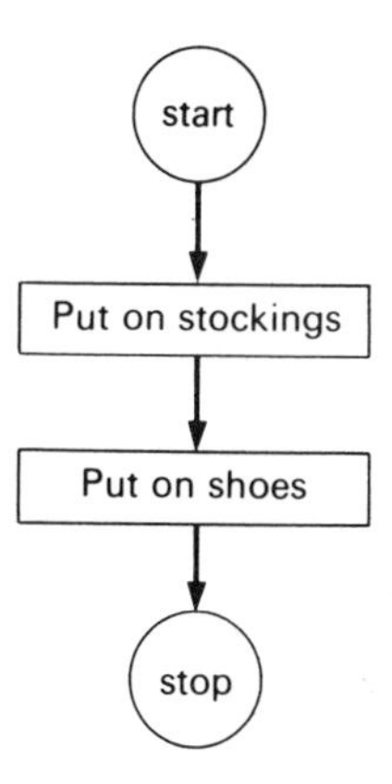

Notice that we use conventions in drawing flowcharts; for example, a different shape is used for the start and stop boxes than for the boxes indicating actions to be performed. Such conventions are rather arbitrary but should be followed consistently, because they emphasize, for the flowchart user, certain important aspects of the process.[3]

Storing Numbers Inside a Computer In this chapter we focus our attention on exclusively numerical computations and on the kinds of operations a computer can perform on numbers. These operations include *storing* numbers generated as intermediate results during the computation, *retrieving* stored numbers for use in later steps of the computation, and *arithmetic* operations on numbers.

A physical process occurs whenever a number is stored in a computer. Storing a number is a real event in space and in time. The space where numbers are stored is divided into *cells* of equal size (Figure 2.2a). Each cell contains enough space for the electromagnetic representation of one number such as 14.37, 356, .000547, or 3.41592. The number stored in a cell is the *contents* of the cell.[4]

---

[3] A slightly different set of conventions is imposed by the American National Standards Institute (ANSI, 1970; see also Chapin, 1970; Bohl, 1971). Our minor deviations from this standard are motivated by typographical considerations.

[4] This definition of a cell is necessarily an oversimplification, because of the wide variety of ways in which numbers are represented in various computing machines (see Maurer, 1972, Chapter 1). All the widely used computer systems, however, have some unit of storage that fits this definition more or less closely, with room for a numerical value consisting of between 5 and 14 (and sometimes a variable number of) significant decimal digits or the equivalent.

14.37	356	.000547	3141592	

2.2a. *A cell is a space inside a computer, just large enough to hold one number.*

A number can be retrieved as often as necessary from the cell where it has been stored, without being destroyed. The contents of a cell does not change until a new number is stored there. When a new number is stored, the old number is lost and can no longer be retrieved from the cell; the contents of the cell from that time on is the new number. Thus, we cannot talk about the contents of a cell unless we specify a point in time. The contents of a cell at any specified time is the last number stored in the cell up to that time.

This dependence on time works the same way for an audio tape recorder. A signal that has been recorded on the tape remains there and can be played back repeatedly without being destroyed. The signal is lost only when a new signal is recorded. A spot on the tape can be compared to a cell in the computer. Only the last signal recorded there can be retrieved at any specified time.

Another analogy, more graphic although not so close to reality, is that of the cells as glass jars (Figure 2.2b). We store a number in the jar by writing it on a piece of paper and putting the paper in the jar. We can read the number through the glass, without changing it. To make the analogy work, we must agree to remove and destroy the old piece of paper whenever we put in a new one. At any time, then, the only number we can read from a jar is the last number that has been put in.

Cell Names Now we need ways to indicate to the computer that it should store a certain number in a cell or should retrieve a number that has been stored there earlier. First, we must be able to identify particular cells—we may want to store several different numbers and then retrieve a particular one. For this purpose, we must *name* the cells that are going to be used in a computation.

The names used to identify cells are *variables*, inasmuch as the number in a cell may vary from time to time. The contents of the cell is the *value* of the variable.

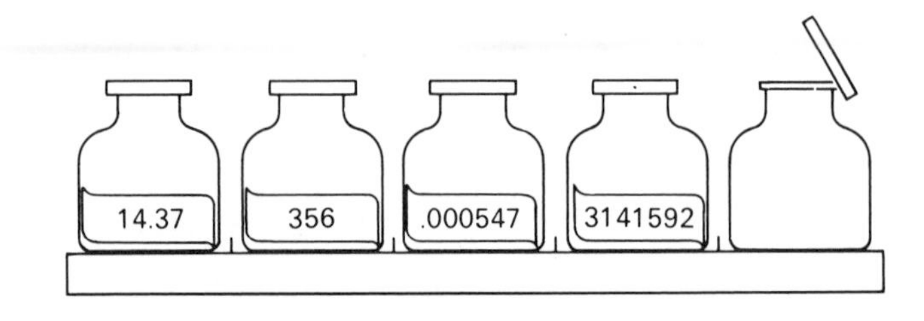

2.2b. *Cells may be compared to glass jars. We can read a number through the glass without changing it.*

Storing a number in the cell *assigns* a value to the variable. When a new value is assigned to a variable, the old value is lost, for the previous contents of the cell can no longer be retrieved. In elementary algebra, a variable is ordinarily a single letter, such as *a* or *x*. Since many computations require more than 26 different variables, programming languages allow combinations of letters such as USA and IBM, as well as combinations of letters and digits such as A1 or X3.

Besides variables, most computations also use some *constants*. A constant has a numerical value which is stored in a cell and is never changed during the computation.

Variables and constants are names (see Tarski, 1965); they are symbols used in referring to the cells where numbers are stored. The names are composed of letters, digits, and certain other characters. A constant appears in the program as an ordinary numeral, composed of one or more digits and perhaps a decimal point. A variable may consist of letters and digits, but the first character must be a letter.

Before a computation begins, a cell inside the computer is allocated to each variable and each constant that appears in the program. The numerical values of the constants are stored in the cells allocated to them, and remain unchanged during the computation. The values of the variables, on the other hand, are initially undefined. This means that no numerical values have been stored in the cells allocated to variables, and any attempt to retrieve numbers from these cells before something is stored there may cause an error. In terms of our analogies, this would correspond to playing a new, unrecorded magnetic tape or attempting to read a number from an empty jar. A constant, on the other hand, might be compared to a sealed jar with a number placed in it ahead of time and not changed during the computation (Figure 2.2c).

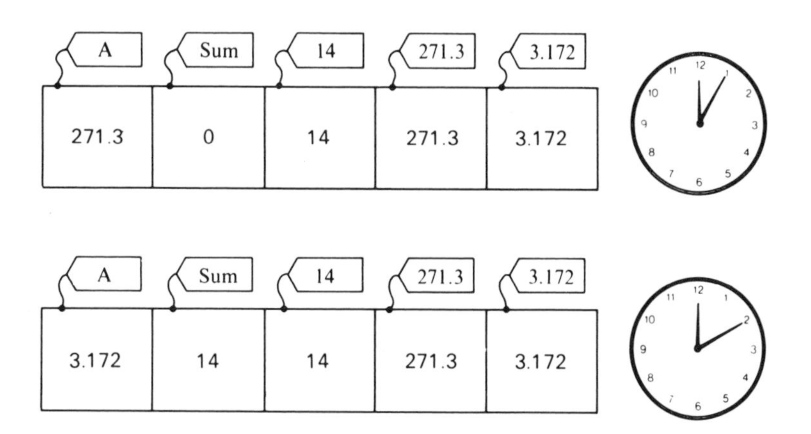

2.2c. *The name of a cell is a* variable *or a* constant. *The number stored inside the cell is the* value *of the variable or constant. The value of a variable may change (vary) during the course of a computation. The value of a constant never changes.*

Flowchart In the flowcharts, we shall use combinations of letters and digits as variables, but the first character of the name will always be a capital letter. Sometimes a variable will be just a single capital letter such as A or X. The remaining characters, if any, may consist of capital letters, lower-case letters, or digits. No blank spaces, commas, or other special characters will be used in variables. Constants will be written simply as strings of digits, with or without a decimal point.

Variables:

X	Sum	Dummy	A12oP3
T	R3	EmpNo	S3Box

Constants:

1	14.37	123456789	19354
2	0.00547	1.98	0

Basic In Basic, a variable must be a single letter or a single letter followed by a single digit. Lower-case letters are not allowed in Basic variables.

Variables:

X	R3	A	T
F1	Q	V1	A9

A constant is a string of digits, with or without a decimal point.

Constants:

1	2	123456789	0
14.37	356	0.00547	1.98

None of the following can be Basic variables or constants:

USA	IBM	AB	AA
Z23	7Q	I2.3	A12B
123,456	Sum	$1.98	I + J

Fortran There are two fundamental numeric *types* in Fortran. *Integer* variables and integer constants are distinguished from *real* variables and real constants. The value of an integer variable or constant must be a positive or negative integer (whole number). Such values are especially useful for counting how many times a calculation is repeated or for indicating a particular item in a sequence of several items. Real variables and constants need not have whole-number values, so they are more useful than integers for most purposes, since the numbers occurring in most calculations are not necessarily integers but may include decimal fractions as well.

In Fortran, a variable consists of up to six characters. The first character must be a letter, and the remaining characters (if any) may be letters or digits. Lower-case letters are not allowed in Fortran. A constant is a string of digits.

Integer variables:

```
IN      MI      JA      L
MAX     I3BOX2  ISAX
```

Real variables:

```
X       R3      XY      YDEV
EMPNO   GCD     TERM1   TEST
A1      XAVG    Q       DUMMY
T       SUM     S3BOX   SUPERC
ALONGS  ABCDEF
```

It is easy to tell by looking at a Fortran variable or constant whether its type is integer or real. The first character of an integer variable must be one of the letters I, J, K, L, M, or N. A real variable must begin with a letter other than these. A real constant must be written with a decimal point; an integer constant must not.

Integer constants:

```
0       1       356     2
123456789
```

Real constants:

```
14.37   .00547
```

None of the following can be Fortran variables or constants:

```
123.4          123,456,789   $1.98    I + J
Sum     7Q                   1AB2
ALONGSEQUENCE                SUPERCALIFRAG
ABCDE12345
```

PL/1 In PL/1 we use numbers of two different *types*, *floating* and *fixed*. Numbers of fixed type are used only for values that must be positive or negative integers (whole numbers). Such values most often occur in indexing operations, such as counting how many times a calculation is repeated or indicating a particular item in a sequence. In the much more common case of values that need not be integers, numbers of floating type are used.

The distinction between numbers of floating and fixed types is rarely of serious concern to a PL/1 programmer, because type conversion is provided automatically when it is needed. Floating numbers should be used in most cases because they can be processed more efficiently. However, fixed numbers must be used for most indexing operations to obtain the required accuracy.

In PL/1 a variable consists of up to 31 characters. The first character must be a letter, and the remaining characters (if any) must be letters or digits.

Variables:

```
IN      MI      JA      L
X       R3      XY      YDEV
EMPNO   GCD     TERM1   TEST
MAX     I3BOX2  T       SUM
ALONGSEQUENCEOFWORDS
SUPERCALIFRAGILISTICEXPIALIDOCI
```

We use *type declarations* to explicitly specify whether the variables in a program are of floating or fixed type:

```
declare (IN, X, EMPNO) float;
declare (MAX, TEST) fixed;
```

In our PL/1 examples, a constant is a string of digits with or without a decimal point. We use constants of this form with variables of either fixed or floating type. (Actually, constants of this form are of fixed type, but the use of floating constants in PL/1 is so unwieldy that we prefer to use fixed constants and to rely on automatic conversion when these are to be combined with variables of floating type.)

Constants:

```
0       1       356     123456789
14.37   2       .00547
```

None of the following can be PL/1 variables or constants:

```
123.4   123,456,789   $1.98   I + J
7Q      1AB2
```

Words to Remember

cell	variable	constant
contents	value	undefined
retrieve	assign	

Exercises

1. For the programming language you are studying, identify each of the following as a valid variable, a valid constant, or neither.

A	TOTALHOURS	4JULY1776	27.28.29
B7	BASERATE	JULY4	Section2.2
X − Y	GROSSPAY	.000000000099	763
1 + 2 + 3 + 4	rdr	if	go to
START	ptr	MCMLXXIII	stop
O	0	o	ZERO
π	Pi	e	i
IF	GO TO	STOP	
ALONGSEQUENCEOFCAPITALLETTERS		123,456,789,876.5	
JawQuizMightVexBlindFrockSpy			

2. Choose *one* of the three programming languages (Basic, Fortran, or PL/1). Invent names for all the numerical quantities mentioned in each of the following simple calculation problems. Each name should be a valid variable or constant for the programming language you have chosen.
   a. The area of a rectangle is determined by multiplying its length by its width.
   b. The area of a triangle is one-half the product of its base and its altitude.
   c. The middle note in a major triad is four semitones above the root.
   d. To find the average of two numbers, add them together and divide by two.

2.3 *How Do We Tell the Computer to Store and Retrieve Numbers?*

An operation that stores a number in a cell is an *assignment operation.* The corresponding instruction is an *assignment instruction* (also sometimes called an assignment statement or a replacement statement). In the simplest case, the value to be assigned to a variable is the value of a constant. A typical assignment instruction in flowchart language has the general form

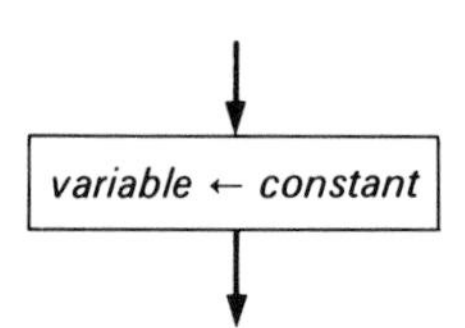

(We show arrows to indicate that this instruction is generally part of a longer sequence of instructions.) When the computer encounters an instruction of this form during a computation, it retrieves (*fetches*) the number that is the value of the constant. It then stores this number as the new value of the variable.[5]

As we have seen, the operations that comprise a calculation are physical events in space and in time. The assignment operation occurs in two phases, each involving a different place and a different time. The first phase relates to the cell named on the right-hand side of the assignment instruction, and the final phase relates to the cell named on the left. This is opposite to our customary manner of reading from left to right. We use the left-pointing arrow as a reminder that the part of the operation designated by the right-hand part of the instruction occurs first, and is followed in time sequence by the change in the contents of the cell named on the left.

Consider, for example, the assignment instruction

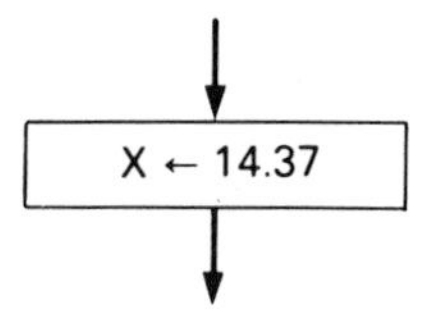

which might appear in a flowchart. We have noted that each variable and each constant in a program has a cell allocated to it. Before the computation begins, then, a cell will have been provided for the variable X and another for the constant 14.37. The numerical value of the constant will have been stored in its cell, but the contents of the cell named X will be initially undefined. The assignment operation prescribed by this instruction causes the value of the constant 14.37 to be assigned to the variable X. Here is a sequence of steps illustrating the use of assignment instructions.

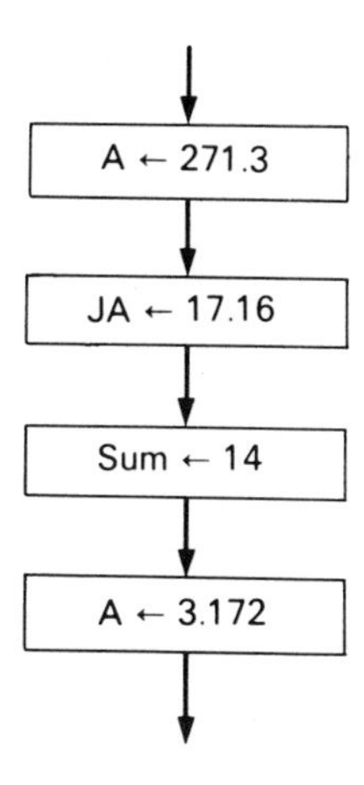

---

[5] We also say that the cell designated by the variable on the left is the *destination* of the number that is the value of the constant on the right. The left-pointing arrow is the assignment operator in our flowchart language. At the end of this section we learn how to write the assignment operator in Basic, Fortran, and PL/1.

The meaning of an assignment of the general form[6]

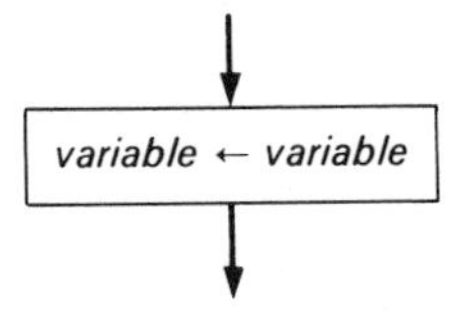

should now be evident. The value of the variable on the right is retrieved from its cell, and is then placed in the cell named by the variable on the left. As before, this operation has two phases. In the first phase, a number is fetched from the cell named on the right; in the second phase, this number is assigned as the new value of the variable on the left. It may happen that no number has ever been stored as the value of the variable on the right; in such a case this instruction would be in error.

Note that the symbol on the right side of an assignment instruction may be a constant or a variable. The symbol on the left side, however, must be a variable, since we are not allowed to write an instruction assigning a value to a constant.

Time Sequence of Assignment Instructions The four assignment instructions in the program

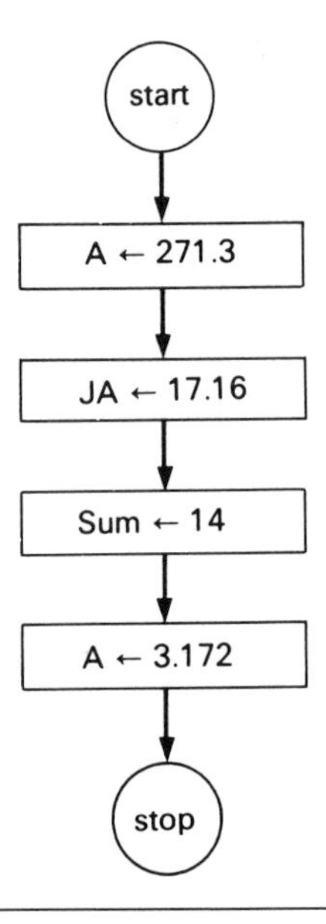

[6] Although the same form *variable* is used twice in this description, it does not mean that an assignment instruction must consist of the same name written two times with an arrow between the two occurrences. A more correct description would be

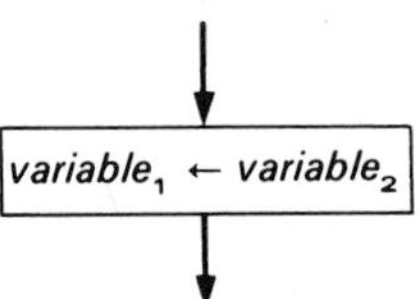

which indicates clearly that the two names may be *different* variables. However, as the typographical complexity of this more correct form may be confusing, we merely stipulate that each occurrence of a form which appears more than once may signify a *different* entity of the designated form.

will be executed by the computer in the sequence indicated by the arrows, beginning with the instruction following start. The variable A will first be assigned the value of the constant 271.3. Values are next assigned to the variables JA and Sum. Finally the variable A is given a different value, so that the final contents of the cell named A will be 3.172. After execution of this last instruction, it is not possible to tell that any other value was previously stored in the cell named A.

This group of assignment instructions illustrates what we mean by the relation between the static program and the dynamic computation. The space relation among the instructions, as specified by the flowchart arrows, corresponds to the time relation among the operations. The instruction immediately following start corresponds to the beginning of the computation. When the sequence of arrows shows one instruction between two others in space, the corresponding operation will occur between the two others in time.

As another example, consider the sequence of instructions

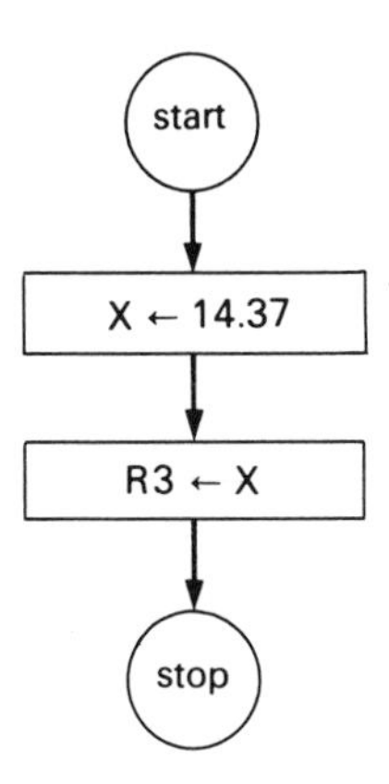

The first instruction results in the value of the constant 14.37 being assigned to the variable X. This value is then fetched from the cell named X and assigned as the new value of the variable R3. Both of these variables will therefore have the same final value.

Conventional programming languages, such as Basic, Fortran, and PL/1, do not use arrows to connect the instructions. Instead, the execution sequence of the instructions is determined from their position on the page. The instruction nearest the top of the page is executed first, and execution proceeds downward. Since execution always begins at the top of the page, it is not necessary to explicitly designate the starting point of the calculation.

Also, in a flowchart we do not always provide a separate box for each instruction; we often combine a group of related instructions into a single box, thus giving added insight into the *structure* of the program. *Within each box*, execution begins at the top and proceeds downward.

Flowchart

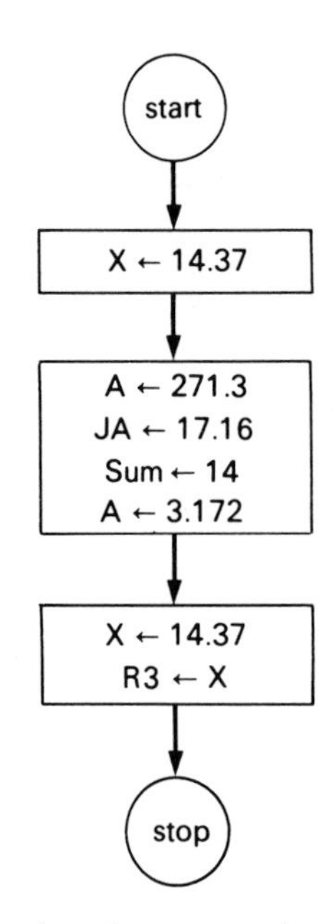

This flowchart is a composite of those displayed in the text. Note that the assignment

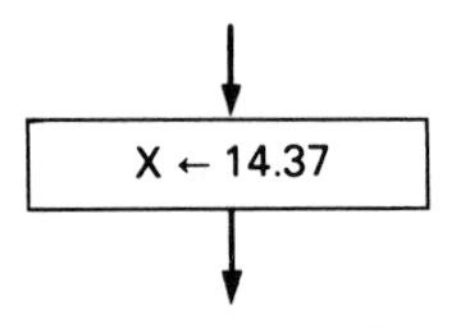

occurs twice in the composite flowchart; in an actual program this would be useless but harmless.

Basic In Basic, each instruction is preceded by a *line number*. An assignment instruction in Basic begins with the word LET and uses an equals sign as the assignment operator.

```
10   LET X = 14.37

110  LET A = 271.3
120  LET J1 = 17.16
130  LET S1 = 14
140  LET A = 3.172

210  LET X = 14.37
220  LET R3 = X
```

Fortran In Fortran an assignment instruction is written with an equals sign as the assignment operator. The *type* of the variable on the left should agree with that of the variable or constant on the right; that is, either they should both be of *real* type, or they should both be of *integer* type.

```
X = 14.37

A = 271.3
AJ = 17.16
MUS = 14
A = 3.172

X = 14.37
R3 = X
```

PL/1 In PL/1 an assignment instruction is written with an equals sign as the assignment operator. Each declaration or instruction in PL/1 ends with a semicolon.

```
declare (X, A, JA, SUM, R3) float;
X = 14.37;

A = 271.3;
JA = 17.16;
SUM = 14;
A = 3.172;

X = 14.37;
R3 = X;
```

Words to Remember

assignment instruction   fetch

Exercises

1. Write assignment instructions to store the values 1.0 through 5.0 in five cells named A through E.
2. Write a sequence of assignment instructions to *interchange* the values of the two variables X and Sum. (Caution: This may be slightly harder than you expect.)
3. Write assignment instructions to store approximate values of your age, your house number, your Social Security number, your bank account balance, your telephone number, and your Zip code, in appropriately named cells.
4. Suppose you have a very slow computer that will execute only one instruction on the first day of January each year. Write the instructions that you would like to have your computer execute, so that the variable Year will have as its value the correct current year number during each of the next five years, beginning at the start of next year.
5. Check your local newspaper to find the call letters of all the television stations that can be received in your area. Using each set of call letters as a variable, assign the channel numbers as values.
6. Look in a cookbook, and find a recipe for angel food cake. Using variables such as Flour and Sugar, assign numerical values indicating the quantity of each ingredient specified in the recipe.

2.4 *How Can Humans Give Data to the Computer and Get Back the Results?*

We have seen how numbers can be moved from one cell to another inside the computer, and how to write instructions to prescribe such operations. To perform computations that are useful to humans, computers must be able to communicate with the "outside world." That is, they must be able to read and store *data* prepared by people, as well as printing *results* in a form that people can read.

An instruction for reading data is a read instruction; an instruction for printing results is a print instruction. In our flowcharts, a read or print instruction for a single number may be written in the same form as an assignment instruction. The only difference is that a read instruction has the special name rdr on the right, to designate the reader (from which data is obtained); and a print instruction uses the

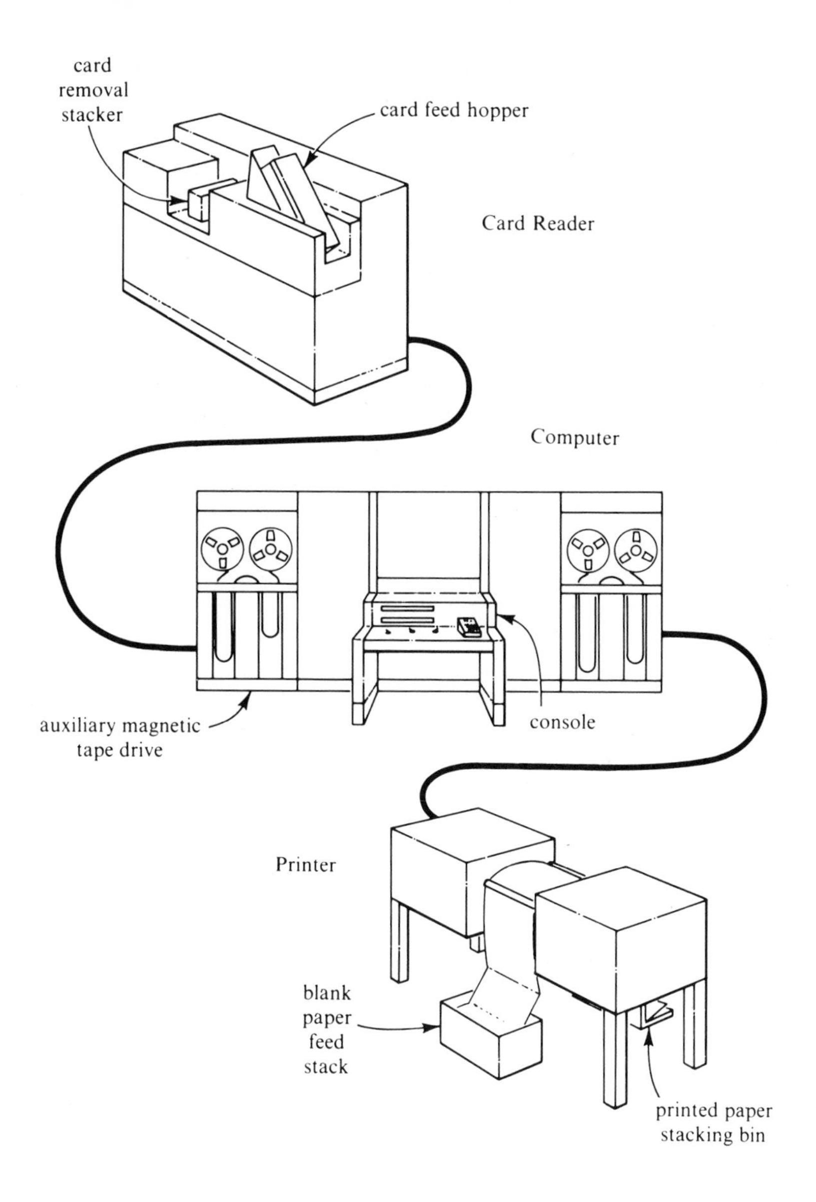

2.4a. Reader, computer, and printer.

special name ptr on the left, to indicate that the destination is not a cell but rather the printer (where results are displayed). For example:

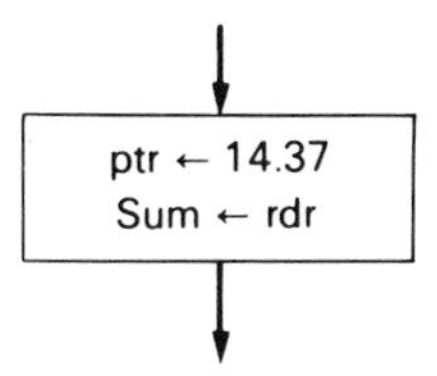

The first of these is a print instruction, which sends the value of the constant 14.37 to the printer, and the second is a read instruction, which assigns to the variable Sum a value obtained from the reader.

The effects of read and print instructions on the values of cells inside the computer are just what we expect if we consider them as assignment instructions. However, the external results are quite different because the reader and the printer bear little physical resemblance to storage cells. We can send a number to the printer but we cannot fetch a number from it. Conversely, we can receive a number from the reader but we cannot send a number to it.

The reader consists of a *queue* or *waiting list* of numbers, prepared by a human, which are presented to the computer one at a time. Each time a read instruction is executed, the next number is removed from the queue and transmitted to the computer. The numbers in the queue are not necessarily in a convenient form for humans to read. Rather, they are coded in some form such as punched holes in cards or tape, or magnetized spots, for the convenience of the computer hardware.

The printer contains a long strip of paper, on which the results of the computation are written in ordinary decimal notation. The print instruction causes each number to be printed on a new line, below the previous number, as on an adding machine tape.

Note that a read instruction changes the contents of a cell inside the computer, but a print instruction does not. The symbol on the left side of a read instruction must therefore be a variable, while the symbol on the right side of a print instruction may be a variable or a constant.

Some Computations Some useful computing tasks can be accomplished using only assignment instructions (including read and print instructions). Table 2.4 lists

Table 2.4 *Population of the World's Five Largest Metropolitan Areas*

	Millions
Tokyo–Yokahama	20.5
New York	16.9
Osaka–Kobe–Kyoto	12.3
London	11.025
Moskva (Moscow)	9.15

the five most populous metropolitan areas of the world (International Atlas, 1969, p. 280) with population in millions. Let us suppose that the numbers in the right-hand column of this list have been prepared and inserted in the reader. We want these numbers to be printed in reverse order, as follows:

```
 9.15
11.025
12.3
16.9
20.5
```

The program for this computation requires five variables. We might use the names TokyoPop, NewYorkPop, OsakaPop, LondonPop, and MoscowPop. The program includes instructions for reading all of the numbers and storing them in the five cells identified by these variables and then for printing the contents of the cells in the desired order:

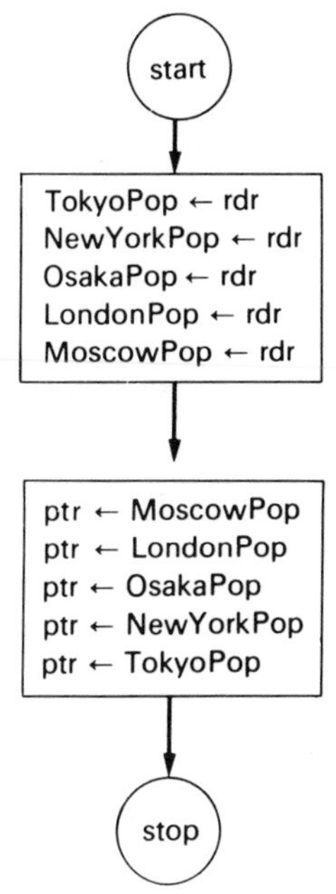

Notice that these operations are performed in sequence, and the printed result depends on the sequence of the data values in the queue of the reader. The first data value is read when the first read instruction is executed and is stored in the cell specified by the variable on the left side of this instruction. The first result, printed by the first print instruction, is the value of the variable on the right side of this instruction.

Read and Print Instructions for Lists Most computations require that several data values be read together, or that several results be printed on the same line. Although the simplified form we have been using emphasizes that reading and printing are closely related to assignment operations, the instructions for handling a group of values become quite unwieldy, as in the previous example. For reading several data values or printing several results, it is more convenient to use a different form. We can prescribe reading or printing of the values of all the variables in a *list* with a single instruction:

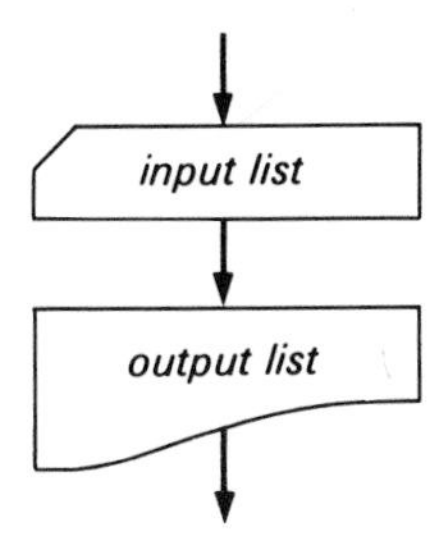

The operation of input or output is indicated by the shape of the box. The box for a read instruction resembles a punched card, which is a common medium for the input queue. The box for print suggests a torn sheet of paper that might have been just removed from the printer.

The *input list* or *output list* may be a single item or a sequence of items separated by commas. Each item in an input list must be a variable; the items in an output list may be constants or variables. For instance, we could rewrite the previous example as:

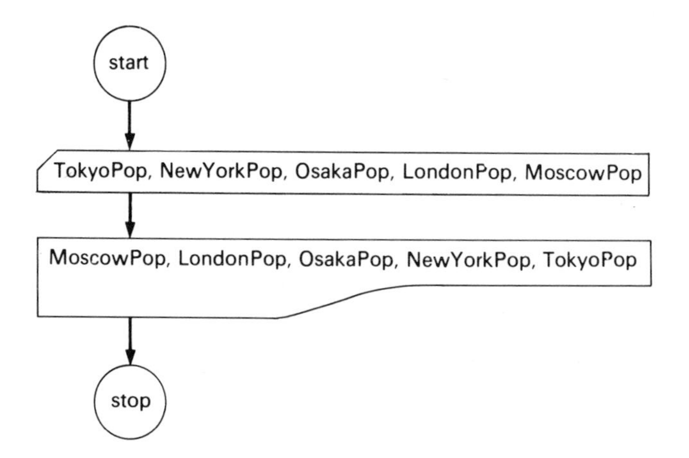

Or we could print the data values in their original order, along with constants to serve as labels:

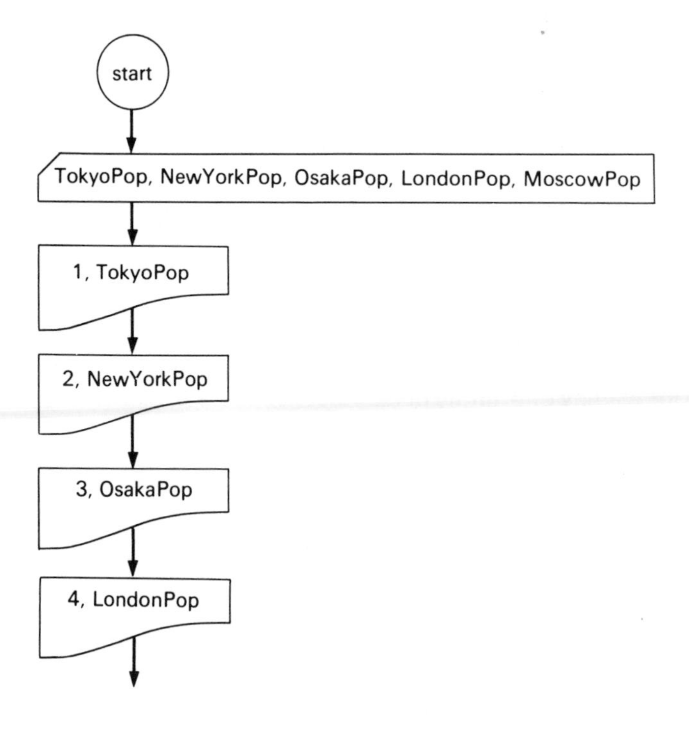

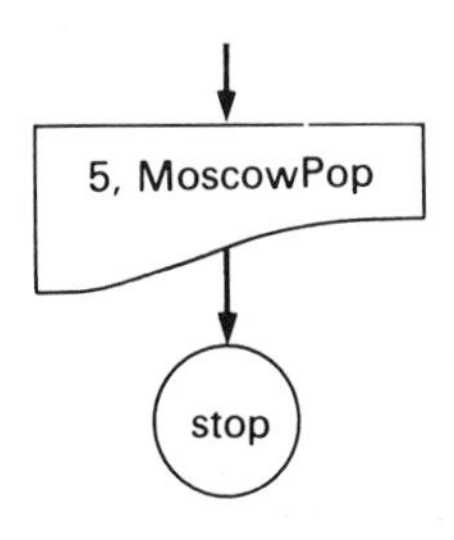

In the first example, the values of all five variables will be printed on the same line:

9.15 11.025 12.3 16.9 20.5

In the second example, the results will be printed on five separate lines because the execution of each print instruction starts a new line on the printer:

```
1    20.5
2    16.9
3    12.3
4    11.025
5     9.15
```

Because the *value* of a variable or constant determines what is printed, it does not matter whether this value was generated by the execution of an assignment instruction or a read instruction. The following program will print the same number twice (assuming that the reader queue contains the number 20.5):

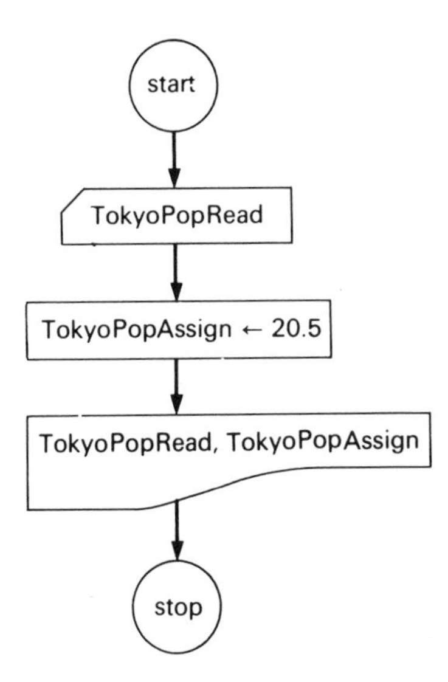

Flowchart For pedagogical purposes, we have first introduced a special category of assignment instructions that can be used to read data or print results:

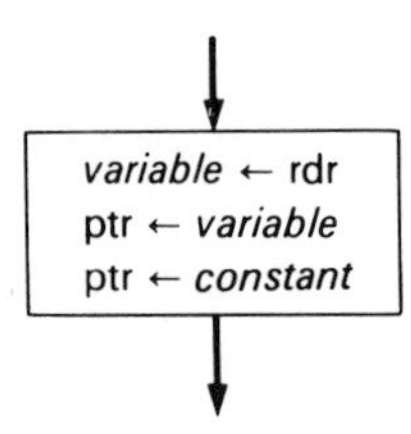

More useful for practical computation problems, on the other hand, are the list forms of the read and print instructions:

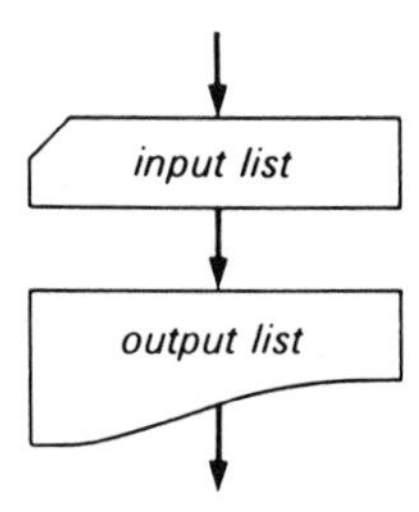

The list consists of a single item, or of a sequence of items separated by commas. Each item in an input list must be a variable; the items in an output list may be variables or constants. For example:

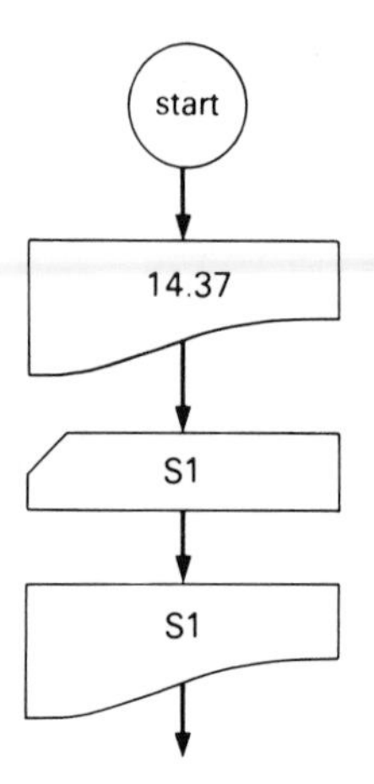

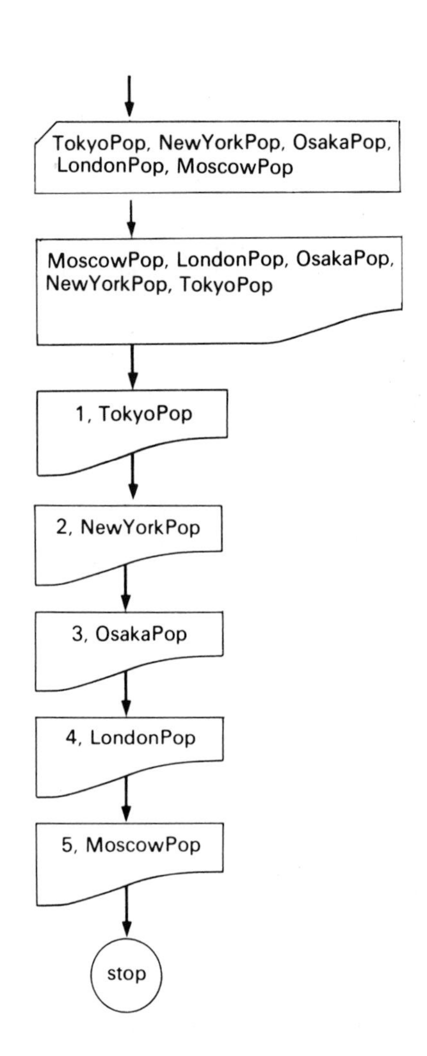

Basic There are two input list forms in Basic, and one output list form:

```
READ input list
INPUT input list
PRINT output list
```

The READ instruction uses data prepared in advance in DATA statements that are combined with the instructions of the program:

```
140 DATA 20.5, 16.9, 12.3, 11.025, 9.15
```

The INPUT instruction is included in Basic on the assumption that a typewriter-like console is connected to the computer during execution. The computer executes the INPUT instruction by typing ? and waiting for data to be entered by the user. The user types a number and ends by striking the carriage return key as a signal for execution to proceed. If several variables are listed in the same INPUT instruction, the computer types ? again and waits for each variable in the list.

The PRINT instruction types the value of each listed variable or constant on the paper at the console.

```
10   PRINT 14.37
20   READ S1
30   INPUT S1

110 READ T, N, O, L, M
120 INPUT T, N, O, L, M
130 PRINT M, L, O, N, T

210 INPUT T, N, O, L, M
220 PRINT 1, T
230 PRINT 2, N
240 PRINT 3, O
250 PRINT 4, L
260 PRINT 5, M
```

Fortran The Fortran READ and WRITE instructions require a *device code* and a *format number*, as well as a list of variables. The device code is an integer assigned at the computer installation. We shall assume that device 5 is the reader and device 6 the printer. The format number tells which *format statement* to use while reading or printing the data. The format statement describes the arrangement of data in the reader queue or of results to be written on the printer.

```
READ (device code, format number)
    input list
WRITE (device code, format number)
    output list
```

The format statement is a *declaration* (or non-executable statement) rather than an instruction. That is, it contains descriptive information and is not actually executed. Included in the format statement is a specification of the type (integer or real) of the variable values to be transmitted. For example:

```
     READ (5, 82) SUM
82   FORMAT (8 F 10.0)
```

The interpretation is: Read from device 5, according to format 82, the value to be assigned to the variable SUM, whose type is real.

Another example:

```
     READ (5, 82) TOKYO, RNYORK,
    1 OSAKA, RLOND, RMOSC
82   FORMAT (8 F 10.0)
```

(When a Fortran statement is too long to fit on one line, the digit 1 is inserted at the left on the continuation line.) Execution of this instruction will read (from device 5, according to format 82) the values of the five real variables in order as they appear in the list. The first number in the reader will be assigned as the value of TOKYO, the next RNYORK, etc. We change the names of the cities to conform to the Fortran type convention, prefixing them with an R if necessary to indicate *real* type, or with an I to indicate *integer* type.

The format statement is not repeated when the same format is used with more than one READ instruction. If the two previous examples were part of the same program, we would write:

```
      READ (5, 82) SUM
82    FORMAT (8 F 10.0)
      READ (5, 82) TOKYO, RNYORK,
     1 OSAKA, RLOND, RMOSC
```

We would *not* repeat 82 FORMAT (8 F 10.0).

Each format statement specifies, by means of the numbers and letters inside the parentheses, the following information concerning the data it will handle:

(a) Maximum number of variables[7] in the corresponding input or output list.

(b) Type (*integer*, I, or *real*, F[8]) of variables in the list.

(c) Maximum number of digits (including, if necessary, the sign and the decimal point).

(d) For the output of *real* variables only, the number of digits to be printed to the right of the decimal point.

Thus (8 F 10.0) may be used with an input list of up to eight real variables; each data value may have up to nine digits plus a decimal point. If the value is negative, the digit space allowance will be decreased by one to provide for the sign.

A different format is required for *integer* variables:

```
      READ (5, 81) ITOKYO,NYORK,IOSAKA,
     1 LONDON,MOSCOW
81    FORMAT (8 I 10)
```

The I in the format specification indicates that the variables are of integer type, and no decimal points are present in the data.

To print results we use a WRITE instruction, such as:

```
      WRITE (6, 81) MOSCOW, LONDON,
     1 IOSAKA, NYORK, ITOKYO
81    FORMAT (8 I 10)
      WRITE (6, 83) RMOSC, RLOND,
     1 OSAKA, RNYORK, TOKYO
83    FORMAT (8 F 15.3)
```

The first WRITE instruction, with format number 81, is used to print integer values. The second WRITE instruction prints real values according to format 83.

The specification (8 F 15.3) indicates that as many as eight variables may appear in the list, that the variables must be of real type, and that the value of each variable may occupy as many as 15 total spaces, including room for a minus sign and decimal point, with three digits printed to the right of the decimal point.

From even this rudimentary description, it is evident that format statements can be constructed to allow great flexibility in input and output. We will not go further into the subject at this point. The three formats used in the previous examples are sufficient for most numerical applications. For input or output of up to eight integer values, we use

```
81    FORMAT (8 I 10)
```

For input of up to eight real values, we use

```
82    FORMAT (8 F 10.0)
```

And for output of up to eight real values, with three digits to the right of the decimal point, we use

```
83    FORMAT (8 F 15.3)
```

[7] Standard Fortran input and output is "record-oriented"; that is, it assumes that the data and results will be arranged in definite units called records. Strictly speaking, the format statement specifies the maximum number of variables *per record* of input or output. A list containing more than this number of variables will cause additional records to be read or printed (and the format statement to be "rescanned"). At the beginning, however, we assume that each READ or WRITE instruction causes just one record to be transmitted.

[8] The letter *F* has historical significance. Numbers of real type were once called "floating-point" numbers. This term pertains to the way numbers are stored internally which, from the point of view of computer hardware design, is responsible for the need to distinguish between the two types of numbers. As programming has developed, however, users have wanted to be concerned as little as possible with hardware details. The term "real," which has greater significance to the user, has replaced the older term in descriptions of Fortran, but the older term remains imbedded in a few places in the language itself.

Constants are not permitted in the list of a Fortran WRITE instruction. Thus we cannot say, WRITE (6, 83) 14.37; instead we must *assign* the value of the constant to a variable, and then use the variable in the list:

```
X = 14.37
WRITE (6, 83) X
```

To print numerical labels, then, we could use five assignment instructions:

```
      X1 = 1.0
      X2 = 2.0
      X3 = 3.0
      X4 = 4.0
      X5 = 5.0
      READ (5, 81) TOKYO, RNYORK,
     1 OSAKA, RLOND, RMOSC
      WRITE (6, 83) X1, TOKYO
      WRITE (6, 83) X2, RNYORK
      WRITE (6, 83) X3, OSAKA
      WRITE (6, 83) X4, RLOND
      WRITE (6, 83) X5, RMOSC
```

PL/1 The instructions get list and put skip list are used in PL/1 to obtain data values and to display results:

```
declare (SUM, TOKYO, NEWYORK,
  OSAKA, LONDON, MOSCOW) float;
put skip list (14.37);
get list (SUM);
put skip list (SUM);
get list (TOKYO, NEWYORK, OSAKA,
  LONDON, MOSCOW);
put skip list (MOSCOW, LONDON,
  OSAKA, NEWYORK, TOKYO);
put skip list (1, TOKYO);
put skip list (2, NEWYORK);
put skip list (3, OSAKA);
put skip list (4, LONDON);
put skip list (5, MOSCOW);
```

Input data in the reader is prepared as a "stream" of constants, separated from each other by a comma or by one or more blanks. Each data item, as it is encountered, is assigned to the next variable in the list. The put skip list instruction will cause the results to be written on the printer in the order listed. The word skip causes the first item in the list to be printed on a new line. Note that PL/1 output lists may include both constants and variables.

Words to Remember

data	printer	print instruction
results	queue	input list
reader	read instruction	output list

Exercises

1. Write a program for reading the values of four variables A1, A2, B1, and B2, in that order, and printing them in the order A1, B1, A2, B2.
2. Will the numbers printed, when the following two program segments are executed, always be identical regardless of the numbers in the queue of the reader?

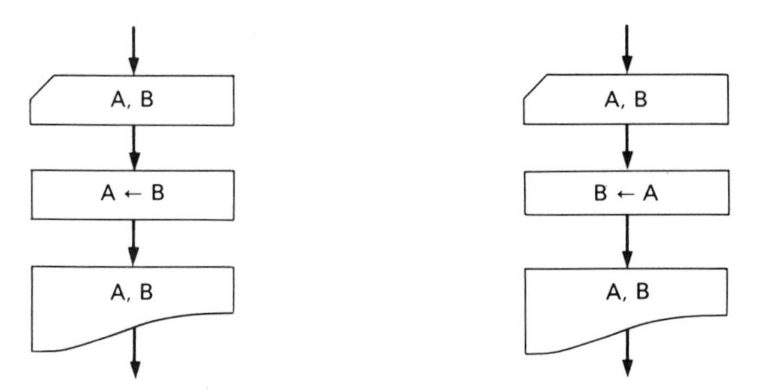

3. Write a program to print the first eight lines of Pascal's triangle, assuming that the numbers have been prepared properly as data in the reader. Run the program on a computer if possible. The results on the printer should appear approximately as follows:

```
1
1   1
1   2   1
1   3   3   1
1   4   6   4   1
1   5  10  10   5   1
1   6  15  20  15   6   1
1   7  21  35  35  21   7   1
```

If we number the lines in this array, starting from zero, and if we number the entries on each line, again starting from zero, we find that entry r on line n is the coefficient of x^r in the binomial expansion of $(1 - x)^n$. Pascal (1623–1662) used these numbers in his studies of probability (Turnbull, 1951), and showed that what we are calling "entry r on line n," usually designated as $\binom{n}{r}$ or as $C[n, r]$, is the number of possible combinations that can be formed when r objects are selected from a group of n objects. Pascal's triangle is also known as the table of binomial coefficients.

2.5 *How Can We Tell the Computer to Do Arithmetic?*

Inside the computer, besides the cells where numbers are stored, there is an *arithmetic unit* that performs the arithmetical manipulations needed in a calculation. This unit has access to all of the cells, as well as to a few *registers*. These registers are similar to cells in that each register can hold one number; however, the instructions written by the programmer cannot refer directly to the registers as they can to the cells.

As part of a calculation, we might want the computer to retrieve numbers from two of its cells and multiply them together, leaving the result in a register. Then we might want it to fetch a number from another cell and add it to this result, depositing the result of the addition in another register.

A programming language needs a way of prescribing that an arithmetic operation be performed in a calculation. In our flowcharting language, we use the ordinary *arithmetic operators* +, −, ×, and ÷ for this purpose. However, conventional programming languages must use symbols that exist on the keyboards of the devices used to enter the programs into the computer, and most such devices do not include the characters × and ÷. Therefore, many conventional programming languages indicate multiplication by the asterisk, *, and division by the slash, /.

For instance, if we want the values of the variables A and B to be multiplied together, we can indicate this in a flowchart as

A × B

If we want the computer to add the constant value 1.74 to this result, we form the longer expression

(A × B) + 1.74

Notice that the new expression contains within it the first expression enclosed in *parentheses*. The value of the first expression, (A × B), will remain in a register while the new result is being calculated and stored in another register.

Here are more examples of expressions that might appear in flowcharts:

(J × K) ÷ L	3 + (KO + Mi)	(X + Y) ÷ 2.0
1 − XY	I + 1	(A1 + B) ÷ C
((A ÷ B) + (C × D)) ÷ (E + (F ÷ G))		

The computer uses the registers and the arithmetic unit in evaluating each of these expressions. To evaluate the long expression at the end of this list of examples, it fetches the values of the variables A and B from their cells, uses the arithmetic unit to compute the quotient, and stores the result in a register (see Figure 2.5a). Similarly, it fetches and multiplies the values of C and D and computes the quotient of the values of F and G; each of these new results is stored in a register. Next the computer combines the values in two of the registers to form the value of the expression ((A ÷ B) + (C × D)), and stores this value in a register. By this time, the machine may be running out of registers; if it recognizes that some of the intermediate values are no longer needed after this point it will reuse one of the registers. Next it combines a register value with the contents of the cell named E to form the value of

(E + (F ÷ G)) in another register (again perhaps reusing one of the registers no longer needed). Finally the values of the two expressions ((A ÷ B) + (C × D)) and (E + (F ÷ G)) are taken from the registers where they have been stored, and their quotient is formed in the arithmetic unit. This quotient is finally stored in a register as the value of the expression ((A ÷ B) + (C × D)) ÷ (E + (F ÷ G)).

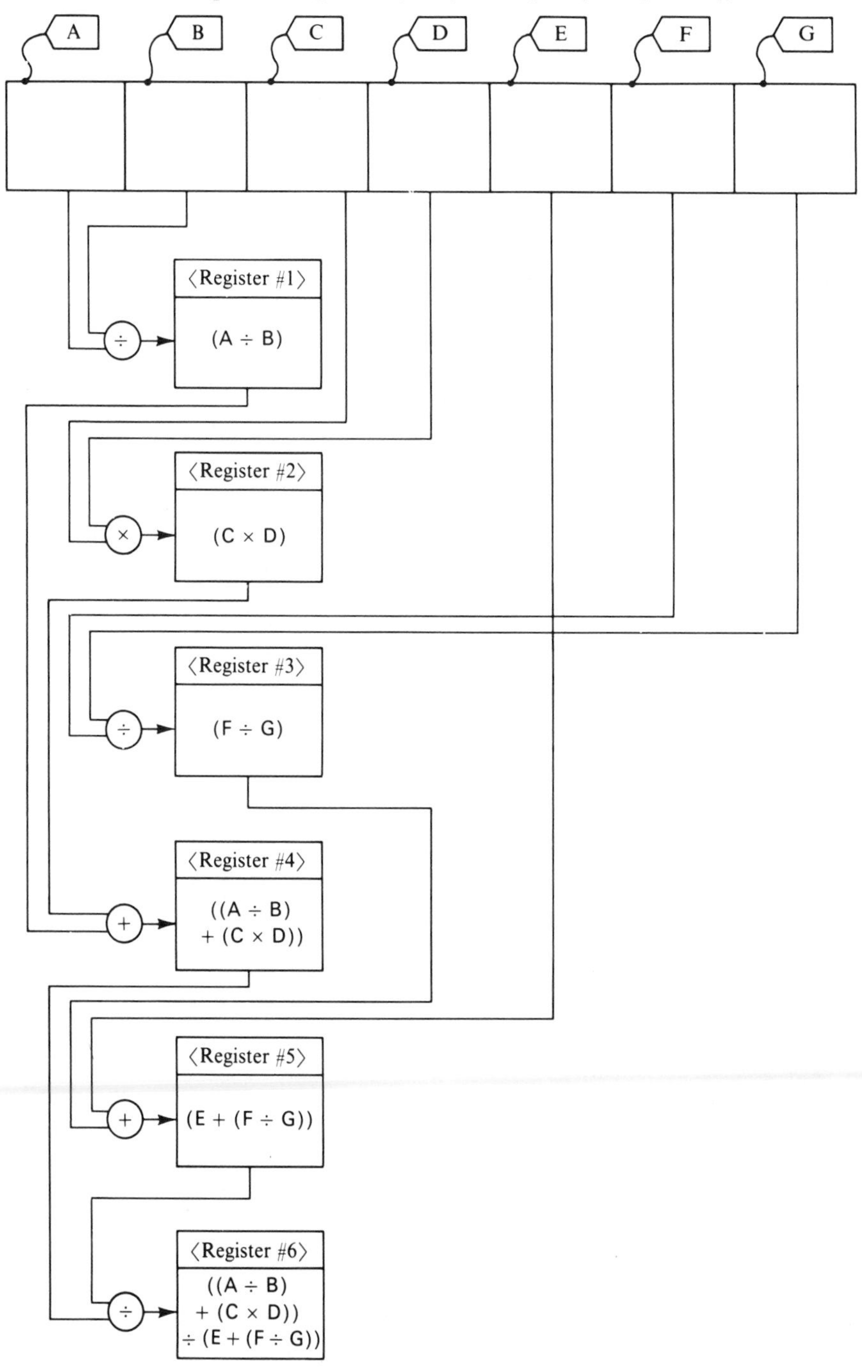

2.5a. *Use of registers in the evaluation of a complicated expression. Some of the registers can be reused when there is no further need to retain the intermediate value.*

The Minus Sign There are actually two ways to use a minus sign. One way, as we have seen, is to write the minus sign between two expressions:

expression − *expression*

The other way is to write the minus sign to the left of an expression:

−*expression*

The operation specified in the first case (joining two expressions) is *subtraction*, and in the second case (with one expression) the operation is *negation*. The value of the result in the second case is the negative of the value of the expression that follows the minus sign.

Examples of Subtraction

```
1 − XY        A − B        ((X + Y) ÷ 2.0) − ((A1 + B) ÷ C)
```

Examples of Negation

```
−(1 + XY)        −(A + B)        −((X + Y) ÷ 2.0)        −((A1 + B) ÷ C)
```

A constant preceded by a minus sign, such as

```
−14.37        −1        −3141592        −.000547
```

can be viewed either as a negative constant (a constant whose value is less than zero) or as an expression in which the operation of negation is applied to a positive constant. We adopt the latter viewpoint.

Parentheses In the previous examples, we have enclosed an expression in parentheses whenever it is used as a part of some larger expression. As we construct more complicated expressions, a great many parentheses would be introduced if we continued to follow this rule strictly. Experience, on the other hand, suggests that not all of these parentheses are really necessary, and that we should not put them in unless they are needed to make the meaning clear. But in some cases, parentheses make a difference in the algebraic meaning of an expression. For instance, in algebra the expression $a - (b + c)$ is not the same as $a - b + c$, and $(b + c) \times d$ does not mean the same thing in algebra as $b + c \times d$. Also we want to include parentheses to separate adjacent operators, as in the expression $a \times (-b)$.

Common sense, along with some careful application of elementary algebra, is usually a reliable guide for deciding whether parentheses are needed in a particular case. And it is always permissible to use parentheses when in doubt.

Words to Remember

arithmetic unit	arithmetic operator	negation
register	expression	

Exercise

For each minus sign in the following examples, tell whether the operation it specifies is *subtraction* or *negation.*

a. (X + Y) − Z
b. −Z + (X + Y)
c. (X + Y) × (−Z)
d. −Z × (X − Y)
e. −Z × (3.27 − X)

2.6 How Do We Tell the Computer to Store the Value of an Arithmetic Expression?

Now that we know how to write expressions and how they are evaluated, we can write assignment instructions of the more general form:

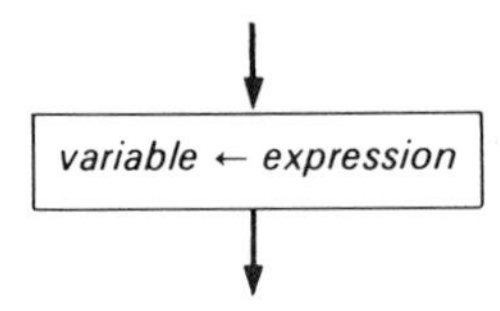

We recall that the operation of executing an assignment instruction has two phases. In the first phase, the value of the expression on the right is determined by the arithmetic unit and is placed in a register. In the second phase, the number in the register is assigned as the contents of the cell named by the variable on the left.

Even if the arithmetic unit has no real work to perform, as when the right-hand side of the assignment instruction consists merely of a single variable or constant, this value is fetched and placed in a register. For this reason, it makes sense to consider a variable or constant, by itself, as a valid expression.

Does it make sense for the variable on the left to appear also in the expression on the right? If we clearly understand what happens during the execution of an assignment instruction, we should be able to answer this question. Let us consider the simple example

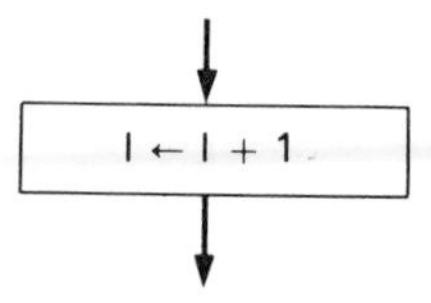

The value of I is fetched from a cell during the first phase of this operation (without changing the value of I) and the expression I + 1 is evaluated. At the end of this first phase, the value of the expression I + 1 is in a register, and the cell named I retains its value (Figure 2.6a). In the second phase, this number is moved from the register and assigned as the new value of I, replacing the number previously stored in that cell (Figure 2.6b). There is no conflict between the appearances of I on the right and those

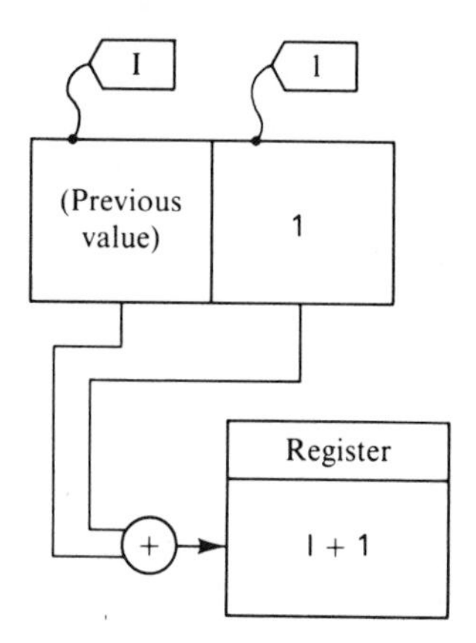

2.6a. *Execution of the assignment instruction* I ← I + 1.
Phase 1: *The value of* I *is fetched from a cell, and* 1 *is added to it. The result is stored in a register. The cell named* I *retains its previous value during this phase.*

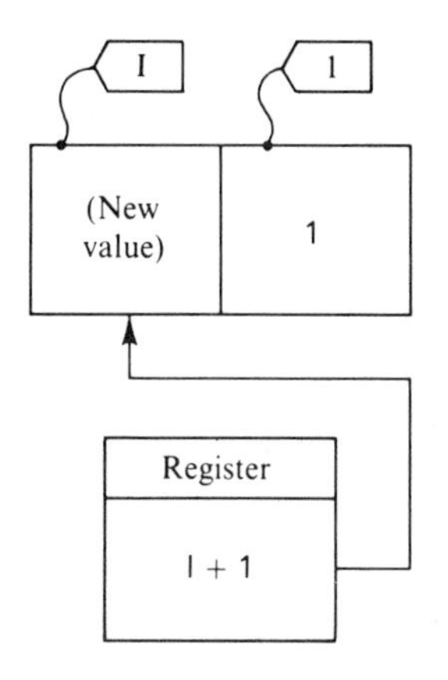

2.6b.
Phase 2: *The value of* I + 1, *computed during the first phase from the previous value of* I, *is moved from the register and assigned as the new value of* I, *replacing the number previously stored in the cell.*

on the left, since the two sides of the instruction refer to the two phases of the assignment operation, which happen at different times. The value of I fetched for use in evaluating the expression on the right is the old value, based on instructions executed before this one. The value stored (during the assignment phase) in the cell named on the left is the new value of I that will be available for operations following this one.

In a similar example, the sum of a sequence of terms is desired. Recalling how we would add the terms together with an adding machine, we write a separate instruction for each term:

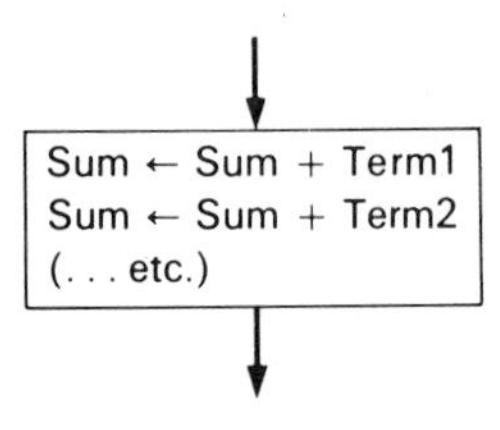

The first instruction adds the first term to the "old" sum, and stores the result as the new sum. The second instruction adds on the second term, and so on.

Undefined Variable Values During the execution of an assignment instruction, values of the variables that appear in the expression on the right are fetched from their cells and combined in the registers to form the value of the expression. But unless values have previously been stored in the cells named by all of these variables, the value of the expression cannot be determined.

A constant is never "undefined," since its value is placed in the cell before the execution of the first instruction that references it. A variable, on the other hand, is undefined until an instruction is executed that defines it by assigning a value to it. This instruction must be an assignment instruction, with the variable on the left, or a read instruction.

When the variable on the left side of an assignment instruction also appears in the expression on the right, we must make sure that this variable has a proper "old" value. In particular, a sum would normally be *initialized*—that is, given a zero value—before the terms are added on.

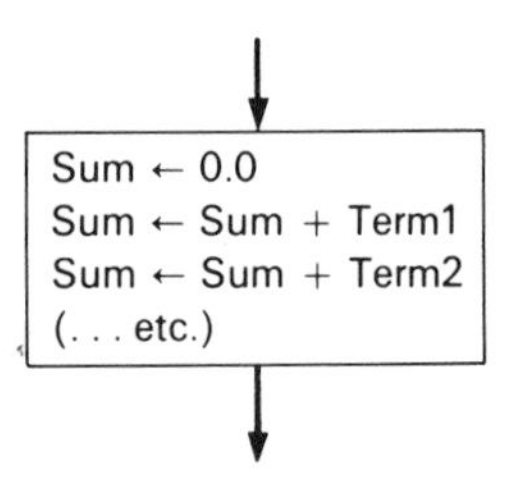

Word to Remember

initialize

Exercises

1. Complete the above example to add together the values of the seven variables Term1, Term2, . . . , Term7 with eight instructions.
2. Show how the same result could be computed in seven instructions, combining the initialization with the incorporation of the first term into the sum.

2.7 *How Can We Combine Instructions into a Complete Program?*

With assignment instructions that include expressions, and read and print instructions, we can compose complete programs to describe some interesting computations.

We can instruct the computer to read the diameter of a circle and print the circumference. Calling the diameter Di and the circumference Ci, we write:

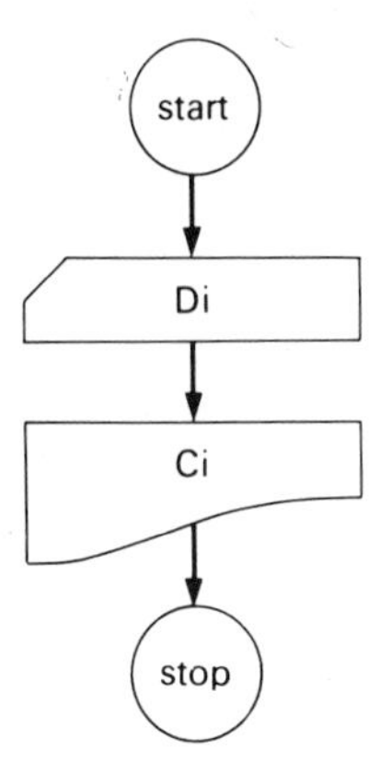

But something is missing here. No value is ever assigned to the variable Ci. We must insert an instruction for muliplying the diameter by Pi to obtain the circumference:

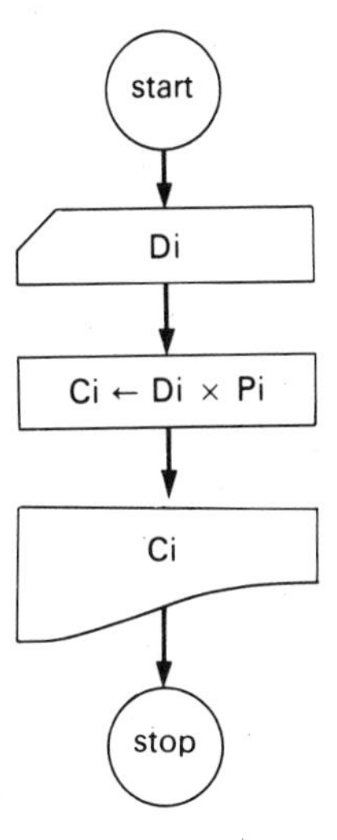

That looks better, but it still won't work. The new instruction involves another variable whose value is undefined. Let's define it:

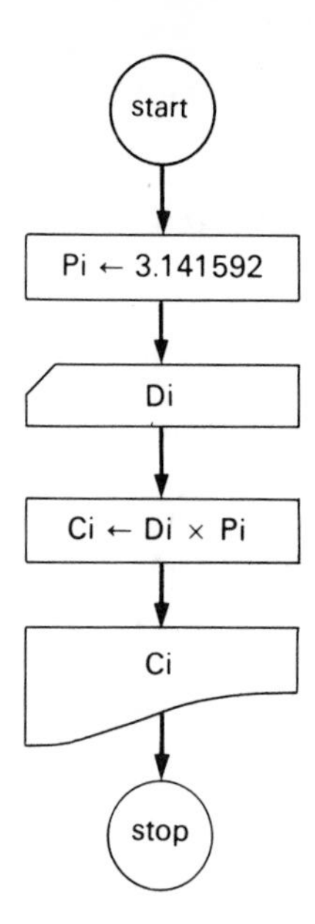

Now we have it. The four instructions form a complete set, each variable being defined before it is used. Or, instead of introducing the variable Pi, we can omit the instruction that assigns its value and use the constant directly:

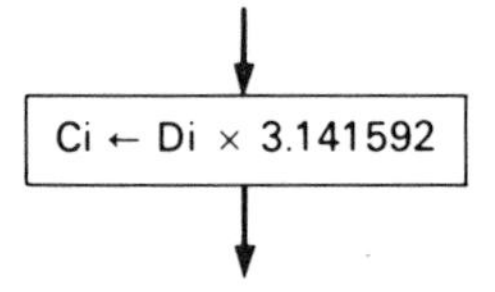

Our choice might depend on how many times the same constant appears in the program. If several other formulas use the constant, we might prefer to use a variable named Pi and assign the value to it just once.

The process of finding the average of two numbers is described by the following program:

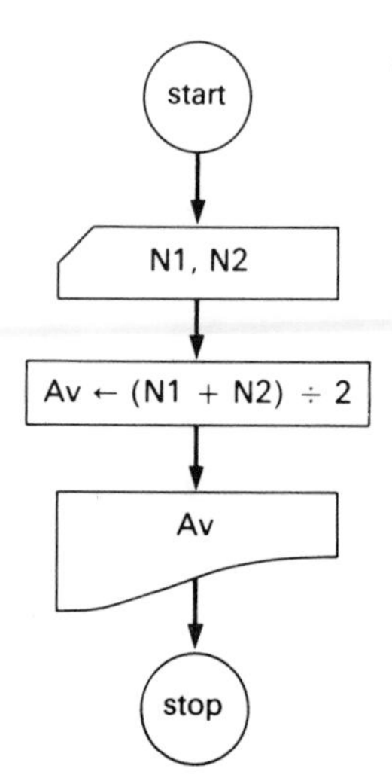

Perhaps it would be well to print the data values for reference:

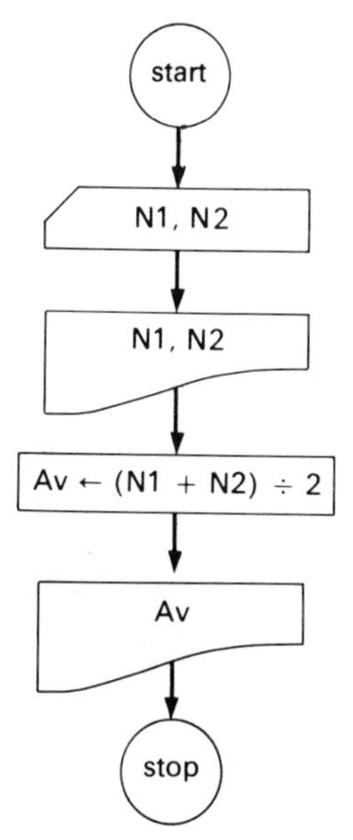

If the average value is not going to be used any further in the calculation, there is no need to store it in a cell; we may as well include the expression directly in the output list:

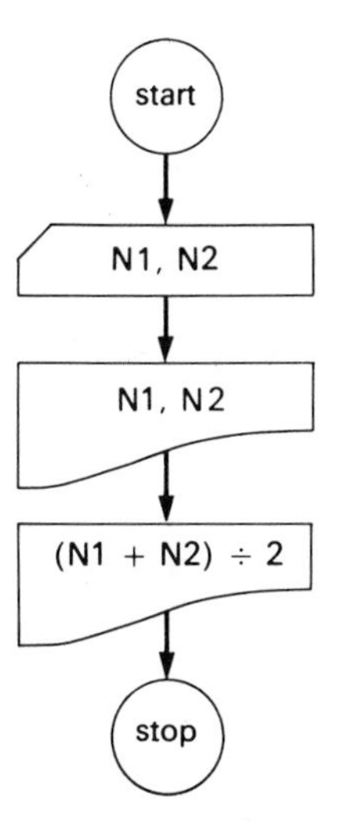

Adding Fractions Let us suppose that we want to write a program for adding two fractions together; for example,

$$\tfrac{1}{2} + \tfrac{1}{3} \qquad \text{or} \qquad \tfrac{35}{51} + \tfrac{49}{72}$$

We want to handle the numerators and denominators as separate integers, and we want to find out that the sum of $\frac{1}{2}$ and $\frac{1}{3}$ is a fraction whose numerator is 5 and whose denominator is 6. The problem does not mean that the values of the fractions expressed as real numbers are to be added together. We are not seeking the answer expressed as:

$$\begin{array}{l} 0.5000 \\ 0.3333\ldots \\ \hline 0.8333\ldots \end{array}$$

In general, to add the fractions a/b and c/d, we use bd as a common denominator and write

$$\frac{a}{b} + \frac{c}{d} = \frac{ad + bc}{bd}$$

Our program must include instructions to find the numerator of the sum and the denominator of the sum, using the numerators and denominators of the given fractions. The first step, as usual, is to choose names for the variables. (This, in itself, may be a rather confusing task for the beginning programmer.) We choose A and C as names for the numerators of the two given fractions, and B and D for the denominators. Then we let NumSum represent the numerator of the sum, and we let DenSum represent the denominator of the sum. We now write the instructions:

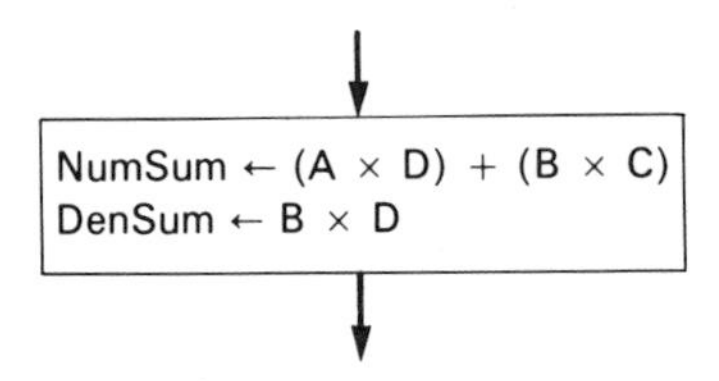

Remember that multiplication signs must not be omitted, and that a variable must begin with a capital letter. The only other instructions we need are those that read the given numerators and denominators and print the results:

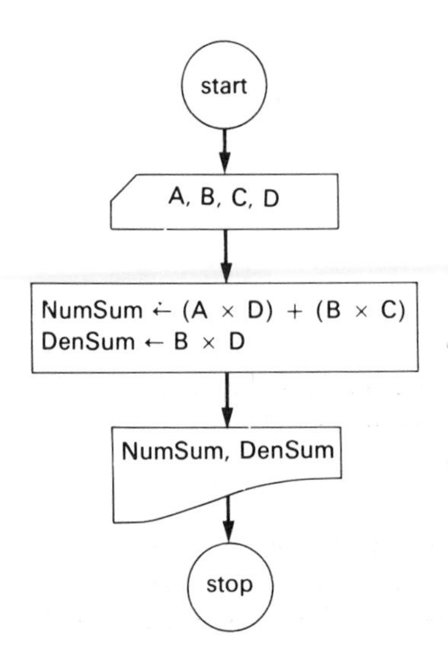

Flowchart

Circle:

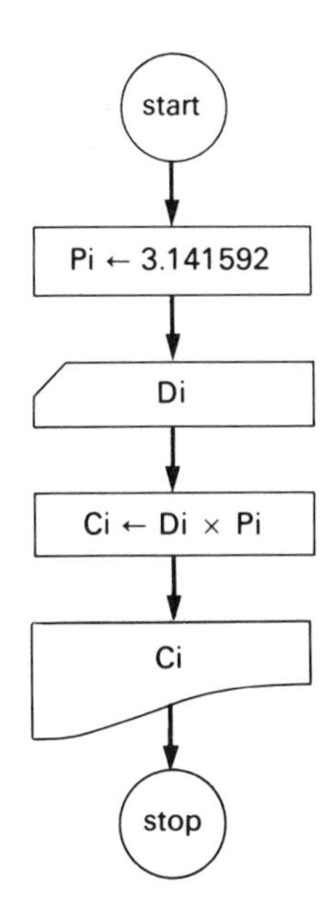

Adding fractions:

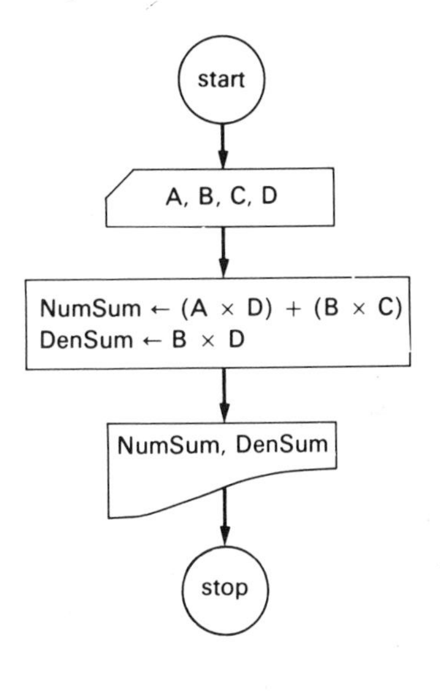

Average:

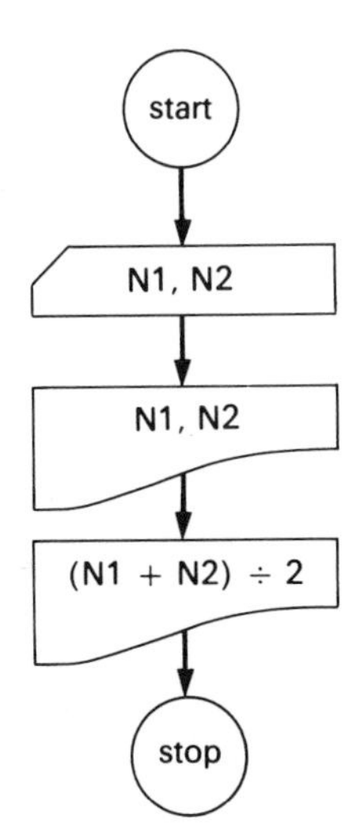

Basic

Circle:

```
10  LET P = 3.141592
20  INPUT D
30  LET C = D * P
40  PRINT C
```

Average:

```
110 INPUT N1, N2
120 PRINT N1, N2
130 PRINT (N1 + N2)/2
```

Adding fractions:

```
210 INPUT A, B, C, D
220 LET N1 = (A * D) + (B * C)
230 LET D1 = B * D
240 PRINT  N1, D1
```

Fortran

Circle:

```
     PI = 3.141592
     READ (5, 82) DI
82   FORMAT (8 F 10.0)
     CI = DI * PI
     WRITE (6, 83) CI
83   FORMAT (8 F 15.3)
```

Average:

Remember that Fortran does not permit expressions in an output list, so we assign the value of the desired expression to the variable AV.

```
     READ (5, 82) RN1, RN2
82   FORMAT (8 F 10.0)
     WRITE (6, 83) RN1, RN2
83   FORMAT (8 F 15.3)
     AV = (RN1 + RN2)/2.0
     WRITE (6, 83) AV
```

Adding fractions:

```
     READ (5, 81) IA, IB, IC, ID
81   FORMAT (8 I 10)
     NS = (IA * ID) + (IB * IC)
     IDS = IB * ID
     WRITE (6, 81) NS, IDS
```

PL/1

Circle:

```
declare (PI, DI, CI) float;
PI = 3.141592;
get list (DI);
CI = DI * PI;
put skip list (CI);
```

Average:

```
declare (N1, N2) float;
get list (N1, N2);
put skip list (N1, N2);
put skip list ((N1 + N2)/2);
```

Adding fractions:

```
declare (A, B, C, D, NUMSUM,
  DENSUM) fixed;
get list (A, B, C, D);
NUMSUM = (A * D) + (B * C);
DENSUM = B * D;
put skip list (NUMSUM, DENSUM);
```

Exercises

1. Modify the program for calculating the circumference so that the data values which are read will also be printed for reference. Use diameters (a) 0.3183099; (b) 1.2732395; (c) 57.2957795.
2. Write a program for calculating the area of a rectangle from the lengths of the sides. The calculation should include the necessary read and print operations. Three sets of data values are given: (a) 6.0, 9.0; (b) 33.1, 24.6; (c) 1200.0, 833.3.
3. Write a program for calculating the sum of the squares of three data values.

	first	second	third
a.	3.0	4.0	5.0
b.	16.0	1.0	−3.0
c.	0.01	0.16	0.03

4. Write a program that will instruct the computer to read three data values, and calculate and print the difference between each possible pair. (There are six pairs, as each of the three numbers can be subtracted from either of the other two.) Also include instructions to print two integers along with each difference, indicating which two of the three data values were used to form the difference. (Use the data in exercise 3, above.)
5. Write a program to multiply Total Hours by Base Rate to determine an employee's Gross Pay. Include the necessary read and print instructions.

Total Hours	40	25	45	40
Base Rate	$1.60	$2.70	$4.31	$9.75

6. Write a program, including read, assignment, and print instructions, to calculate simple interest according to the formula *interest* = *principal* × *rate* × *time.*

	principal	rate	time (years)
a.	$100.00	6%	0.25
b.	$1,500.00	$7\frac{1}{4}$%	4
c.	$700.00	20%	2
d.	$100.00	6%	12

7. Write a program to convert temperature readings from degrees Fahrenheit to degrees centigrade. Use Fahrenheit values (a) −40; (b) 0; (c) 32; (d) 98.6; (e) 212; (f) 1000.
8. Write a program for multiplying two fractions, handling the numerators and denominators as separate integers. Include appropriate read and print instructions.

	first numerator	first denominator	second numerator	second denominator
a.	1	2	2	3
b.	35	51	49	72
c.	2	3	4	1
d.	34	55	89	144

9. Write a program with eight variables, P1 to P8, which will compute and print the first eight lines of Pascal's triangle (see exercise 3, Section 2.4). After printing each line, determine the values of as many of the variables as necessary for the next line according to the following rules:
   a. Each line involves one more variable than the previous line.
   b. The value of the first and of the last variable on any line is 1.
   c. The value of each of the other variables (beginning with the third line) *between* the first and the last variable on the line is obtained by adding the previous value of the *same* variable to the previous value of the *next* variable on its left. For example, on the sixth line, the value 10 for P3 is obtained by adding the previous value (6) of P3 to the previous value (4) of P2.

   Write a separate assignment instruction for each variable to be printed on each line, and write a separate print instruction for each line. (Hint: For each line, you may wish to compute and assign the variable values in reverse order, that is, from right to left. Why?)

2.8 *How Can We Use Functions in Forming Expressions?*

Some important computations with numbers require operations that are slightly more complicated than the five arithmetic operations discussed in Section 2.5.

Computations in applied mathematics (including statistics and geometry) frequently use *square roots*. In statistics, the standard deviation of a sample is computed as the square root of the variance. Another example, from geometry, is the Pythagorean theorem: the length of the hypotenuse of a right-angle triangle (or the length of the diagonal of a rectangle) is the square root of the sum of the squares of the lengths of the sides. In our flowcharts, the operation of finding the square root of an expression is designated by:

$$\sqrt{expression}$$

The value of the expression must not be negative.

A useful operation whose applications are not quite so familiar is that of converting a real number to an integer. For example, a life insurance premium is often based on "age at last birthday" of the insured at the time the policy is purchased. If a person's age is 22.75 (22 years 9 months) the insurance company uses the table labeled 22 to find the amount of the premium. In fact, all ages from 22.00 to 22.99 (and even up to $22^{364}/_{365}$) use the same table of premiums. If the value of the expression is an exact integer, the result is the same as the value of the expression. Otherwise, it is the next lower integer. The conversion operation gives the integer that is equal to, or slightly less than, the value of a given expression. This is called the *integer part* of the given value, and in our flowcharts we indicate it as follows:

$$\lfloor expression \rfloor$$

Another operation, finding the *absolute value* of an expression, is done by computing its numerical value and then simply ignoring the sign. The absolute value of an expression is written in our flowcharts as

$$|expression|$$

An application of this operation occurs when we have two numbers designating the positions of two objects, measured along a line. The distance between the objects is obtained by subtracting the position measurements and taking the absolute value of the difference.

An expression whose value is the square root, integer part, or absolute value of a subexpression is a *function.* The subexpression is the *argument* of the function.

A function can be incorporated into any expression, wherever a variable or a constant can be used. Any expression may be used as the argument of a function, including an expression with arithmetic operators or functions.

Examples

$$ISAX \leftarrow \lfloor\sqrt{|X|} + 0.05\rfloor$$

$$HYP \leftarrow \sqrt{A \times A + B \times B}$$

$$X1 \leftarrow (-B + \sqrt{B \times B - 4 \times A \times C}) \div (2 \times A)$$

$$X2 \leftarrow (-B - \sqrt{B \times B - 4 \times A \times C}) \div (2 \times A)$$

$$DIST \leftarrow |A1 - A2|$$

$$Tabl \leftarrow \lfloor Age\rfloor$$

$$UX \leftarrow \lfloor X\rfloor - 10 \times \lfloor\lfloor X\rfloor \div 10\rfloor$$

Rounding The value of $\lfloor X\rfloor$ is the closest integer not larger than the value of X. Thus the value of $\lfloor 3.999\rfloor$ is 3. In many applications, we need to *round* the value of X to the closest integer. But the closest integer may be either larger or smaller than the value of X. However, in all cases, the closest integer is the integer part of the expression X + 0.5:

$$\lfloor X + 0.5\rfloor$$

For example, if the value of X is at least 3 but less than 3.5, then the value of X + 0.5 will be at least 3.5 but less than 4.0, and the value of $\lfloor X + 0.5\rfloor$ will be 3. On the other hand, if the value of X is at least 3.5 but less than 4.0, then the value of X + 0.5 will be at least 4.0 but less than 4.5, and the value of $\lfloor X + 0.5\rfloor$ will be 4. In either case, we see that the expression gives the properly rounded result. Notice that a number exactly half way between two integers will be rounded *upward.* If the value of X is exactly 3.5, the rounded result will be 4.

We would like to be able to use the same expression for rounding a negative expression as for rounding a positive one. There will be no difficulty if we properly understand the definition of the integer part function[9] for negative values of its

[9] All three languages provide functions (Basic, INT; Fortran, AINT; PL/1, INT and FLOOR) that produce the integer part of a *positive* argument. However, the Fortran function AINT and the PL/1 function INT, when applied to a *negative* argument, produce a function whose *absolute value* is the integer part of the absolute value of the argument. In Fortran there is also a function called INT, which converts an argument of real type to a function value of integer type.

argument. When the value of A is negative but is not an exact integer, the value of $\lfloor A \rfloor$ is the next integer *more negative* than the value of A. The following algebraic formula will be valid for all values of A:

$$A - 1 < \lfloor A \rfloor \leq A$$

For instance, when the argument has the value −3.1, the value of the integer part function is −4. Thus, when X has the value −3.6 in the expression $\lfloor X + 0.5 \rfloor$, the value of the argument is −3.1 and the function value is −4 as desired.

Integer Division with Remainder This is one of the most important applications of the integer part function. In elementary arithmetic, we learn to express the result of division as an *integer quotient* and a *remainder*. For instance, we say that 23 divided by 4 gives 5 as the quotient and 3 as the remainder. The desired integer quotient, 5, is simply the integer part of the complete quotient, 5.75. In general, let us designate the dividend or numerator by N and the divisor or denominator by D. We then obtain the integer quotient as the value of the expression

$$\lfloor N \div D \rfloor$$

Recalling the process of long division from elementary arithmetic, we see that the remainder is expressed by

$$N - D \times \lfloor N \div D \rfloor$$

Notice carefully the sequence of operations here. $D \times \lfloor N \div D \rfloor$ is *not* the same as $\lfloor D \times (N \div D) \rfloor$, for the value of the latter expression is in fact identical to that of $\lfloor N \rfloor$. In finding the remainder, we first calculate the integer quotient, then (as in long division) we multiply the integer quotient by the value of D and subtract the product from the value of N.

If D divides evenly into N, the remainder is zero. This fact is used in many applications where the divisors (or factors) of an integer need to be found.

It is important to realize that this remainder calculation $N - D \times \lfloor N \div D \rfloor$ still makes sense when the numerator (dividend) is smaller than the denominator (divisor). For in this case the integer quotient is zero and the remainder is the same as the numerator.

A special use of the remainder calculation is in isolating the various individual decimal digits of a number. For example, the expression

$$\lfloor X \rfloor - 10 \times \lfloor \lfloor X \rfloor \div 10 \rfloor$$

gives the first digit to the left of the decimal point, that is, the digit in the "ones" or units place. To see this, consider an example such as 123.45 for the value of X. The value of $\lfloor X \rfloor$ is 123. Using the remainder formula, with $\lfloor X \rfloor$ instead of N and 10 instead of D, we see that the value of the entire expression is the remainder when 123 is divided by 10. This number is the desired units digit, 3.

Cyclic Number Systems If we look only at the units place of a counter, such as the mileage indicator on an automobile, we see that the digits from 0 to 9 appear in cyclic order. When the last digit, 9, is reached, the sequence starts over from 0 and repeats in the same order as before. Similar cyclic sequences, with cycle lengths of 2,

3, or some other number instead of 9, occur frequently in computations. Problems based on time use a cycle of 60 minutes.

In many of these problems some other value changes at the point where the sequence returns to the beginning. On a mileage indicator, the tens digit increases when the units indicator goes from 9 back to 0. On a clock, when the number of minutes goes from 59 to 0 the hour increases by one.

Programs to keep track of these cyclic number systems often use integer division. The remainder forms the lowest cycle, and the integer quotient produces the higher sequence. As an example, let us write instructions for modifying a cycle of hours and minutes, on the basis of elapsed time.

If the time is 8:59 and we wait 2 minutes, it will be 9:01. Let us use H and M to express the time in hours and minutes, and let E be the elapsed time in minutes. We can perform the following computation:

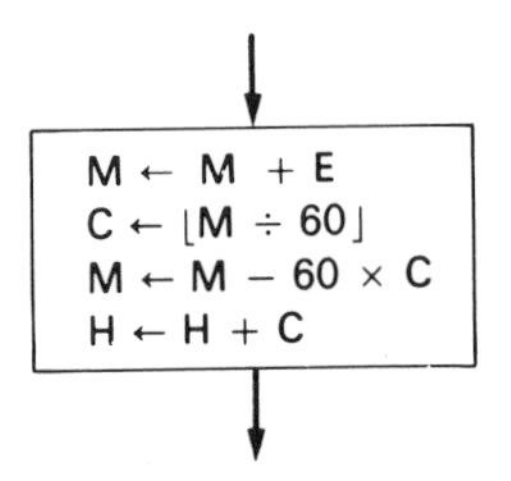

We first add the elapsed time to the value of M. We have noted that the integer quotient is zero when the numerator is less than the denominator. Hence, if the result is less than 60, the value of C will be zero and the last two instructions will accomplish nothing. However, if the result of the addition is 60 or more, the third instruction will give M a new value that is the remainder after division by 60. Also, in this case, the value of C is the number of extra hours to be carried over and added onto the value of H.

Flowchart

$SAX \leftarrow \lfloor \sqrt{|X|} + 0.5 \rfloor$

$HYP \leftarrow \sqrt{A \times A + B \times B}$

$X1 \leftarrow (-B + \sqrt{B \times B - 4 \times A \times C}) \div (2 \times A)$

$X2 \leftarrow (-B - \sqrt{B \times B - 4 \times A \times C}) \div (2 \times A)$

$DIST \leftarrow |A1 - A2|$

$Tabl \leftarrow \lfloor Age \rfloor$

$UX \leftarrow \lfloor X \rfloor - 10 \times \lfloor \lfloor X \rfloor \div 10 \rfloor$

Elapsed time:

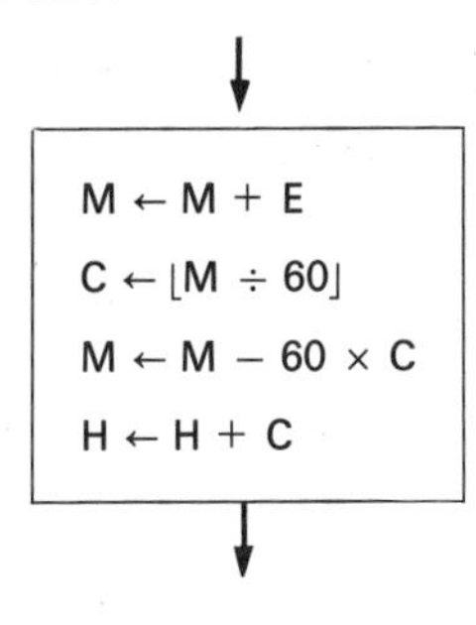

Basic The functions SQR, INT, and ABS in Basic produce the square root, integer part, and absolute value, respectively. The arguments of a function in Basic are enclosed in parentheses.

```
10  LET I1 = INT (SQR (ABS (X)) + 0.5)
20  LET H1 = SQR (A * A + B * B)
30  LET R = SQR (B * B - 4 * A * C)
35  LET X1 = (-B + R)/(2 * A)
40  LET X2 = (-B - R)/(2 * A)
50  LET D1 = ABS (A1 - A2)
60  LET T1 = INT (A1)
70  LET Q = INT (INT (X)/10)
75  LET U1 = INT (X) - 10 * Q
```

Elapsed time:

```
110 LET M = M + E
120 LET C = INT (M/60)
130 LET M = M - 60 * C
140 LET H = H + C
```

Fortran In Fortran, the arguments of a function are enclosed in parentheses and written after the function name. The type (real or integer) of a number must be taken into account when using functions in Fortran.

For real numbers, the Fortran functions SQRT, AINT, and ABS produce the square root, integer part, and absolute value, respectively. However, AINT, when applied to an argument whose value is negative but is not an exact integer, gives a slightly different result—the absolute value of the result is the integer part of the absolute value of the argument, making the function value closer to zero than the argument value. In other words, the function has the effect of deleting everything to the right of the decimal point.

For integers, the Fortran function IABS gives the absolute value.

There are two other important functions, FLOAT and INT, which perform *type conversions.* An expression used as the argument of FLOAT must be of integer type; the value of the function will be of real type. The function INT is applied to a real expression to give a value of integer type. In effect, the function INT first applies AINT and then deletes the decimal point. (This is true for both negative and positive argument values.)

In Fortran, when the quotient of two expressions of integer type is computed, the result must be an integer. Therefore truncation of the true quotient occurs. The value of the quotient I1 / I2 is the same as the value that would be obtained from the expression

```
INT (FLOAT (I1)/FLOAT (I2))
```

In other words, the decimal point and everything to its right has been deleted from the true quotient.

When one of the type conversion functions (INT or FLOAT) is used, the expression used as the argument will, of course, form a sub-expression of different type. ANSI Standard FORTRAN provides "implicit" type conversion automatically when the left and right sides of an assignment instruction are of different types. Many systems include implicit conversion of subexpressions as well. When the INT function is invoked implicitly in this way, truncation of course takes place. The novice programmer is advised to avoid this implicit conversion and to write the INT and FLOAT functions explicitly.

The remainder from integer division is given directly by the Fortran MOD function. Thus, instead of writing N − M * (N / M), we may use the shorter form, MOD (N, M) to obtain the remainder that results from integer division. Note that, even in the longer form, we do not write INT (N / M) in Fortran because truncation takes place automatically with division of integers. Furthermore, the argument of the INT function must be of real type.

If the value of N is less than that of M, the value of the integer quotient will be zero and the value of MOD (N, M) will be the same as the value of N.

```
      ISAX = INT (SQRT (ABS (X)) + 0.5)
      HYP = SQRT (A * A + B * B)
      X1 = (-B + SQRT (B * B
    1  - 4.0 * A * C))/(2.0 * A)
      X2 = (-B - SQRT (B * B
    1  - 4.0 * A * C))/(2.0 * A)
      DIST = ABS (A1 - A2)
      TABL = AINT (AGE)
      ITABL = INT (AGE)
      IUX = MOD (INT (X), 10)
```

Elapsed time:

```
M = M + IE
IC = M/60
M = MOD (M, 60)
IH = IH + IC
```

PL/1 The PL/1 functions SQRT, FLOOR, and ABS produce the square root, integer part, and absolute value, respectively. PL/1 also has a ROUND function. The arguments of a PL/1 function are enclosed in parentheses and written after the function name.

```
declare (ISAX, X, HYP, A, B, X1, X2, C,
   DIST, A1, A2, TABL, AGE, UX) float;
ISAX = ROUND (SQRT (ABS (X)));
HYP = SQRT (A * A + B * B);
X1 = (-B + SQRT (B * B
  - 4 * A * C))/(2 * A);
X2 = (-B - SQRT (B * B
  - 4 * A * C))/(2 * A);
DIST = ABS (A1 - A2);
TABL = FLOOR (AGE);
UX = MOD (FLOOR (X), 10);
```

The remainder from integer division is given directly by the MOD function. Thus, instead of writing N − M * FLOOR (N / M), we may use the shorter form, MOD (N, M) to obtain the remainder that results from integer division.

If the value of N is less than that of M, the value of the integer quotient will be zero and the value of MOD (N, M) is the same as the value of N.

Elapsed time:

```
M = M + E;
C = FLOOR (M/60);
M = MOD (M, 60);
H = H + C;
```

Words to Remember

function	absolute value	integer quotient
square root	argument	remainder
integer part	rounding	

Exercises

1. Write a program to read the value of X and print
   a. the tens digit;
   b. the tenths digit (the first digit to the right of the decimal point);
   c. a combination of the tens and hundreds digits—that is, it should produce the number 23 when 1234.5 is the value of X.
   Use the following values for X: (a) 4046.873; (b) 2150.42; (c) 283.1701; (d) 16387.162; (e) 35.314445.
2. A certain game is played on a board shaped like an equilateral triangle. The corners of the triangle are numbered 0, 1, and 2 in clockwise order. As part of the game, a marker is moved one step at a time around the board in the clockwise direction.
   a. Write a single expression that will give the result of a move from any corner to the next corner in the clockwise direction.
   b. Write a program to print a sequence of numbers, giving the position of the marker at each of the first seven steps, if it starts at corner 0.
3. Modify the expression in exercise 2a for the case in which the marker is moved counterclockwise.
4. Write a program to make change. Given an amount of money less than $1.00, find the number of half-dollars it contains; then find the number of quarters in the remainder; then the number of dimes, nickels, and pennies. (Hint: First convert the original amount to cents.) Data: (a) 1¢; (b) 99¢; (c) 50¢; (d) 34¢; (e) 55¢; (f) 51¢.

2.9 *What Happens to a Program after It Is Written?*

We have been emphasizing the difference between the dynamic *process* of a computation and the static *program* that we write to prescribe the process. The program, as we have seen, is composed of letters, digits, and other symbols written on paper. But so long as it is merely written on paper, the program cannot be executed. It must first be transmitted to the computer.

A program that exists only as a flowchart must be translated into a conventional programming language. As we have pointed out, flowcharts are excellent for communicating the steps of a calculation from one person to another, or to the same person at a later time; however, they are not suitable for transmitting instructions to the computer. After the program has been prepared according to the rules of the language, the instructions are punched into cards or paper tape, or in some way converted to a form that the computer can directly accept, and are entered into the computer.

The flow of information inside the computer is controlled by devices that can perform only very simple operations. These devices are not capable of directly interpreting and executing instructions written in a conventional programming language such as Basic, Fortran, or PL/1. Although such instructions are capable of being *transmitted* directly to the computer and are thus more machine-oriented than the steps of a flowchart, still (from the standpoint of the limited information handling devices inside the computer) the conventional programming languages are *human-oriented.*

In summary, the steps of a flowchart are intended to be written by humans and read by humans; the steps of a program written in a conventional programming language are to be written by humans and transmitted to machines. These latter instructions, before they can be executed, must be transformed into a totally machine-oriented language, the steps of which can be interpreted directly by the unsophisticated information-handling devices inside the computer.

Some Characteristics of a Machine-Oriented Language The machine-oriented form of a program differs from the human-oriented form in two major ways. First, because the information-handling devices and the arithmetic unit inside the computer are so limited, the instructions of the original program have to be subdivided into very elementary steps, each of which involves only a small machine operation, such as the transmission of a single number from a cell to a register or a single arithmetic step.

Furthermore, the symbolic cell names (variables and constants) used in the human-oriented version of the program cannot be interpreted by the cell selection mechanism. These devices must be told, in a more direct way, where to find the desired cell. For this reason, the cells inside the computer are numbered consecutively; the consecutive number of a cell is its *address.* During the transformation of an instruction into the internal machine language, the symbolic name of a cell must be replaced by this address, that is, by the consecutive position number of the cell among all the cells in the computer. (Thus, the address becomes the machine-language *name* of the cell.) The selection mechanism uses the address to identify a cell when it stores or fetches a number.

As an example, we will show how an assignment instruction involving a complicated expression might appear after conversion to machine language. Consider the assignment instruction

$$X \leftarrow (((A \div B) + (C \times D)) \div (E + (F \div G)))$$

First, cells must be allocated for each variable that appears in this instruction. Suppose that cells numbered 4221 to 4227 are allocated to the variables A through G, and that cell number 4228 is to be used for the variable X.

A single machine instruction is just powerful enough to do any one of the following:

fetch the contents of a cell;
store the contents of a register in a cell;
add—compute the contents of a register plus the contents of a cell;
subtract—compute the contents of a register minus the contents of a cell;
multiply—compute the contents of a register times the contents of a cell;
divide—compute the contents of a register divided by the contents of a cell.

To evaluate the complicated expression, we begin by calculating the value of the expression (A ÷ B). The following pair of machine instructions will prescribe that calculation:

```
1.          fetch cell 4221 to reg 1
2. reg 1  ÷  cell 4222 to reg 2
```

A single machine instruction never involves more than one cell, for that is the limit of the capacity of the cell selection mechanism. The hundreds or thousands of cells in a computer are relatively remote from its control and arithmetic operational portions, so it is a rather complicated task for the limited devices inside the computer to find a cell and transmit a number to it or from it. On the other hand, the registers are more central within the computer, making it a much simpler task to transmit numbers to and from the registers. Therefore, to perform an arithmetic operation involving two cells, two machine instructions must be used: the first fetches a number from one cell to a register, and the second combines the number from the other cell with the contents of that register. (The numbers that enter into an arithmetic operation are called the *operands*. Thus we may say that the computer cannot fetch both operands from cells; when there are two operands, one of them must come from a register.)

After execution of the first of the above machine instructions, the contents of the cell named A or 4221 has been brought to register 1. After the second instruction, this number has been divided by the contents of the cell named B or 4222, and the result has been stored in register 2. Remember that the numbers 4221 and 4222 are not the numbers stored in the cells named A and B; rather, these numbers are the addresses or machine-language names of the cells, corresponding to the human-oriented names A and B (Figure 2.9a).

The following machine instructions continue the calculation of the long expression:

```
3.          fetch cell 4223 to reg 1
4. reg 1  ×  cell 4224 to reg 3
5. reg 2  +  reg      3 to reg 1
```

Instruction 3 brings the value of C to register 1. Note that we want to be as economical as possible, because the number of registers is limited. The value of A, brought

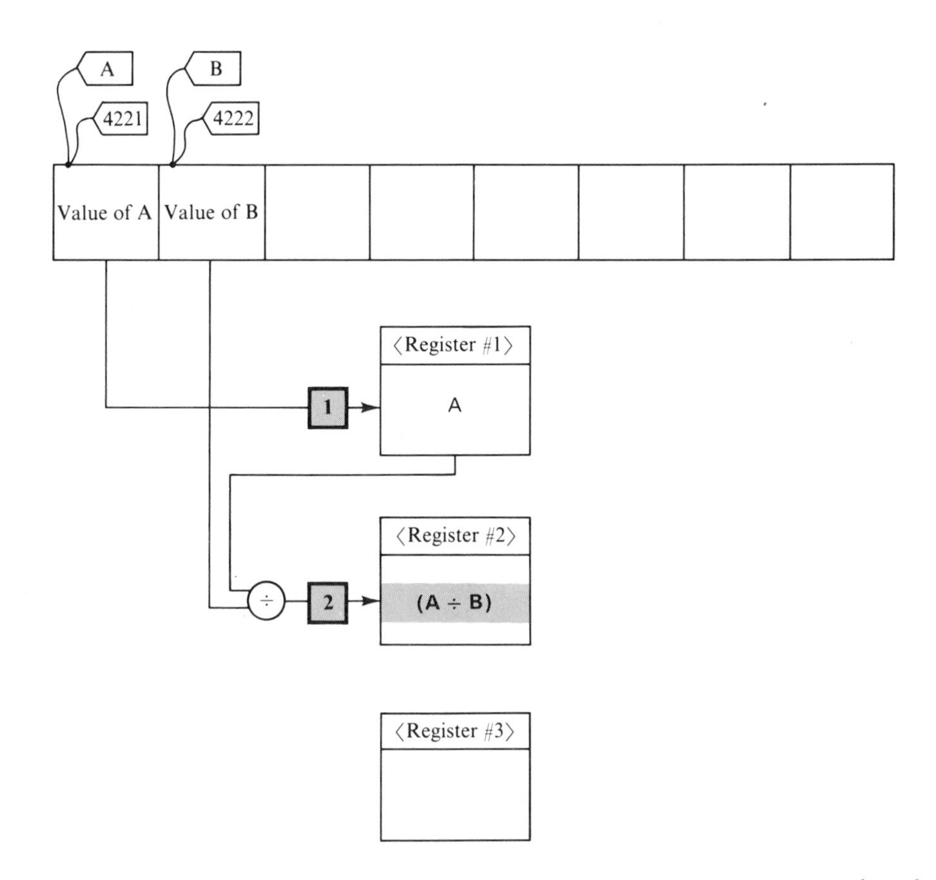

2.9a. *As a result of the execution of the first two machine-oriented instructions, the value of* (A ÷ B) *has been stored in Register #2.*

to register 1 by instruction 1, is no longer needed after (A ÷ B) has been calculated, so we use this register again for the value of C in instruction 3. Instruction 5 combines the values of two registers, rather than a register and a cell. Because a register is more accessible than a cell, it may be used at any place in an instruction where a cell would normally be used. However, it is not possible to use a cell where a register is expected (Figure 2.9b).

Continuing the calculation (Figure 2.9c):

```
 6.         fetch   cell 4226 to reg 2
 7. reg 2     ÷     cell 4227 to reg 3
 8. reg 3     +     cell 4225 to reg 2
 9. reg 1     ÷     reg       2 to reg 3
10. reg 3 store     cell 4228
```

After instruction 7, register 3 contains the value of the expression (F ÷ G). Instruction 8 adds the value of E to this, forming ((F ÷ G) + E). Since the operation is addition, it makes no difference which operand is in the register and which is brought from

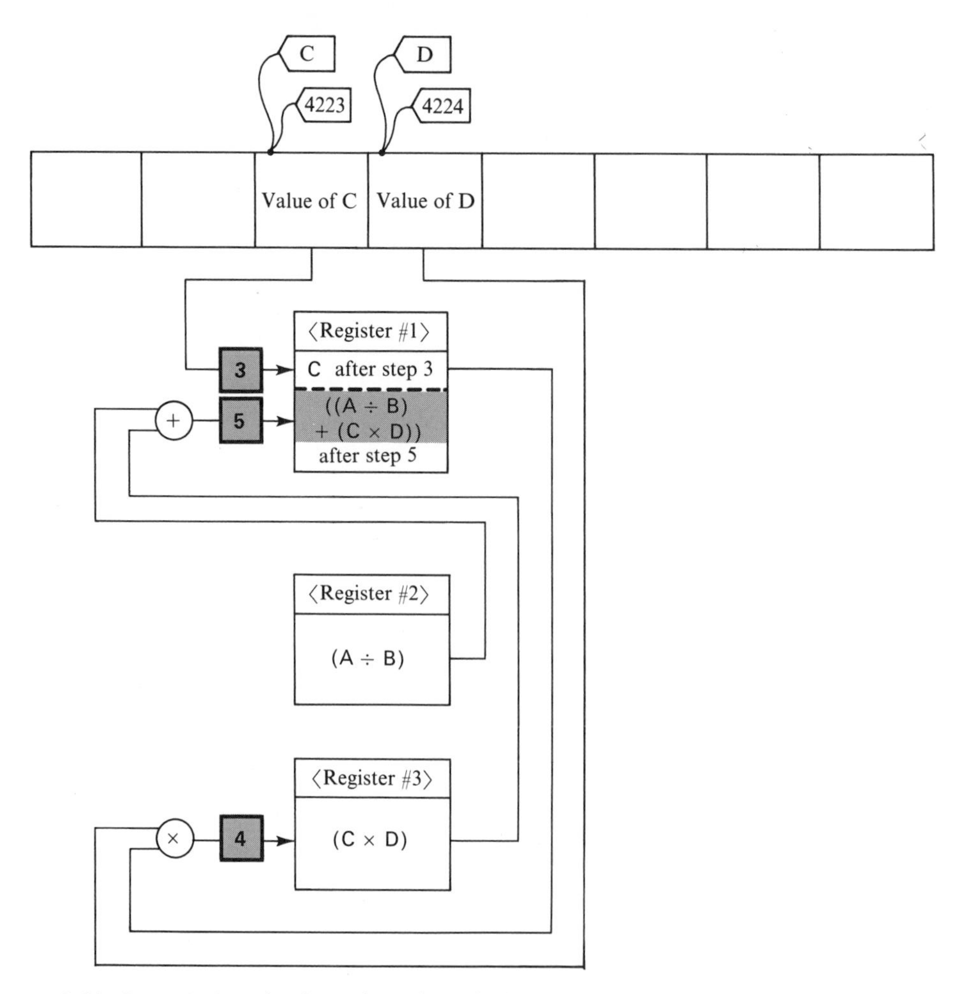

2.9b. *Steps 3, 4, and 5 form the value of* ((A ÷ B) + (C × D)) *in Register #1.*

the cell. For division or subtraction, on the other hand, it would not be possible to reverse the operands in this way. In such a case, it might be necessary to use one more register to perform the operations in the correct sequence. However, for addition, a change in the sequence of operations is permissible and results in the use of fewer registers, so we make the choice that will be most economical.

Reviewing these instructions as a group, we see that the most common pattern consists of an arithmetic operation involving first a register and second a cell, and that the result of the operation ends up in a register. Several variations to this pattern occur: the second operand in an arithmetic instruction may be a register instead of a cell; the fetch instruction omits the first operand; and the store instruction uses a cell as its destination.

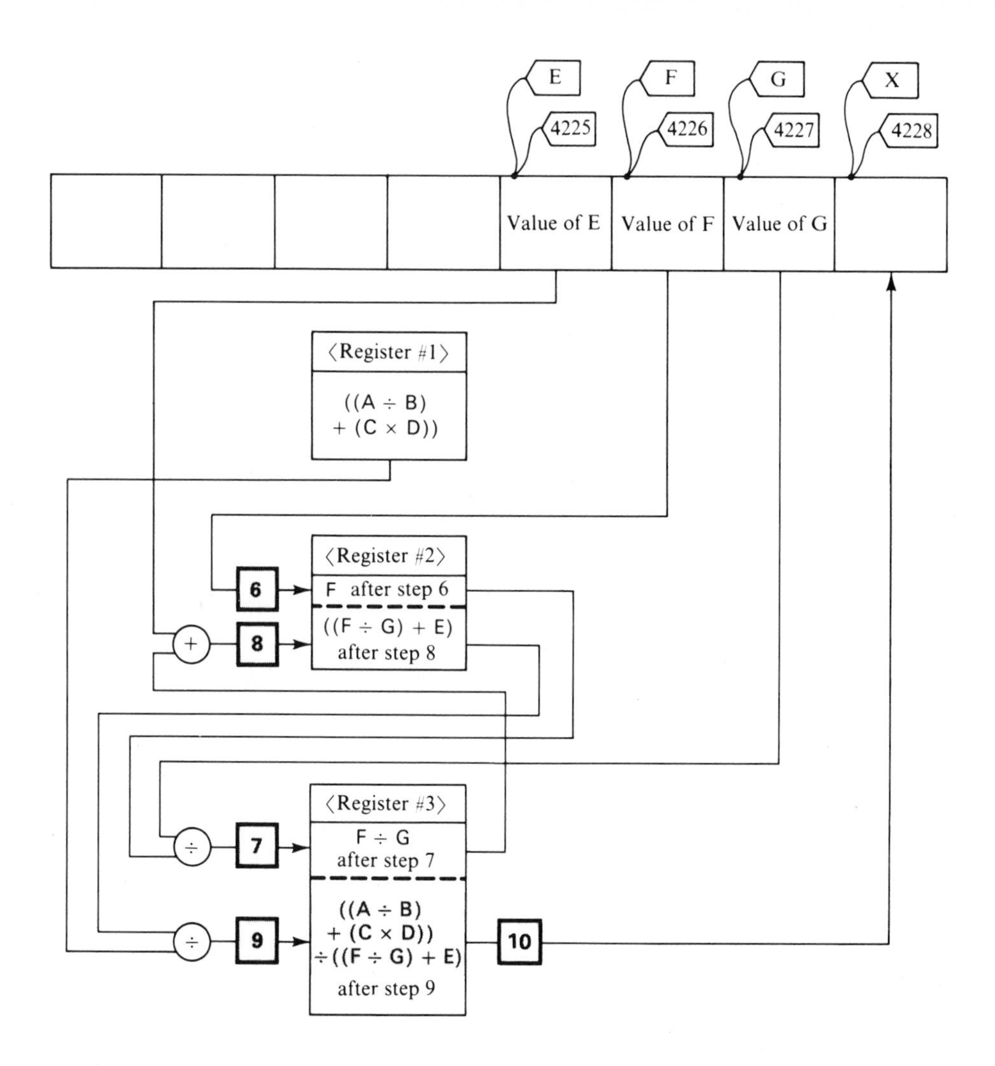

2.9c. *Steps 6 through 10 finish the evaluation of* ((A ÷ B) + (C × D)) ÷ (F ÷ G) + E) *and store the result at the cell named* X *or* 4228.

The Stored Program Concept Only one more step remains in the conversion of an instruction from human-oriented to machine-oriented form. To complete the dehumanization of the program, we must rewrite the machine instructions in completely numeric form. In that form, we can store the machine instructions *as numbers* inside the computer.[10]

---

[10] This concept seems to have been originated by Von Neumann, about 1949 (see Knuth, 1970).

The identification of each register requires a one-digit decimal number; cell addresses are four-digit numbers. It only remains to convert the operation code (fetch, store, add, or whatever) to a number. Every computer has some simple scheme of numeric operation codes such as the following:

fetch	1
add	2
subtract	3
multiply	4
divide	5
store	6

Using this code, we can convert each machine instruction to a seven-digit decimal number. Instruction 1, for example, becomes:

0 1 4221 2

The 1 in the second digit position is the operation code for fetch; the zero at the beginning is just a filler since a fetch instruction does not designate any register in this position. The four digits 4221 are the cell number, and the final 2 designates the destination register.

Thus, we have seen how to convert a human-oriented assignment instruction, involving a complicated expression, into machine-oriented form, by (a) subdividing the evaluation of the expression into very simple steps and (b) changing the symbolic names of cells to numeric cell addresses. The result of this complete dehumanization of the assignment instruction is the following list of ten seven-digit numbers:

1.	0	1	4221	2
2.	1	5	4222	2
3.	0	1	4223	2
4.	1	4	4224	3
5.	2	2	0003	1
6.	0	1	4226	2
7.	2	5	4227	3
8.	3	2	4225	2
9.	1	5	0002	3
10.	3	6	4228	0

These seven-digit numbers can be stored inside the computer. The simple devices that control the operation of the machine are just powerful enough to interpret these numbers as machine-language instructions and thus to execute the equivalent of the original human-oriented instruction.

The Compiler During the "dark ages" of computing (before about 1957), the transformation to internal machine language had to be performed manually by the programmer himself, or by an assistant, who was usually called a *coder*. However, well before 1957 *assembler programs* were used in the later stages of this task, to put in the numerical operation codes and even to substitute cell numbers for symbolic variables; the subdivision of expressions was still done by people up to that time. In effect, a coder was a slave to the machine—he converted expressions from the form desired by the human programmer into the form demanded by the machine. People eventually realized that this conversion task was just another manipulation of the sort that a machine could be instructed to perform (see Backus, 1957; Rosen, 1967; Sammet, 1969, especially pp. 1–5).

The instructions for performing this particular kind of manipulation, converting human-oriented instructions into machine-oriented instructions, are more sophisticated than those we have seen in the previous programming examples. The input data to be processed is a *source program* written in some human-oriented language such as Basic, Fortran, or PL/1. This data does not consist purely of numbers, but also includes letters and other symbols such as parentheses, plus signs, and decimal points. The results form the *object program* in numeric form, suitable for storing within the machine. This sophisticated program, which converts a source program into an object program, is called a *compiler*.

Two detailed features of compilers are worth mentioning. First, to keep track of the substitution of addresses for variables and constants (as cell names), the compiler uses a *symbol table* (Table 2.9). At the beginning of its operation, the compiler has a "pool" of cell numbers that it can allocate to variables and constants that appear in the source program. Whenever the compiler encounters a new variable or constant, it adds the symbolic name to the symbol table. The compiler also allocates a cell from the pool and enters the cell number in the table opposite the symbolic name. Thus it can locate this cell number again whenever the same variable or constant appears in the source program. For a constant, the compiler also ensures that the proper value will be stored in the cell before the object program is executed.

Table 2.9 *Symbol Table*

Variable or Constant (Human-Oriented Cell Name)	*Address (Machine-Oriented Cell Name)*
A	4221
B	4222
C	4223
D	4224
E	4225
F	4226
G	4227
X	4228

Another particularly interesting feature of the compiler is its *parser*. This is perhaps the most complicated part of the compiler. It consists, in effect, of a programmed version of the rules for writing an expression in the conventional programming language. It applies these rules to properly generate the machine instructions of the object program. An important auxiliary benefit occurs if the instruction cannot be recognized according to the rules describing the language. Then, the compiler

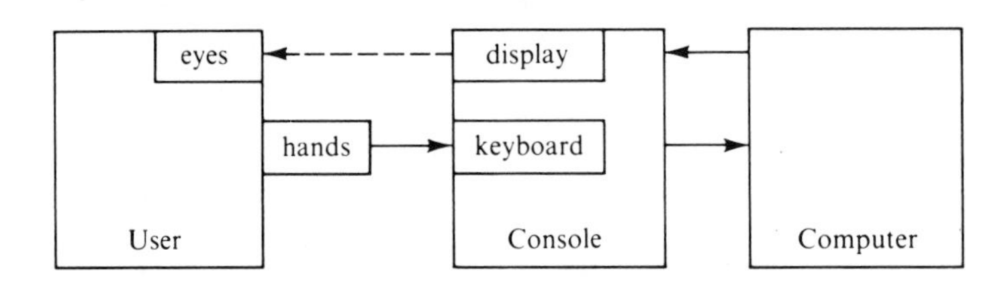

Flow of information in an
"on-line" computing system

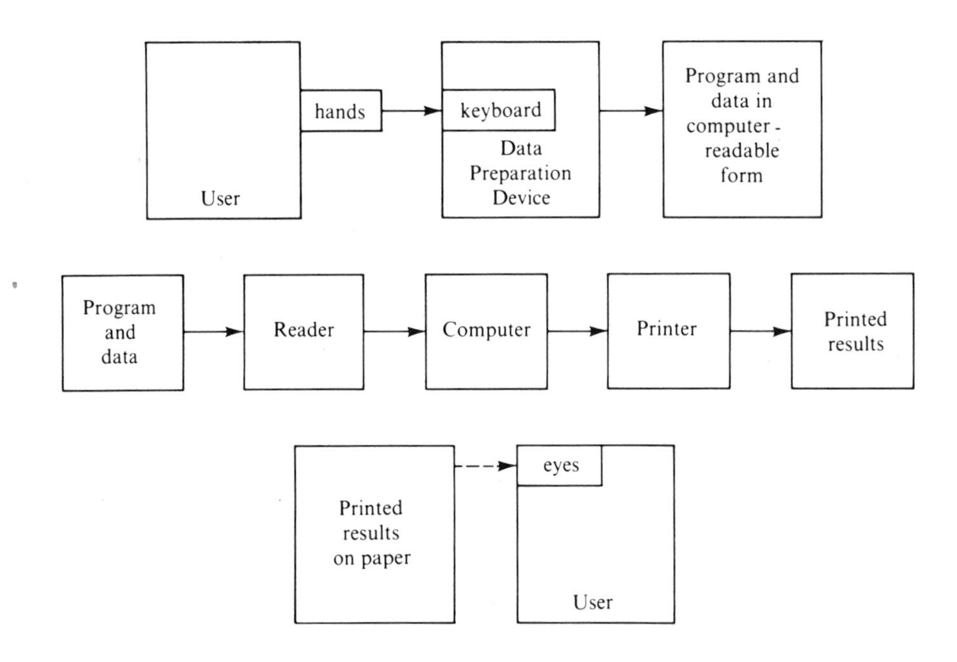

Flow of information in an
"off-line" computing system

2.9d. *On-line and off-line computing.*

has detected an error in the source program. In most such cases, the compiler will not generate an object program but will instead produce an error message for the programmer, giving as much information as possible about the location and nature of the error in the source program.

Preparing a Source Program Computers can be used in two quite different ways: *on-line* or *off-line.* These terms describe the relation of the user to the computer while his program is being compiled and executed (Figure 2.9d). When the user is on-line, he interacts with the computer. Typically, he sits at a typewriter-like console connected to the computer. Some very small computers have only one on-line user, but

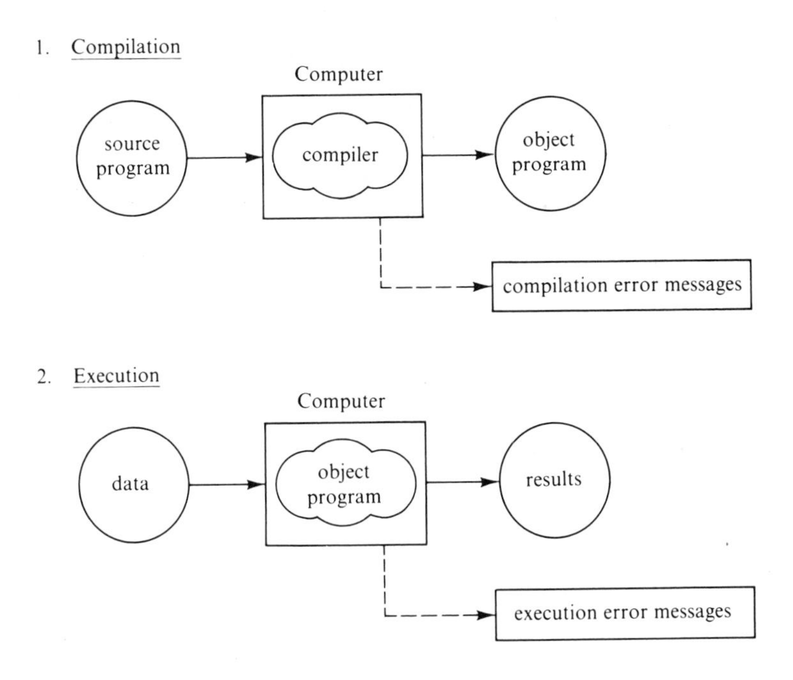

***2.9e.** Off-line computing systems usually perform compilation and execution in two separate stages.*

in most cases the computer is far faster than the user and is too expensive to be allowed to sit idle while waiting for him to complete his next entry. Thus, on-line computing is almost always time-shared computing—several people are using different terminals connected to the computer at the same time. Modern computers are fast enough to respond to all users so effectively that none of them notices any significant delay. On-line operation is usually associated with simple programs or those that must be modified frequently, or with jobs that intrinsically involve interaction with a user (such as computer-controlled drill in arithmetic). Large programs, during their developmental stages, may also benefit from on-line computer use.

Off-line operation, on the other hand, is used for larger programs (especially those in later stages of development) or for programs whose efficiency of execution is particularly important. The user of an off-line system prepares his program and his data carefully in advance, submits it to the computer, and goes away. Most off-line systems use punched cards as the input medium, both for the instructions of the program and for the data. The jobs enter a queue as they are submitted and are processed in turn as computer facilities become available. The results of the computation are written on a printer and returned to the user.

Compilers for on-line systems usually do very little work on the source program before execution begins. Most of the processing of each source program instruction is performed after the previous instruction has been executed. An instruction that is executed more than once must be compiled each time it is executed. Although this seems inefficient, it provides a close relation between the execution process and the instructions of the source program. When a source program error is detected, execution is suspended and a message to the user is printed on the console. Because the user is at the console, he can correct the error and try again. After the compiler has processed an instruction that requires input from the user, the computer executes this instruction. A signal appears on the console, indicating that the user should enter some data. After the data are entered, the computer compiles and executes the subsequent instructions, which tell how the data should be processed. Results are printed on the console as they are computed. Serious data errors cause a message to be printed at the console, after which execution is suspended pending further user action.

In off-line systems, compilation and execution usually take place in separate stages (Figure 2.9e). In the first stage, the compiler program resides in the computer and is executed. The compiler uses the source program as its data and produces the object program as its result. If the compiler detects any serious source program errors, it stops work on the object program and produces only the information necessary for the programmer to locate and correct the errors. If the compilation is successful, the object program will be loaded and executed, replacing the compiler as the resident program inside the computer. A copy of the object program may also be preserved, so that the same program can be executed at some other time without repeating the compilaton process. The object program uses the data prepared by the user along with the source program to produce results that are written on the printer and eventually delivered to the user. Serious data errors will stop execution of the object program with an indicative message. After normal or abnormal termination of execution, the computer facilities are released by the job so that they may be allocated for compiling and executing the next job in the queue.

We have described two different ways in which compilers are commonly used. A system that compiles and executes each source-language instruction individually (as most on-line systems do) is called an *interpreter*. The term *compiler* is often reserved for systems that generate the complete object program before starting to execute it (as most off-line systems do). It has been pointed out (Maurer, 1970; Glass, 1969) that many systems actually use a combination of these methods, modifying the whole source program to some extent before execution begins and finishing the job on one instruction at a time.

Flowchart We may include comments or remarks along with flowcharts, to explain the meaning or intent of certain steps, or for other purposes. Such information will be enclosed in a cloud-shaped outline and written alongside the regular boxes, out of the normal line of flow. For example:

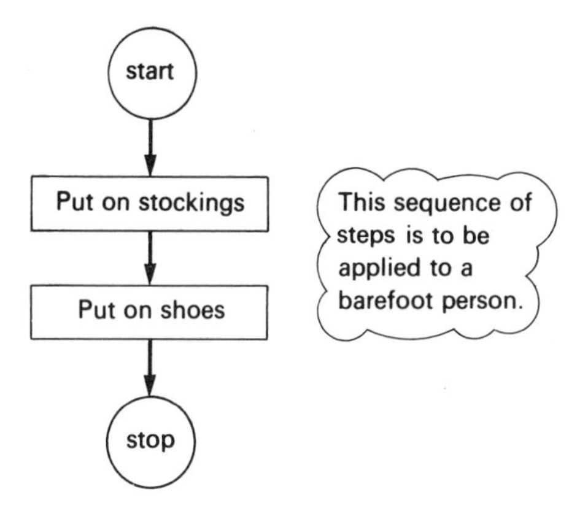

Basic Programs are usually entered in Basic by the user at a console connected directly to the computer. The user begins by typing some control and identification information, as required by the computer system he is using. The identification information is used for allocation of computer costs, and the control information is used to prepare the system to accept Basic instructions.

Then the instructions are typed, one per line, with line numbers. When all the instructions have been entered, the user indicates this fact by typing an END statement. This is just a line number followed by the word END. Then the user may type LIST to obtain a copy of the program as it is stored in the computer.

After examining the program, he may decide to insert more instructions. If he has used nonconsecutive line numbers (as has been done above) he can insert an instruction in the middle of the program by giving it a line number that will place the new instruction in the desired sequence. If he types LIST again, the program will be typed in order by line number. He can replace an instruction by simply typing a new line with the same number as the line to be replaced. Note that the END statement must have a line number that is larger than that of any instruction in the program.

When the user types RUN, the computer will begin executing the program. Compilers in Basic systems are (almost invariably) nearly pure interpreters, although some systems partially analyze the instructions as they are entered to detect certain kinds of program bugs. Other errors are often detected during execution. The run may end normally, or it may end when the compiler reaches an impasse due to a program bug, or it may be stopped by the user at the console. He may then run the program again immediately, or after listing it and replacing or inserting instructions.

A *remark* may be included anywhere among the instructions of a program. A line number, followed by REM and any desired sequence of characters, will be included in the program listing as an aid to the programmer, but will have no effect when the program is executed.

Blank spaces may be interspersed at will among the characters on any line, and will be ignored by the compiler. Thus, they should be included as desired to improve the readability of the instructions.

Fortran Fortran programs are usually prepared on an off-line key punch—one that is not directly connected to the computer. The key punch produces cards, each having 80 columns into which 80 characters may be punched. The instructions (and other statements such as format statements) are punched in the cards in a fixed format. The format number of a format statement is punched in columns 1–5. The main portion of a statement (the instruction or declaration) is punched in columns 7–72. Column 6 is blank except when a statement is so long that it must be continued on more than one card. Columns 73–80 are not used by the compiler.

Note that this fixed format applies only to the cards containing the statements of the program, which are read by the compiler. The data cards read by the program itself during execution are governed by the format specifications in the program (such as 8 F 10.0 or 8 I 10).

The cards comprising the source program, along with cards containing control and identification information (used to allocate computer costs, and to prepare the computer to compile and execute the program) and data cards to be read by the program during execution, are transmitted to the computer via a card reader. If the compiler detects any serious mistakes in the source program, execution is suppressed. The source program listing, with indications of mistakes found by the compiler, is returned to the user along with the results (if any) printed by the program during execution. The user may replace some of the source program cards or data cards and repeat the process.

A *comment* card may be inserted anywhere among the statements of a program. This is a card with a C in column one, and any desired sequence of characters in the remaining columns. Such a card will be included in the source program listing as an aid to the programmer, but will be ignored by the compiler.

Blank columns may be interspersed at will among the characters in columns 7–72 of any *program* card and will be ignored by the compiler. Thus, they should be included as desired to improve the readability of the instructions. Blank columns in a numeric field on a *data* card, however, will *not* be ignored but will be treated as zeros.

An END statement must be included after the last statement of the program. This is punched on a card with the characters END beginning in column 7.

PL/1 This language is used mostly in off-line systems, in which the user does not interact directly with the computer but prepares the instructions and data ahead of time on punched cards. The instructions (and other statements, such as declarations) are punched in columns 2–72, and are separated by semicolons. It is not necessary to use a separate card for each statement, or to insert any special indication when a statement continues from one card to the next.

A complete program must have a *name*. A program name consists of a letter followed by a string of letters or digits. In contrast to a variable, however, a program name is limited in length to seven characters. This name appears at the beginning of the program, and is followed by a colon and then by the words procedure options (main);. The last statement of the program consists of the word end followed by the program name and, of course, a semicolon.

The data is also punched into cards in columns 1–80. The get list instruction will read data items that are separated from each other by a comma or blank.

The cards comprising the source program, along with cards containing control and identification information (used to allocate computer costs and to prepare the computer to compile and execute the program) and data cards to be read by the program during execution, are transmitted to the computer via a card reader. If the compiler detects any serious mistakes in the source program, execution is suppressed. The source program listing, with indications of mistakes found by the compiler, is returned to the user along with the results (if any) printed by the program during execution. The user may replace some of the source program cards or data cards and repeat the process.

Blanks may be interspersed at will among the characters of the source program, except that variables, constants, and other groups of characters considered as a unit by the compiler may not contain imbedded blanks.

A comment, which is a string of characters beginning with the character pair /* and ending with the pair */, may be inserted at any point in the source program where a blank is permitted. A comment will be treated as a blank by the compiler and hence will have no effect except to be included in the source program listing for the convenience of the programmer.

Words to Remember

machine-oriented language	source program	on-line
address	object program	off-line
operation code	symbol table	interpreter
compiler	parser	operand

Exercise

Using the assumptions applied in the example in this section, write a sequence of machine instructions corresponding to the assignment instruction

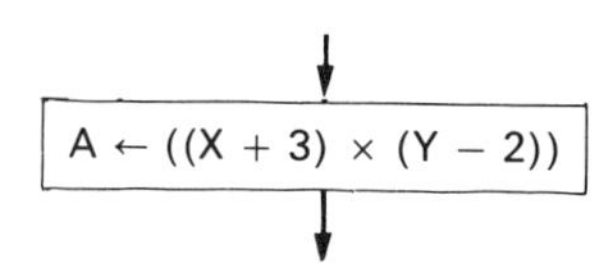

given the symbol table:

Variable or Constant	*Address*
X	4763
Y	4764
A	4765
3	4766
2	4767

Assume that the values of the variables X and Y and of the constants 3 and 2 have already been stored in the proper cells.

2.10 How Can the Rules of a Programming Language Be Stated More Formally?

A programmer soon learns that the computer does what he tells it to do and not necessarily what he wants it to do. For instance, while designing an expression for computing

$$\frac{A}{B - C}$$

the programmer may write A ÷ B − C, which is interpreted by the compiler as (A ÷ B) − C. The computer, having no information about the programmer's intentions, must carry out his instructions according to its rules as implemented in the compiler. If chaos is to be avoided, these rules must be rigid—they must be capable of producing the same computation each time the compiler is presented with the same sequence of instructions.

The rules must also be explicit. In other words, the programmer should have access to a detailed and unambiguous statement of the rules. This will make it possible for him to tell exactly how to write an instruction specifying a desired task, and to tell exactly how an instruction that he writes will be executed.

A compiler itself provides a complete specification of the rules of the programming language it will accept. In fact, for about ten years the only complete definition of Fortran in existence was in this form, and users had to learn the more obscure rules from experience. At the other extreme, we could write a rather long description in ordinary language of the rules followed by a compiler. The document defining Standard Fortran (ANSI, 1966) is such a description. Between these extremes, and often more useful than either, are the formal language descriptions, which are more accessible than the compiler itself and more precise than ordinary language. The archetypal formal language description is the Algol-60 report (Naur, 1963). Formal descriptions are used in the development of new compilers; in fact, it is possible to produce a formal language description in machine-readable form, and thence to produce a compiler automatically (see Gries, 1971).

Let us see how to make some practical use of a formal description of one small portion of a language. If we want to answer the question, "Are the parentheses in the expression (B + C) × D optional, or are they required for the expression to be interpreted correctly?" we must discover whether removal of the parentheses will have any effect. In other words, we must compare the compiler's interpretation of the two expressions (B + C) × D and B + C × D.

The compiler will try to analyze each of these expressions by matching it against its list of rules. If the compiler finds a way of subdividing the expression in accordance with the rules, it can then proceed to generate the corresponding machine-oriented instructions; otherwise, it will have detected an error. A typical set of rules for analyzing an expression follows. (The concepts *variable* and *constant* are supposed to have been formally defined already, by rules not listed here; the intermediate concepts *term*, *factor*, and *simple expression*, are introduced as aids to the definition of *expression*.)

Rule 1. An *expression* is a *term* optionally preceded by −, and followed by any number of cases of + or − followed by another *term*.
Rule 2. A *term* is a *factor* followed by any number of cases of × or ÷ followed by another *factor*.
Rule 3. A *factor* is a *simple expression* or an *expression* enclosed in parentheses.
Rule 4. A *simple expression* is a *variable* or a *constant*.
In Rules 1 and 2, "any number of cases" includes zero cases. Thus an expression may consist of a single term, and a term may consist of a single factor (see Gries, 1971, p. 269).

If we analyze the expression B + C × D according to these rules, we see that it consists of two terms: B and C × D; the term B consists of a single factor which happens to be a simple expression, and the term C × D contains two factors, each of which is a simple expression. On the other hand, the expression (B + C) × D consists of a single term composed of two factors, (B + C) and D; the factor (B + C) is an expression enclosed in parentheses, and the factor D is a simple expression.

It follows that the parentheses in the expression (B + C) × D are required and are not optional; their removal will change the compiler's interpretation of the expression. If the parentheses are removed, C will be multiplied by D but B will not be, while if the parentheses are retained the sum (B + C) will be multiplied by D.

This analysis can be arranged in the following tabular form:

```
B + C × D        (B + C) × D

t + t            t
f   f × f        f        × f
s   s   s        (e     )   s
v   v   v         t + t     v
                  f   f
                  s   s
                  v   v
```

In this analysis, we first write the expression at the top. We then apply Rule 1; thus on the second line we must have one or more terms separated by + or − signs. On the third line, we use Rule 2, analyzing each term into factors. On the fourth line, each factor is reduced to a simple expression or an expression enclosed in parentheses, in accordance with Rule 3. Thus if the original expression contains parentheses, they will reappear on the fourth line.

When an expression enclosed in parentheses appears, we start applying the rules to the expression inside the parentheses, just as we applied them to the entire expression originally. The fifth line shows that the parenthetical expression consists of two terms, and the following lines indicate that each of these terms is a factor, which is a simple expression, which is a variable.

Any expression constructed according to these rules can be analyzed in an analogous way and reduced to its elementary components, which are either variables or constants.

Basic, Fortran, and PL/1 Rules 1 through 4 apply to all three languages, except for some slight differences in typography.

In Fortran, variables of a single type (all real or all integer) should be used throughout an expression.

In Basic, Fortran, and PL/1, multiplication is indicated by * and division by /. All three of these languages also provide another arithmetic operation, exponentiation, raising an expression to a power denoted by the value of another expression.

In Basic, exponentiation is denoted by an "up arrow":

expression ↑ *expression*

Fortran and PL/1 use a double asterisk to denote exponentiation:

expression ** *expression*

In Fortran, a real expression may be raised to either a *real* or an *integer* power, but an integer expression may be used only with an integer exponent.

All languages are, of course, subject to mathematical restrictions on exponentiation. For instance, a negative value raised to a noninteger power does not make much sense mathematically.

We can modify Rule 3 to cover exponentiation. In stating this rule, we use ↑; for Fortran and PL/1, ** would, of course, be used instead.

Rule 3a. A *factor* is a *primary* followed by any number of cases of ↑ followed by another *primary*.

Rule 3b. A *primary* is a *simple expression*, or an *expression* enclosed in parentheses.

The effect of these rules is to interpret the expressions A + B ↑ C and D × E ↑ F as if they were written A + (B ↑ C) and D × (E ↑ F). In the expression X ↑ Y ↑ Z, application of Rule 3a requires that we consider Z as the exponent since Y ↑ Z cannot be a primary. Hence, this expression is equivalent to (X ↑ Y) ↑ Z, contrary to the usual mathematical interpretation. We may group the subexpressions in other ways by using parentheses. In the expression X ↑ (Y ↑ Z) the subexpression (Y ↑ Z) may be a primary since it is a factor (hence a term and an expression) enclosed in parentheses.

In this text we include very few examples of exponentiation, partly to increase the nonnumerical emphasis. It should be pointed out that polynomial evaluation is possible and, in fact, more efficient without the use of exponentiation:

A + X × (B + X × (C + X × (...)))

is a much better way, from the computing standpoint, to write a polynomial. For example, instead of

1 + 2 × X + 3 × X ↑ 2 + 4 × X ↑ 3 + 5 × X ↑ 4

it would be better to write

1 + X × (2 + X × (3 + X × (4 + X × 5)))

Words to Remember

factor term simple expression

Exercises

1. Analyze each of the following expressions according to Rules 1 through 4 above.
   a. A + 3 b. −A + B c. −A × B d. 3.1416 × R × R
2. Analyze the following pairs of expressions. In each case, tell whether the two expressions in the pair have the same meaning.
   a. A − (B + C), A − B + C b. (A − B) + C, A − B + C
   c. (B + C) × D, B + C × D d. B + (C × D), B + C × D
3. Analyze the expression A × (−B). Explain where Rules 1 through 4 fail when we try to apply them to the expression A × − B.
4. Analyze the expression ((A ÷ B) + (C × D)) ÷ (E + (F ÷ G)).
5. Rewrite Rules 1 through 4 for Basic, Fortran, or PL/1, making the necessary typographic changes. For Fortran, define *simple real expression* and *simple integer expression*, and use these to define *real factor*, *real term*, *integer factor*, *integer term*, etc.
6. What is the value of the expression X ↑ 0.5 when X has the value −1?
7. Modify the rules defining an *expression* to include *functions*.
8. Rewrite the rules defining an *expression* to take into account the type conversion functions of Fortran. First define *simple real expression*, *real factor*, *real term*, and *real expression* in terms of *real constant* and *real variable*; do the same thing for the corresponding integer concepts. Introduce also the fact that a real factor may have the form FLOAT (*integer expression*) and an integer factor may have the form INT (*real expression*).

References

American National Standards Institute, *ANSI Standard FORTRAN*, ANSI X3.9-1966. New York: American National Standards Institute, 1966.

American National Standards Institute, *Standard Flowchart Symbols and Their Use in Information Processing*, ANSI X3.5-1970. New York: American National Standards Institute, 1970.

Backus, J. W., et al., "The FORTRAN Automatic Coding System," *Proceedings of the Western Joint Computer Conference*, Vol. II. New York: Institute of Radio Engineers, 1957. Pp. 188–198.

Bohl, M., *Flowcharting Techniques*. Chicago: Science Research Associates, 1971.

Chapin, N., "Flowcharting with the ANSI Standard: A Tutorial." *Computing Surveys* 2: 119–146 (Jun 1970).

Glass, R. L., "An Elementary Discussion of Compiler and Interpreter Writing." *Computing Surveys* 1: 55–77 (Mar 1969).

Gries, D., *Compiler Construction for Digital Computers*. New York: Wiley, 1971.

The International Atlas. Chicago: Rand McNally, 1969.

Knuth, D. E., "Von Neumann's First Computer Program," *Computing Surveys* 2: 247–260 (Dec 1970).

Maurer, W. D., "Generalized Interpretation and Compilation," *Software Engineering*, Vol. I. New York: Academic Press, 1970. Pp. 139–150.

Maurer, W. D., *Programming: An Introduction to Computer Techniques*. San Francisco: Holden-Day, 1972.

Naur, P., ed., "Revised Report on the Algorithmic Language ALGOL 60." *Communications of the Association for Computing Machinery* 6: 1–17 (Jan 1963).

Rosen, S., ed., *Programming Systems and Languages*. New York: McGraw-Hill, 1967.

Sammet, J. E., *Programming Languages: History and Fundamentals*. Englewood Cliffs, N.J.: Prentice-Hall, 1969.

Tarski, A., *Introduction to Logic and to the Methodology of Deductive Sciences*. New York: Oxford University Press, 1965.

Turnbull, H. W., *The Great Mathematicians*. London: Methuen, 1951.

3 Algorithms and Program Structures

3.1 What Is an Algorithm?

Many different definitions and descriptions of the word *algorithm* have been proposed in recent years. Some are:

"A set of operations reduced to a uniform procedure for solving a specific type of problem." (*Encyclopaedia Britannica*, 1970, Vol. I, p. 630)

"An exact prescription, defining a computational process, leading from various initial data to the desired result." (Markov, 1961, p. 1)

"A list of instructions specifying a sequence of operations which will give the answer to any problem of a given type." (Trakhtenbrot, 1963, p. 3)

"A list of instructions for carrying out some process step by step." (Forsythe et al., 1969, p. 3)

"Any unambiguous plan telling how to carry out a process in a finite number of steps." (School Mathematics Study Group, 1965, p. 25)

"An unambiguous, complete procedure for solving a specified problem in a finite number of steps." (Maisel, 1969, p. 60)

"Any sequence of unambiguously defined steps leading to the solution of a given problem." (Cole, 1969, p. 57)

"A step-by-step problem solving procedure that can be carried out on a machine." (Hull and Day, 1970, p. 3)

"A procedure which is utilized to solve a problem in a finite number of steps where the procedure consists of an ordered set of unambiguous rules." (Walker and Cotterman, 1970, p. 9)

"A list of instructions for such a [completely deterministic] machine or mechanical process." (Rice and Rice, 1969, p. 20)

"A procedure that a particular computer can perform." (Sterling and Pollack, 1970, p. 1)

"A prescribed set of well-defined rules or processes for the solution of a problem in a finite number of steps." (American National Standards Institute, 1966)

"A computational method which terminates in finitely many steps." (Knuth, 1968, p. 8)

This last quotation is part of a longer and more formal definition, which we shall discuss. The complete definition is based on the concepts of *state* and *transition*.

States and Transitions A woman goes to a diagnostic clinic for a complete physical checkup to learn the *state* of her health. Every facet of her condition that can be measured is recorded on a chart. Her physician reviews the recorded data, makes other observations of the patient, and decides on a course of action. He may declare that the patient's health is satisfactory and that she needs no further treatment. But if he decides that the state of her health is unsatisfactory, he sets out to improve it. From the possible courses of action he selects the *one* that he thinks will do the most good and applies it to the patient. Then he asks her to return in, say, one month. At that time, he tries to find out what new state of health has been produced by the treatment, and he again decides whether to continue the treatment, to modify it, or to release the patient from his immediate care. The patient finally gets well or dies, and in either case she reaches a state in which the doctor (for a time at least) is no longer concerned about the state of her health.

Or consider a twelve-year-old-boy who is putting together a jigsaw puzzle. We can describe the state of the puzzle at a particular time by telling which pieces have been put into their proper positions. The boy changes the state every time he fits a new piece into its place. He has a number of options available to him at each stage. He chooses *one* of the available options and thus makes a *transition* to a new state. Then he chooses again, making a series of transitions until some final state is reached. This occurs when he either completes the puzzle or gives up, takes it apart again, and puts the whole thing away.

The "system" of a jigsaw puzzle or of a patient in a health care center resembles a computer system in some ways. Each system possesses a number of states, and from some of these states one or more transitions can be made to other states. Also, some states can be recognized as potential final states, meaning that one possible option is to leave the system in such a state indefinitely.

In each example above, the number of possible states is rather large. Furthermore, the "systems" have few systematic features, so it seems difficult to reduce their apparent complexity or to adequately analyze them. On the other hand, some very elegant and rather simple systems arise in mathematical recreations (see Ball, 1960; Bellman, Cooke, and Lockett, 1970; Banerji, 1969). These problems often have an underlying pattern that makes it possible to extend the analysis of simple cases to more complex related puzzles. As an example, we shall consider in detail the Tower of Hanoi puzzle (Ball, 1960, p. 303).

> In the great temple at Benares, beneath the dome which marks the center of the world, rests a brass plate in which are fixed three diamond needles, each a cubit high and as thick as the body of a bee. On one of these needles, at the creation, God placed sixty-four discs of pure gold, the largest disc resting on the brass plate, and the others getting smaller up to the top one. This is the tower of Bramah. Day and night unceasingly the priests transfer the discs from one diamond needle to another according to the fixed and immutable laws of Bramah, which require

> that the priest on duty must not move more than one disc at a time and that he must place this disc on a needle so that there is no smaller disc below it. When the sixty-four discs shall have been thus transferred from the needle on which at Creation God placed them to one of the other needles, tower, temple, and Bramins alike will crumble into dust, and with a thunderclap the world will vanish.

Fortunately, most systems do not have such a dramatic final state!

Rather than considering all the possible states of the system of 64 discs, we shall examine a simple puzzle with three pegs and two discs. We start with both discs on the peg shown at the upper part of Figure 3.1a. The problem is to move both discs to either of the other pegs, moving one disc at a time and never placing the large disc on top of the small one.

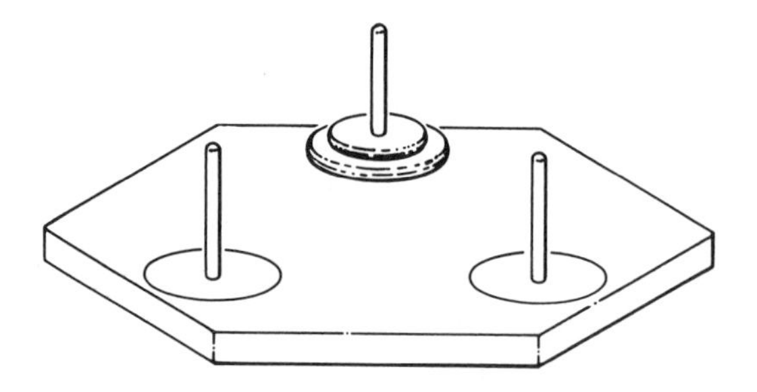

3.1a. *Tower of Hanoi puzzle with two discs.*

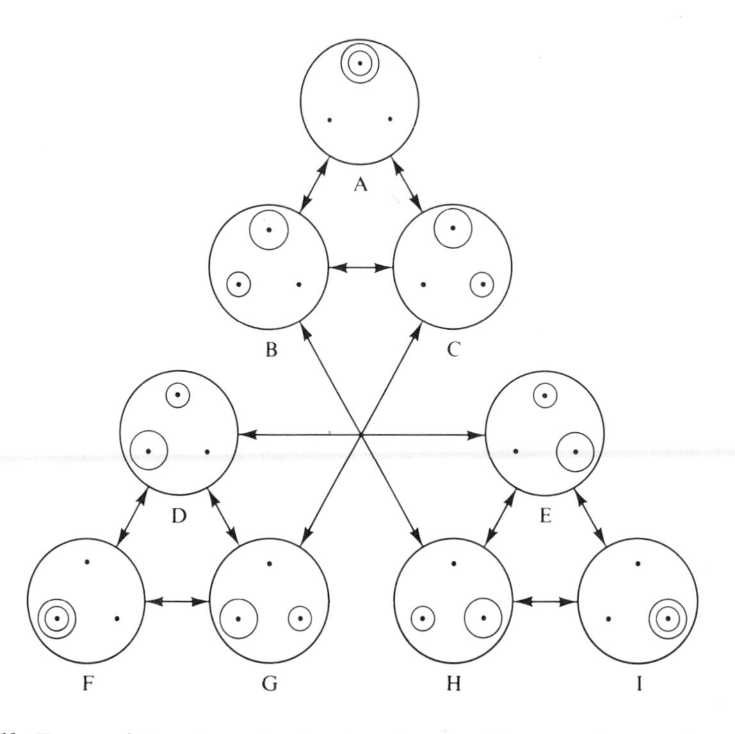

3.1b. *Tower of Hanoi puzzle, showing nine possible states for two discs.*

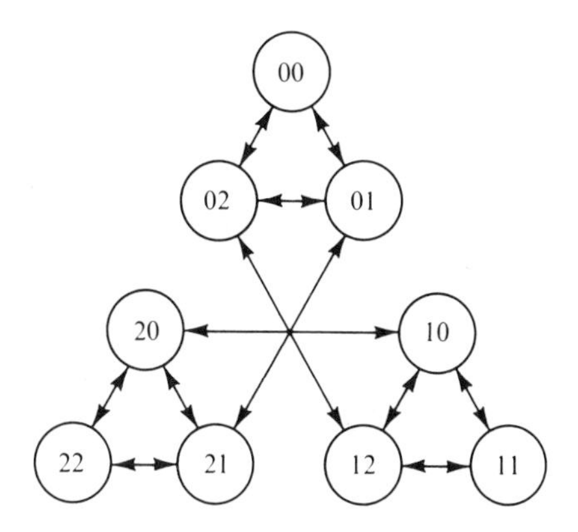

3.1c. *Tower of Hanoi puzzle with two discs, showing a numeric representation of the states. The digit on the right indicates the position of the small disc; the digit on the left indicates the position of the large disc.*

We distinguish nine different arrangements of the two discs on the three pegs (observing the rule that the large disc must not cover the small one): thus we have nine *states* of the system of two discs and three pegs. These nine states are designated by letters A through I in Figure 3.1b and are systematically numbered in Figure 3.1c.

The problem is to make a sequence of transitions leading from state A either to state F or to state I. The arrows show transitions that we can make if we move only one disc at a time. (A curious feature of this diagram is that the short arrows represent moves of the small disc and the long arrows represent moves of the large disc; furthermore, the direction of the arrow is the same as the direction in which the disc moves.) The shortest path from A to I is the sequence of transitions A, B, H, I. What is the shortest path from A to F?

Algorithms Let us consider some system with a finite number of states and with specific permissible transitions from some states to other states. Let us also suppose that there are particular states where we may remain without making any transition; that is, it is permissible to make a *null* transition from one of these states to itself.

For instance, the two-disc Tower of Hanoi puzzle (see Figure 3.1c) has nine states. From each of the three states in the corners of the diagram (states 00, 11, and 22), we can choose between two permissible transitions without violating the rules; from each of the other six states there is a choice of three permissible transitions. If we agree not to give up and abandon the puzzle until we have both discs on the same peg, then we say that the null transition is permissible only from each of the three corner states.

In the general case, one state is designated as the *initial state*, and a *terminal state* is chosen from among those states from which the null transition is permissible. For example, 00 is the initial state, and we may choose 11 as the terminal state in the two-disc Tower of Hanoi problem.

From the initial state we must construct a *path* leading to the terminal state. To construct a path means to designate exactly *one* of the permissible transitions from each state lying on the path. From the terminal state, the chosen transition must be the null transition, so that once a path reaches the terminal state it never leaves that state. For example, we might choose the sequence of transitions leading from state 00 to 02, and thence to 12 and finally to 11. It happens that this sequence obeys the rule that from each state we choose the transition to the next state with the highest possible number. It is not necessary to have any such rule in mind, however. Such a path, leading from the initial state to the terminal state, is an *algorithm.*

We must generalize this definition a bit more, however, for cases in which there is more than one initial state or more than one terminal state. In this more general case, instead of a single path we will have a *network* of paths, leading from each initial state to some terminal state. The same requirement holds as before, that we must designate exactly one transition from each state lying on any of the paths. However, it may happen that transitions from two different states lead to the same state. That is, two paths may join and continue as one path, but a single path may not fork into two separate paths. For example, in Figure 3.1d we designate 01, 10, 20, and 21 as additional states, and 22 as another terminal state. Again choosing the transition to the next state with the highest possible number, we obtain the following additional paths:

01, 21, 22
10, 20, 22
20, 22
21, 22

The requirements that paths may not fork and that each path must lead to a terminal state guarantee that no "loops" may occur in the network of paths. Because

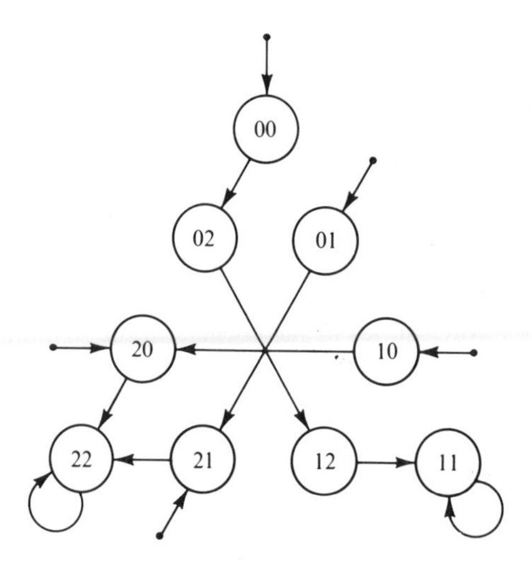

3.1d. *Algorithm for the Tower of Hanoi with two discs. In this algorithm there are five initial states and two terminal states.*

the total number of states is finite, the absence of loops insures that each path is finite in length. That is, *an algorithm must actually lead to a solution* after a finite number of steps from any initial state. (Compare Knuth, 1968, pp. 1–9; Minsky, 1967, p. 153; Birkhoff and Bartree, 1970, pp. 63–95.)

Words to Remember

state	initial state	path
transition	terminal state	algorithm
null transition		

Exercises

1. Remove state 11 from the list of terminal states (see Figure 3.1e). From each state on a path, select the transition that goes to the next state with the highest possible number. Do the resulting paths satisfy the requirements for an algorithm? Explain.

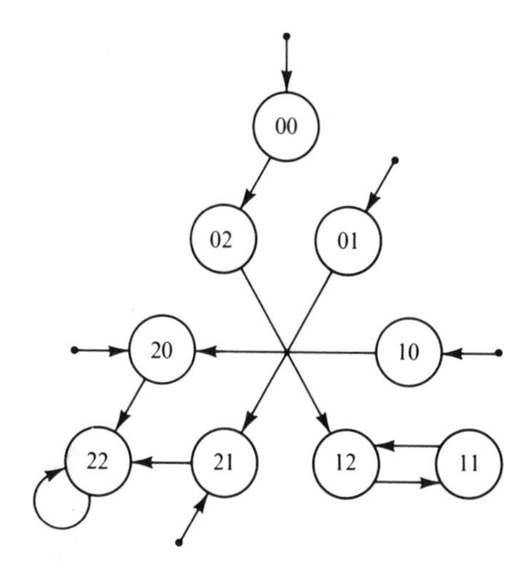

3.1e. *Diagram with five initial states and one terminal state.*

2. After removing state 11 from the list of terminal states, also remove 00 from the list of initial states (see Figure 3.1f). If we now construct paths in the same way as for exercise 1, will we obtain an algorithm? Explain.
3. Consider the version of the Hanoi puzzle shown in Figure 3.1g, with three discs and 27 states. Let 000 be an initial state and 111 and 222 terminal states.
   a. As before, from any nonterminal state select a permissible transition that goes next to the state with the highest possible number. Show that this rule does not produce an algorithm.
   b. Find a path from 000 to 222.

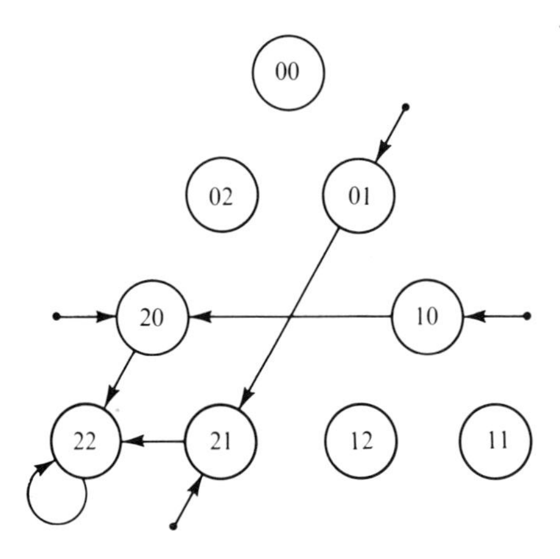

3.1f. *Diagram with four initial states and one terminal state.*

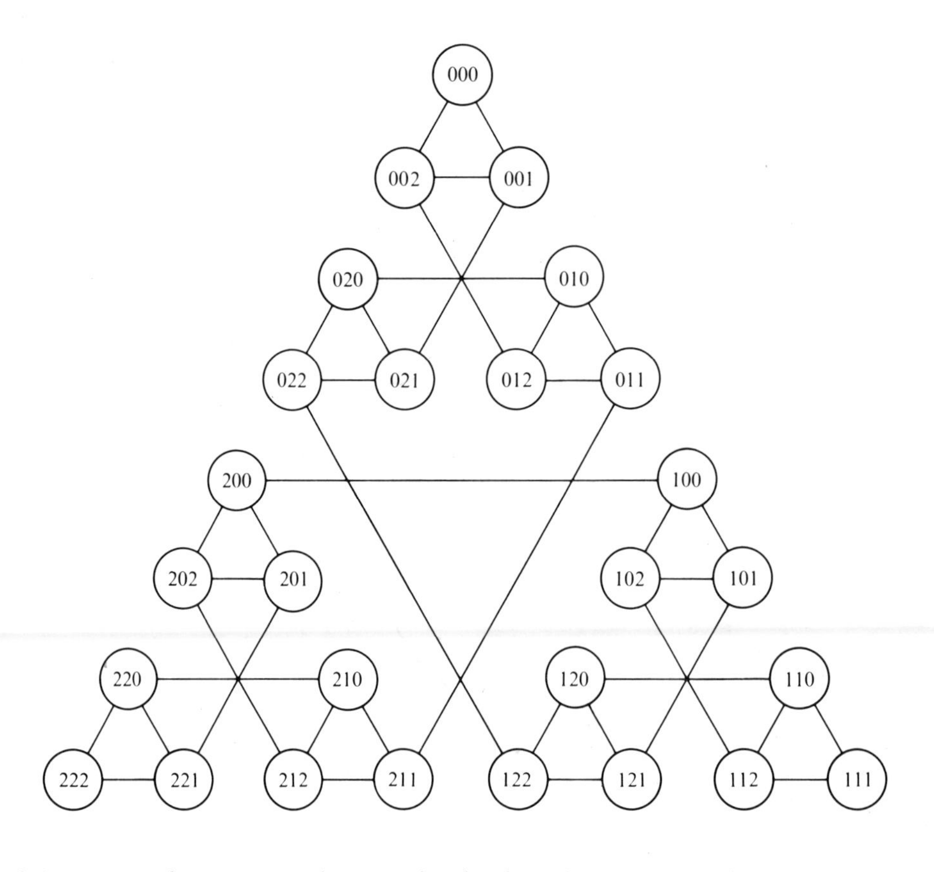

3.1g. *Tower of Hanoi state diagram, for the three-disc version of the puzzle. Can you find the path from 000 to 222 with only seven transitions?*

c. Can you find a *systematic pattern* of states forming a path from 000 to 222? If so, describe the pattern in words. Then modify the pattern to find an algorithm leading to the terminal state 111.

d. Construct a model of the Tower of Hanoi puzzle with three discs. You need not use discs and pegs: you can use three cards labeled 1, 2, and 3, and three saucers, or squares drawn on paper, for the pegs. Try to move the "discs" from one "peg" to another without placing a larger number on top of a smaller one. Relate your moves to the algorithm in parts (b) and (c) above.

e. Try to do the same thing with four, five, or more discs. Can you find a systematic pattern?

4. Figure 3.1h shows a skeleton of the Tower of Hanoi state diagram for five discs. Number the states so that 00000 is at the top, 11111 is at the lower right, and 22222 is at the lower left. Find the shortest path from 00000 to 22222. (It can be done in 31 transitions.) Translate this state-diagram solution to moves in the physical puzzle.

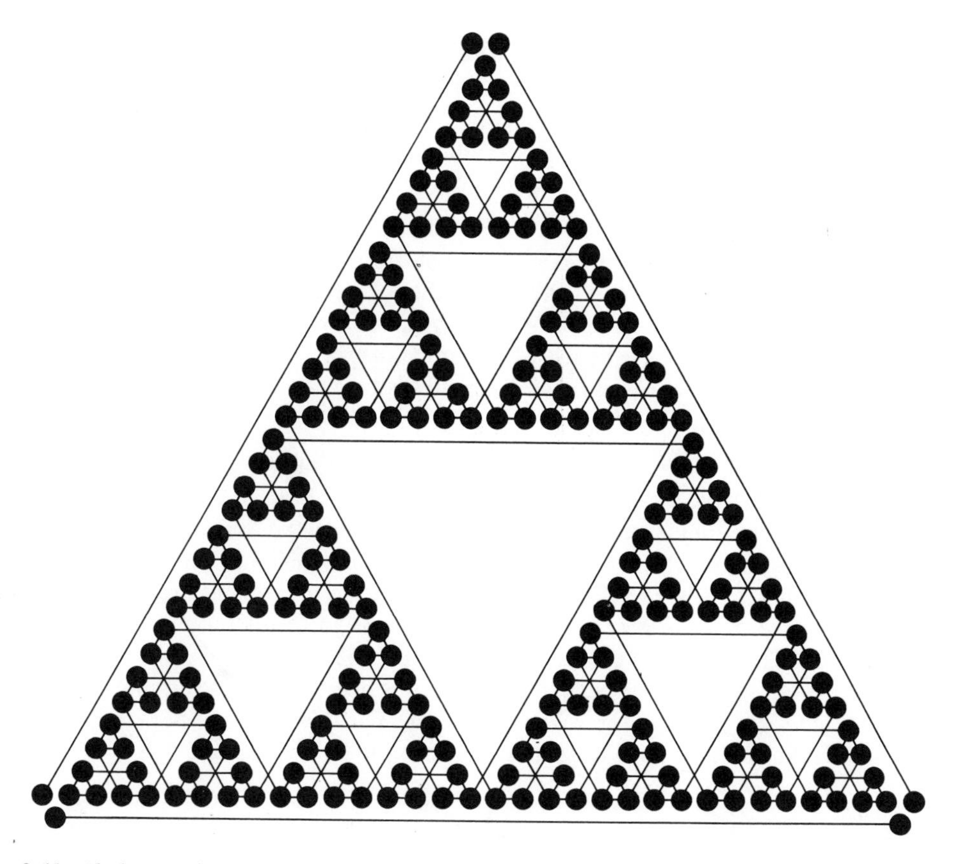

3.1h. *Skeleton of Tower of Hanoi state diagram for five discs. The pattern shown here can be repeated indefinitely.*

3.2 How Is an Algorithm Created?

There is a great deal of difference between carrying out an algorithm (following a clearly defined path from an initial state to a terminal state) and discovering an algorithm in the first place. The process of following algorithms requires careful attention to detail, but no imagination. (In fact, imagination may be a handicap.) It is essentially a mechanical process, as some of the quotations at the beginning of this chapter point out. The *creation* of an algorithm, on the other hand, is an inventive process, which is hard to describe (see Simon, 1962; Minsky, 1967; Banerji, 1969). The suggestions given by Polya (1945) are of some value; the following points may also help.

1. Be sure you understand the system.
   What are the initial and terminal states? Precisely what data are given, and what results are wanted? What are the permissible transitions? What tools are available? What transformations are you allowed to perform? What will each permissible transition accomplish?
2. Look for intermediate goals. (Simon, 1962)
   If you could see an obvious path from each initial state to a terminal state, the problem would be solved. But usually you have to set an intermediate goal, and subdivide the problem into two parts—finding a path from the initial state to the intermediate state, and finding a path from the intermediate state to the final state.

For instance, in the three-disc Tower of Hanoi puzzle (Figure 3.1g), we see that it is necessary to go through either 011 or 100 to get from 000 to 222. So we try to find a path from 000 to either 011 or 100 and from either of these intermediate states to 222. In general there is no guarantee that we will be able to solve either or both parts of the subdivided problem, but at least we can hope that one of the parts is easier than the original problem. Here are some further suggestions about the choice of intermediate goals.

2a. Look especially for intermediate goals near an initial state or near a terminal state, as these have a good chance of being on some path in the final solution.
2b. Look for "mountain passes." Some systems consist of several loosely connected subsystems. If the initial state is in one subsystem and the terminal state is in another, then the few connections between the subsystems are good candidates as intermediate goals, since any path must go through one of these connections. (This situation can be observed in Figure 3.1h.)
2c. Consider an indirect approach. Sometimes it is necessary to use an intermediate goal that seems to be in the wrong direction. When the direct approach leads to a dead end, you may have to retreat or go around an obstruction.

Polya gives an example of an animal in a three-sided enclosure (Figure 3.2a) with food outside. A dog, a chimpanzee, or a four-year-old child (according to Polya) will quickly discover how to reach the goal by going in the opposite direction, but a hen will usually fail.

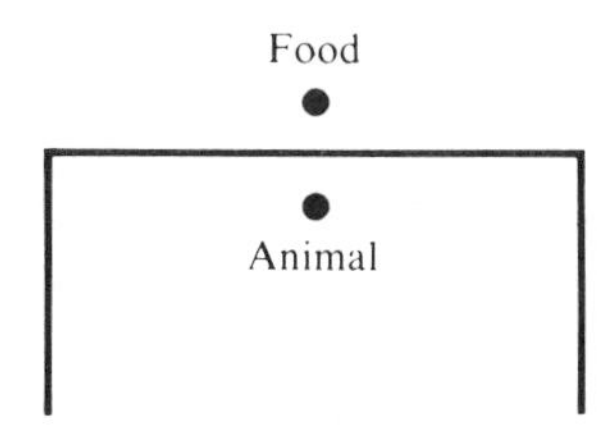

3.2a. A dog, a chimpanzee, or a four-year-old child will quickly discover how to reach the goal by going in the opposite direction, but a hen will usually fail. (Adapted from Polya, 1945.)

2d. Sometimes you must use a *pair* of intermediate goals. You can't see how to construct a path connecting either one of them to an initial or terminal state, but you can see how to get from one to the other. So you construct a tentative path, in hopes that it may be of some use later, or that it may at least provide some helpful experience.

A good example here is a jigsaw puzzle, in which you may find a group of pieces that go together without knowing how the group fits into the final puzzle.

Pattern Recognition Perhaps the most valuable concept of all in the invention of algorithms is that of recognizing *patterns* of states of the system. The five-disc Tower of Hanoi problem (Figure 3.1h) has a clearly visible repetitive pattern structure —the simple nine-state group of the two-disc puzzle keeps repeating over and over. An even more important discovery, however, is that there is a larger pattern in the diagram. The complete five-disc system contains nine groups (of 27 states each) that are related in *exactly* the same way as the nine individual states of the two-disc puzzle! This realization makes it possible to solve a six-disc, seven-disc, or theoretically even a 64-disc puzzle.

As a more familiar example of a pattern within a pattern, let us consider the problem of teaching a Martian to count. If you don't have any little green men in your circle of acquaintances, you could use a seven-year-old Earthling instead.

We use a counting frame (Fig. 3.2b) with several horizontal rods, each holding nine beads. The beads on each rod are divided into two groups: some beads are at the right end, and the others are at the left end. Either group may be empty. We insist that the two groups must each be compact (every bead must be as close as possible to the other beads in its group) and that the two groups must be separated as far as possible. If one of the groups of beads on a rod is empty, all nine beads on that rod will be in the other group and will be pushed as close as possible to the end of the rod.

Each rod represents one decimal digit; the numerical value of this digit is the number of beads in the group at the left end of the rod. The bottom rod is the units place, the next higher rod is tens, the next hundreds, and so forth. Figure 3.2b shows how the number 1974 is represented on the counting frame.

A frame with ten rods has ten billion states, corresponding to the numbers from 0 to 9,999,999,999. A transition from any state to any other state is permissible— that is, we can move any number of beads from either end of any rod to the other end of the same rod, and we can do this simultaneously on as many rods as we wish.

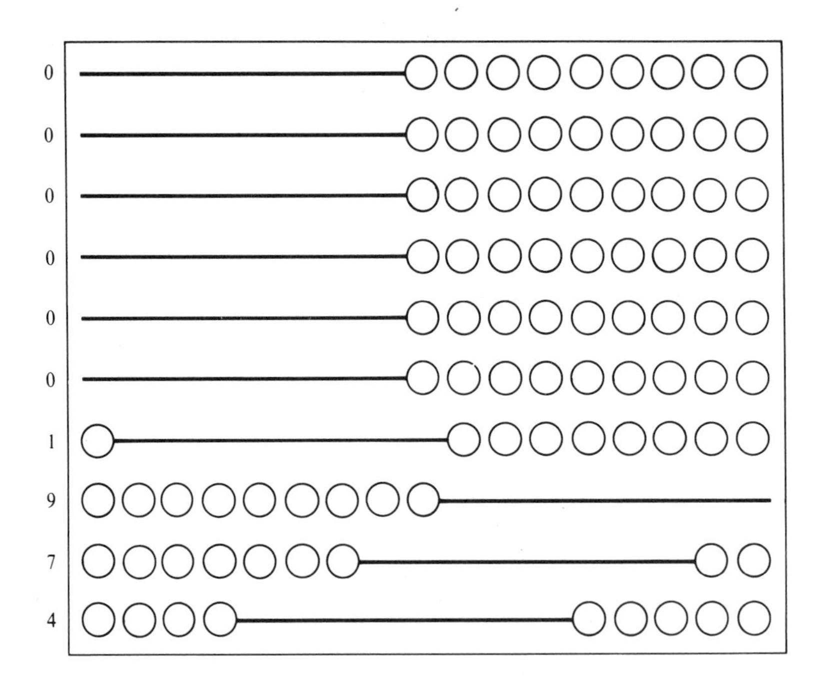

3.2b. *Counting frame with ten rods. Each rod represents one decimal digit. The numerical value of this digit is the number of beads at the left end of the rod. The frame in this illustration represents the number 1974.*

We want the Martian (or substitute Earthling) to learn how to go from any state to the *next* state, that is, to the state corresponding to the next number in normal counting sequence. In other words, we want him to construct a single path from the initial state 0 to the terminal state 9,999,999,999.

Our teaching technique will be to use a pair of identical counting frames, which we might label "current state" and "next state." We show the Martian a number of simple examples; then we ask him to state in words what he has learned.

At first he makes statements like "If the current state is 1,974 the next state will be 1,975" or "If the current state is 47,301 the next state will be 47,302." According to our definition, a list of ten billion rules like this would comprise an algorithm. But instead of waiting for our Martian to compose such a list, we ask him if he can detect any *pattern.* He comes up with the following:

Rule 1: To go from any state to the next state, move one bead on the bottom rod from right to left.

We now proceed in a more systematic way from the initial state, and begin teaching him to count from 0 to 1, 2, and so on. Rule 1 works for a while, but when we come to 9 a new problem arises, since there is no bead to move as required by the rule. We stop him and show him that 9 is followed by 10. After a little difficulty, he learns *not* to proceed directly from 10 to 20 and 30 but to go back to Rule 1 and progress to 11 and 12. As he continues, we show him what to do after he reaches 19 and 29, and he soon grasps this new pattern. He now restates Rule 1 and adds Rule 2:

Rule 1. To go from any state to the next state, *if you can*, move one bead on the first rod from right to left and thus complete the transition. Otherwise, proceed to Rule 2.

Rule 2. Return all the beads on the first rod to the right side. Move one bead on the second rod from right to left and thus complete the transition.

This rule works satisfactorily until the Martian reaches 99. With some further training, he learns what to do next. He restates Rule 2 and adds Rule 3.

Rule 2. Return all the beads on the first rod to the right side. *If you can*, move one bead on the second rod from right to left and thus complete the transition. Otherwise, proceed to Rule 3.

Rule 3. Return all beads on the second rod to the right side.Move one bead on the third rod from right to left and thus complete the transition.

All is well until he reaches 999. After some more training, he then restates Rule 3 and adds Rule 4. Later, he revises Rule 4 and adds Rule 5; somewhat later he revises Rule 5 and adds Rule 6. At this point we end the day's training session, having taught the Martian how to count up to 999,999.

Our Martian, a noble descendant of a race of creatures intelligent enough to build canals on a planet totally lacking in water, achieves the ultimate act of insight. Having dreamed all night that he was moving beads and writing rules, he notices that the first and last rules are always a bit special, but the rules in between are identical except for a couple of words. By morning he is ready to enunciate the following final set of rules.

Rule 1. To go from any state to the next state, *if you can*, move one bead on the first rod from right to left and thus complete the transition. (Return to Rule 1 to start the next transition.) Otherwise, remember which rod you just found empty, and proceed to Rule 2.

Rule 2. Return all the beads, on the rod you just found empty, to the right side. *If you can*, move one bead on the *next* rod from right to left and thus complete the transition. (Return to Rule 1 to start the next transition.) Otherwise, remember which rod you just found empty, and proceed to Rule 3.

Rule 3. If the rod you just found empty is the *last* rod, you have reached the Terminal State. Stop. Otherwise, go back to Rule 2.

Words to Remember

intermediate goal pattern

Exercises

1. Write similar rules for counting by 2s or by 5s.
2. (harder) Write similar rules for counting by 3s or by 7s.
3. Use some other number base, such as 8, 5, 3, or 2. Compare base 3 counting with Tower of Hanoi puzzle states.

3.3 *Is a Computer Program the Same Thing as an Algorithm?*

Although our definition of algorithm can be applied to systems of many different kinds (such as hospital patients, puzzles, and counting boards), the term is most often used in connection with computers and other *finite state automata.*[1] An algorithm expressed as a sequence of computer instructions is a *program.* Not all programs are algorithms, however, for it is possible to write programs that never terminate.

For completeness, we should perhaps give an explicit definition of program. Let us say that a program is any sequence of instructions written in accordance with the rules of some (human-oriented or machine-oriented) computer language. Thus, the *definition* of neither a program nor an algorithm demands performing any useful work. A program must be related to a particular computer language, while an algorithm (strictly speaking) need not have any connection with a computer. An algorithm, by definition, must reach a terminal state after a finite number of steps, while a program (in principle) could repeat the same group of steps forever.

We take the point of view that a (suitable) program is a representation of an algorithm, and that an algorithm is independent of the program used to represent it. In particular, a given algorithm may be represented by programs in several different computer languages. In this book, we frequently present an algorithm as a flowchart and then as a program in Basic, Fortran, and PL/1.

Computer States and Transitions A counting frame with ten rows of beads has ten billion states. One cell in a computer has about the same number of states. How, then, can we keep track of all the different possible states of a computer that has thousands of cells?[2] Fortunately, an algorithm does not have to specify transitions for all the possible states of the system, but only for the states that lie on a path—that is, states that can occur during execution of the algorithm.

And in all useful algorithms, these paths form patterns. As in the case of the counting frame, we generally specify an algorithm as a set of rules based on patterns of states and transitions. We never actually list all states except in simple algorithms like the two-disc Hanoi puzzle.

For computer algorithms, these rules are the instructions of a program. Each such rule in fact defines a very large family of computer state transitions. For example, the instruction

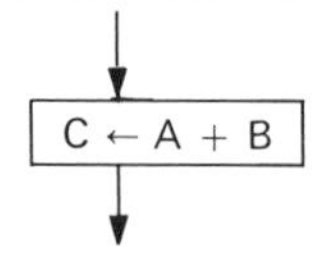

[1] A finite state automaton is simply a machine with a finite number of states. The subject of automata theory, as developed by Turing, Von Neumann, Minsky, and others, is discussed in Minsky (1967).

[2] Strictly speaking, we ought to consider states of the larger system, including the data values in the queue of the reader as well as the internal portion of the computer.

prescribes a transition to a new state in which only the contents of the cell named C has changed. This new value of C depends on the contents of the cells named A and B —that is, on the state of the system just before this transition. But in telling how the new value of C is determined, we need not consider the state of any part of the system except the cells named A and B. Furthermore, the rules of addition built into the arithmetic unit of the computer permit us to describe a very large number of transitions with this single instruction. These transitions include "If the value of A is 1 and the value of B is 1, make a transition to the state in which everything is unchanged except that the value of C, no matter what it was previously, is now set to 2" as well as all other cases in which the numbers (1, 1, 2) in the above statement are replaced by other appropriate sets of numbers such as (1, 2, 3) or (100, 200, 300) or (377, 610, 987).

Furthermore, as we have seen, an algorithm often contains a larger pattern. Similar transitions applied to similar states in different parts of an algorithm may often be expressed by the same rules. In solving the five-disc Tower of Hanoi puzzle, the same instructions can be used whenever a certain pattern of states occurs. In the counting algorithm in Section 3.2, we applied the same rule repeatedly after we had used the larger pattern to combine several similar rules into a single more general rule. In computer programs, we may wish to return to the same instruction when a similar pattern occurs repeatedly.

This does not mean that the same state occurs more than once. In fact, if the system were to reach exactly the same state more than once it would have to repeat the same cycle of steps forever, and this would violate the requirement that an algorithm must eventually terminate. A return to the same instruction, then, does not mean a return to the same state. It merely means that the pattern of transitions applied before is now to be applied to the new state.

A Historical Algorithm: Fibonacci Leonard of Pisa, surnamed Fibonacci (c. 1170–1230), was the son of a business and government official, "a representative of the new class produced by the Commercial Revolution" (Gies and Gies, 1969). In his youth, during his travels around the Mediterranean, Leonard learned about the "new" Hindu numerals. These had been introduced into Arabia about four centuries earlier and were described in a book by the mathematician al Khwarizmi of Baghdad (from whose name the word *algorithm* was later derived), but were not yet in general use except among scholars. Merchants and others did their calculations mainly on the abacus, recording only the final results in written form.

Leonard wrote a book called *Liber abaci*, explaining how the new number system, by using zero as a placeholder, permitted computations to be performed with the numerals directly. The operational techniques still taught in the schools are those he introduced to the Western world.

Leonard's book included abundant examples and illustrations of the reckoning techniques, as well as many practical and recreational problems demonstrating how the numerals might be used in commerce and mathematics. Among these problems is the famous Fibonacci rabbit problem:

> A certain man put a pair of rabbits in a place surrounded by a wall. How many pairs of rabbits can be produced from that pair in a year, if it is supposed that every month each pair begets a new pair which from the second month on becomes productive? (Gies and Gies, 1969, p. 77).

The solution is obtained by observing that the number of pairs, from the third month on, is obtained by adding together the number of pairs for the two previous months. Assuming (see also Weland, 1967, and Hoggatt, 1969) that the original pair was newborn when placed in the cage, so that no offspring are produced until the second month (see Table 3.3a), we obtain the classical Fibonacci sequence

$$1, 1, 2, 3, 5, 8, 13, 21, 34, 55, 89, 144, \ldots$$

This sequence of numbers (see Knuth, 1968, pp. 78–83; Newman, 1956, pp. 23, 98, 718; Alfred, 1963) was studied by J. Kepler (1571–1630), A. Girard (1595–1632), A. de Moivre (1667–1754), N. Bernoulli (1687–1759), G. Lamé (1795–1870), and extensively by E. Lucas (1842–1891).

Table 3.3a *Fibonacci Rabbit Problem*

Month	*Newborn Pairs*	*Month-Old Pairs*	*New Adult Pairs*	*Total Adult Pairs*	*Total Pairs*
1	1	0	0	0	1
2	0	1	0	0	1
3	1	0	1	1	2
4	1	1	0	1	3
5	2	1	1	2	5
6	3	2	1	3	8
7	5	3	2	5	13
8	8	5	3	8	21
9	13	8	5	13	34
10	21	13	8	21	55
11	34	21	13	34	89
12	55	34	21	55	144

The Fibonacci sequence is of interest in modern mathematics because of several remarkable properties—including the fact that the *ratio* of two consecutive numbers in the sequence approaches the "golden ratio," 1.6180339.... This number was considered mystically significant by Pythagoreans in ancient Greece, because it is the ratio between two lengths, A and B, such that $A/B = (A + B)/A$. Botanists have found the Fibonacci sequence of numbers in the spiral patterns of sunflower seeds, pine cone scales, and similar arrangements (Land, 1963, Chapter 13).

A computer algorithm for generating the first few terms of the Fibonacci sequence is very simple. Here are the steps required to generate and print the first five terms. Note that the first two steps differ somewhat from the remaining ones.

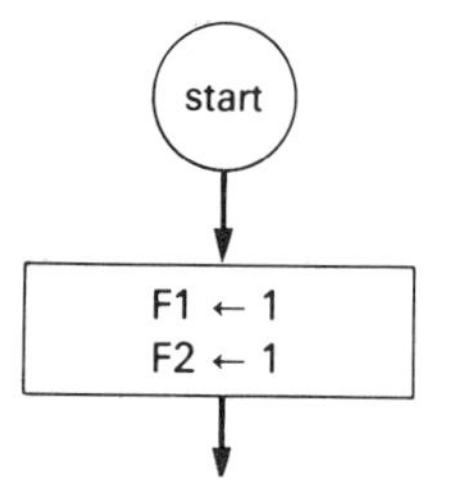

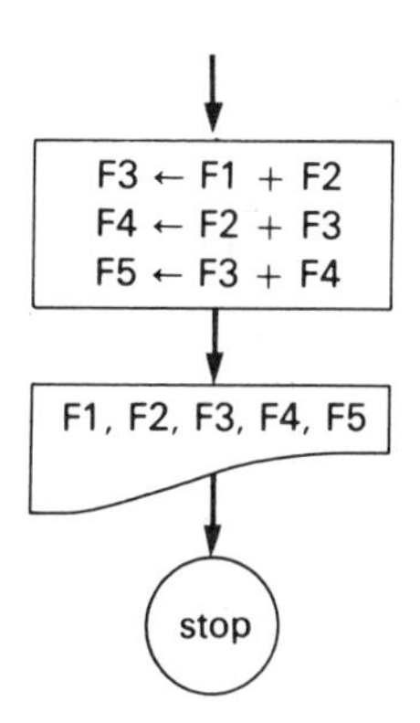

Another Historical Algorithm: Euclid About 300 B.C., Ptolemy (the successor of Alexander the Great in his African dominions) founded the great library at Alexandria, reputed to hold 700,000 volumes.

> He had in effect established a University; and among the earliest of the teachers was Euclid. We know little of his life and character, but he most probably passed his years of tuition at Athens before accepting the invitation of Ptolemy to settle in Alexandria. For twenty or thirty years he taught, writing his well-known *Elements* and many other works of importance. This teaching bore notable fruit in the achievements of Archimedes and Apollonius, two of the greatest members of the University.
>
> The picture has been handed down of a genial man of learning, modest and scrupulously fair, always ready to acknowledge the original work of others, and conspicuously kind and patient. Someone who had begun to read geometry with Euclid, on learning the first theorem asked, "What shall I get by learning these things?" Euclid called his slave and said, "Give him threepence, since he must make gain out of what he learns." (Turnbull, 1951)

Euclid (? 330–275 B.C.) described a procedure, now known as the *Euclidean algorithm*, for finding the *greatest common divisor* of two positive integers. That is, it finds the largest integer that divides evenly into both of the given numbers. For instance, the greatest common divisor of 21 and 24 is 3; the greatest common divisor of 30 and 45 is 15. A familiar problem where this algorithm can be applied is that of reducing a fraction to its lowest terms, by dividing the numerator and the denominator by their greatest common divisor. For example, $^{21}/_{24}$ becomes $^{7}/_{8}$ (dividing numerator and denominator by 3), and $^{30}/_{45}$ becomes $^{2}/_{3}$ (dividing by 15). If a fraction is already in lowest terms, then 1 is the greatest common divisor of its numerator and denominator.

To execute the Euclidean algorithm, we begin by dividing the smaller of the two given positive integers into the larger, ignoring the quotient and saving the remainder:

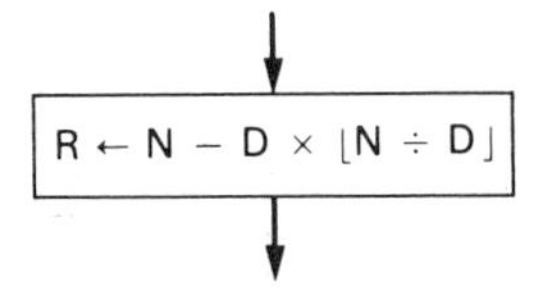

If R is zero, then D is a divisor of N and is the greatest common divisor of N and D. In this case we are finished. Otherwise, we divide R into D, again saving the remainder and ignoring the quotient. Actually, to keep our notation straight, instead of dividing R into D we assign the value of D to N, and then we assign the value of R as the new value of D. We now divide the new value of D into the new value of N, ignoring the quotient and calling the new remainder R:

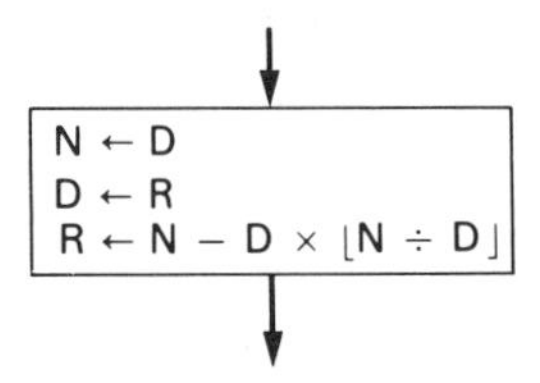

As we repeat this process, the value of R becomes smaller until, eventually, we find zero as a remainder. When that happens, the number which is the current value of D is a divisor of the current value of N. As Euclid pointed out (see McCoy, 1960, pp. 54–58), it is easy to show that this value of D is also the greatest common divisor of the two original numbers.

Words to Remember

finite state automaton	program	Fibonacci sequence
Euclidean algorithm	greatest common divisor	

3.4 *How Can the Sequencing of Instructions Be Specified?*

As we saw in Chapter 2, the instructions of a program are normally executed in sequence. That is, the time sequence of operations corresponds to the space sequence of instructions.

However, we have seen that the description of an algorithm often contains phrases like "repeat this step" or "return to the first step," which require interruptions in the sequence of execution. Only by means of these interruptions in sequence are we able to take advantage of the systematic patterns in a computing process. These patterns, in turn, allow us to describe rather complicated paths with a few rules. In Section 3.2, for example, we used three rules containing less than two hundred words to describe a procedure for counting through ten billion states. This was possible because the same pattern of transitions is applied repeatedly to similar patterns of states.

We can gain insight by expressing these patterns in flowchart form. A break in the normal execution sequence is indicated by an arrow from the bottom of some box, that does not go to the top of the box just below it but goes to some other box on the page instead.

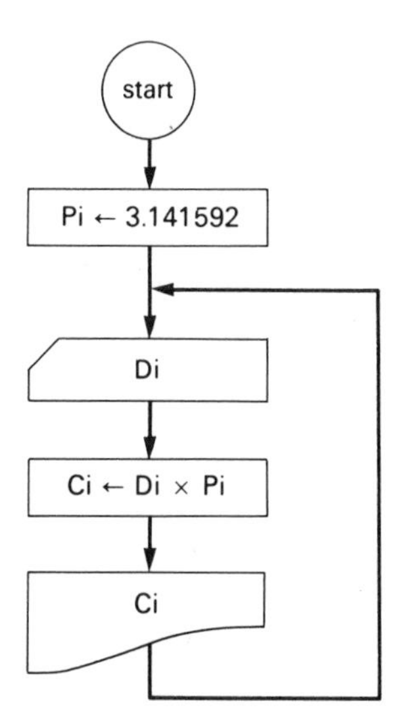

The flowchart above represents a simple calculation where a formula is to be applied to a number of sets of input data. Instructions are first executed for reading a diameter and computing and printing a circumference. After these operations are performed in normal sequence, the sequence is interrupted. As the flowchart arrow indicates, the next instruction to be executed after the circumference is printed is the read instruction.

This kind of closed path on a flowchart, indicating a sequence that returns to an instruction that has been executed previously, is called a *loop*. Notice that the instruction assigning a constant value to Pi need not be repeated and is not included within the loop.

Read and Print Instructions in Loops A loop may include read instructions for several different variables. For instance, in the addition of fractions problem of Chapter 2, the values of A, B, C, and D enter into the calculation. In Chapter 2 we saw how to read a set of values for these four variables, how to use them to compute the sum of the two fractions, and how to print the numerator and denominator of the sum. Now suppose we want to repeat this entire calculation, each time reading four values and printing two values:

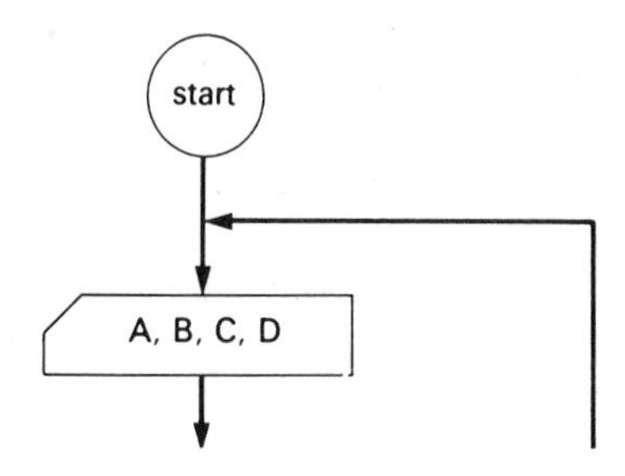

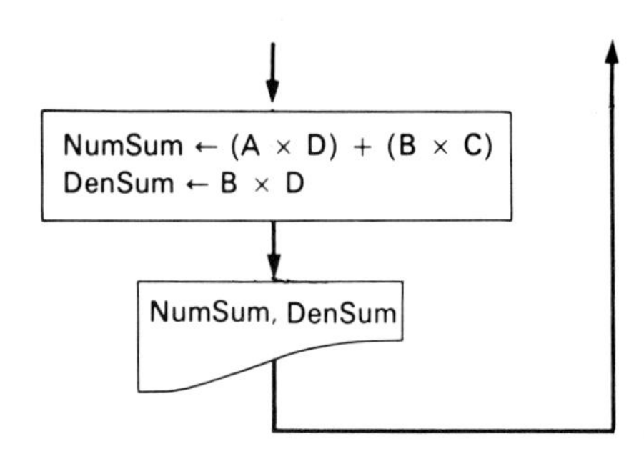

The sequence of numbers in the reader queue must be prepared with great care; the first four numbers placed there will be used the first time the computation is performed, and the next four numbers will be used the next time. If a number is omitted or added, the input portion of the problem will get out of step, and some unexpected answers may result.

The programs represented by the two flowcharts above will continue indefinitely reading data and printing results. These flowcharts represent programs but not algorithms, because they do not provide for terminal states.

Unconditional Control Instructions In a flowchart, we use lines to indicate which instruction is to be executed next after a break in the normal sequence. When we translate the flowchart into a conventional program in Basic, Fortran, or PL/1, however, we cannot connect the instructions with lines and arrows in this way. A control instruction must therefore have some other way of referring to an instruction elsewhere in the program. Accordingly, we provide names for instructions as well as for cells. The name of a cell is a variable or a constant; the name of an instruction is a *label.* An instruction does not need a label unless it is referred to by some control instruction. We sometimes also use labels in our flowcharts, to indicate how the individual instructions correspond to those in the translated version of the program. We write a label above and to the left of the box that it identifies, and we end the label with a colon.

An instruction for an unconditional break in sequence is a go to instruction. The precise form of this instruction varies slightly among the three languages; in general, it consists of the words "go to" followed by a label. Such an instruction specifies that the normal sequence of instructions is to be interrupted, and the next instruction to be executed in time sequence is the one whose label is specified.

Conditions and Branching The instructions in a rectangular flowchart box correspond to state transitions of the computer. However, the definition of an algorithm requires that when the sequence of instructions returns to the same box, the computer must not return to precisely the same state. Each time a loop is repeated, some portion of the state must change. The repetition must end eventually, so the appearance of some condition must be detected within the changing portion of the state.

The counting algorithm of Section 3.2, for example, reaches the terminal state when the topmost rod is found empty. Therefore, each time the rules are applied, a test is made to detect this condition. Figure 3.4a is a flowchart for the counting algorithm.

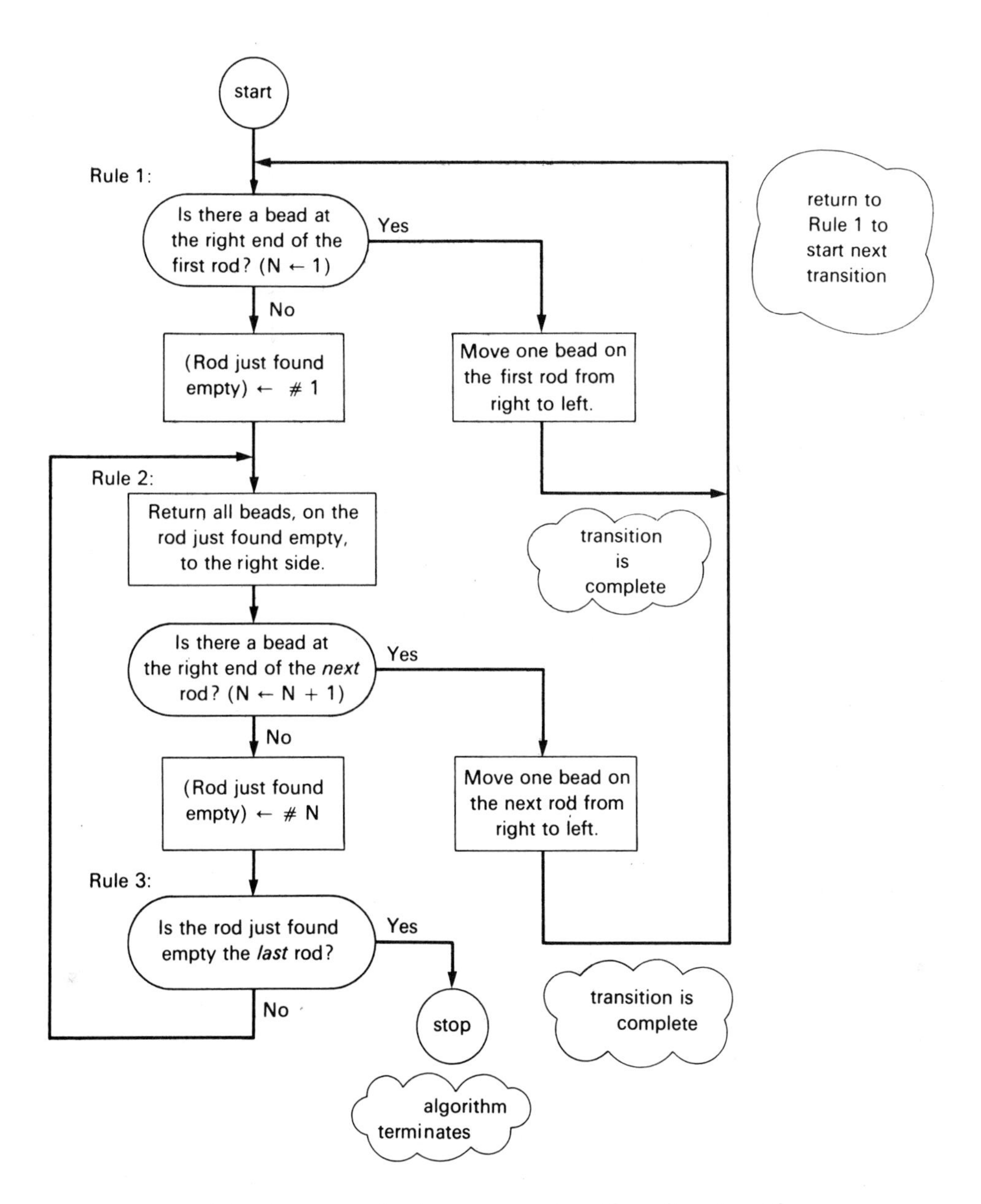

3.4a. *Flowchart for the counting algorithm of Section 3.2.*

In our flowcharts, we write a condition inside an oval box. The use of conditional sequence breaks is called *branching*, because two lines emerge from an oval box although only one arrow leads into the box.

An instruction for branching must involve one of certain specific conditions that the computer is able to test. The condition is designated in the form of an *assertion*, which may involve any two expressions and one of six relations. The effect is to assert that the value of the first expression is related in the specified way to the value of the second expression. The following six relations are available:

$=$	equal	$\neq$	not equal
$>$	greater	$\leq$	not greater
$<$	less	$\geq$	not less

Note that the meaning of "not greater" is "less or equal," while "not less" means "greater or equal."

The assertion in a conditional control instruction may be either true or false. If the assertion is true, a break in the execution sequence occurs, and execution continues with the designated instruction rather than with the next instruction in the normal sequence. We may say that the break in execution sequence is *contingent* on the truth of the assertion: the break occurs if the assertion is true but is ignored if the assertion is false. Thus, when the assertion is false the effect is the same as if the entire conditional instruction were omitted.

A Payroll Example Mr. W. R. Hearst works for the Blank Manufacturing Company. His weekly gross pay, including overtime, is determined from the total number of hours he has worked and from his basic wage rate. If he works more than 40 hours, he is paid for all additional hours at a bonus rate that is 1.5 times his basic rate. The payroll program used by the Blank Manufacturing Company first multiplies the basic rate by the total hours worked, and then, if Mr. Hearst has worked more than 40 hours, adds the overtime bonus of 0.5 times the basic rate for all additional hours beyond 40.

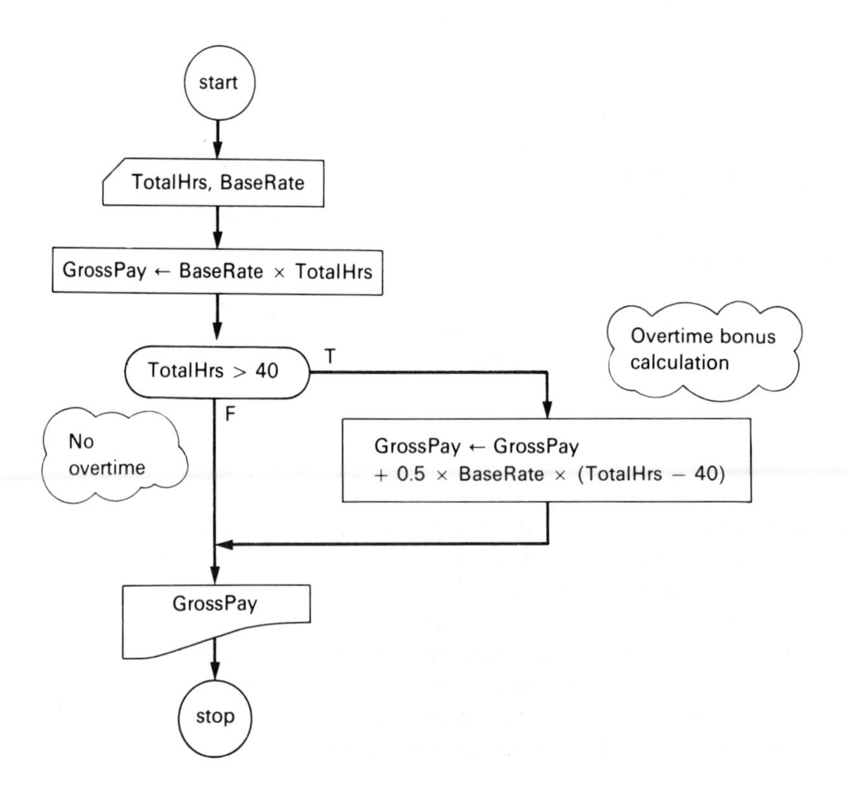

Reverse Effect of the Conditional* go to *Instruction It will be noted, in the following language comparison, that a conditional go to instruction causes a *break* in the sequence of instructions when the assertion is true. Using such an instruction, therefore, we cannot write the equivalent of a step like:

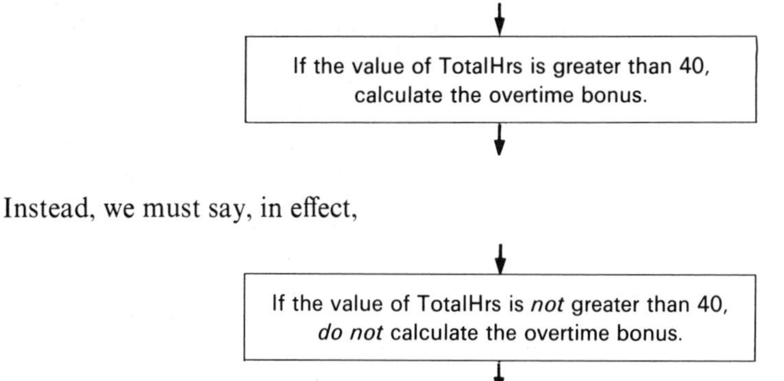
If the value of TotalHrs is greater than 40,
calculate the overtime bonus.
If the value of TotalHrs is not greater than 40,
do not calculate the overtime bonus.

Instead, we must say, in effect,

This somewhat confusing double-negative form is imposed by the rather negative effect of the go to instruction itself: it can be used to skip over a calculation step, but it cannot be used directly to perform a calculation step.

Flowchart

Payroll:

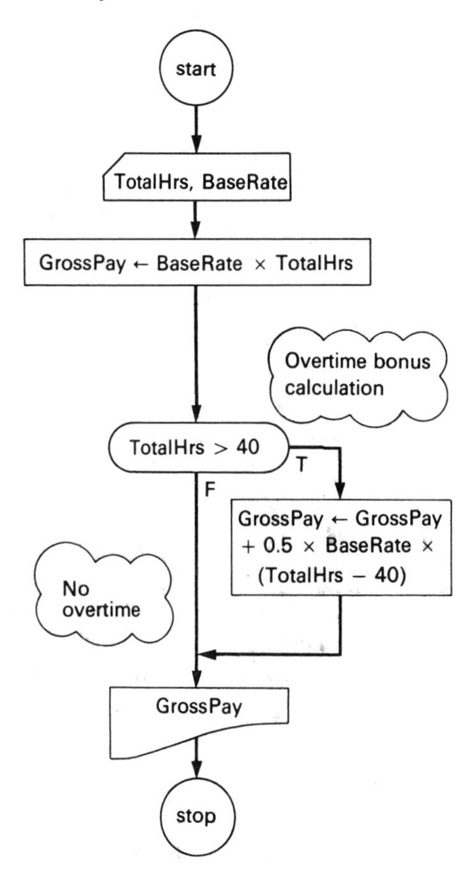

Basic In Basic, each instruction has a *line number*. This number is used as the instruction label.

The form GO TO *label* is used for unconditional control in Basic.

The six relations are used in Basic; however, most console keyboards do not have the symbols ≠, ≤, or ≥. Instead, the following *symbol pairs* are used:

< >	not equal
< =	less or equal (not greater)
> =	greater or equal (not less)

The conditional instruction uses the word THEN:

IF (*assertion*) THEN *label*

Payroll:

```
110 INPUT H, R
120 LET G = R * H
130 IF H < = 40 THEN 160
140 REM: OVERTIME BONUS
150 LET G = G + 0.5 * R * (H - 40)
160 PRINT G
```

Fortran In Fortran, format statements and instructions may have labels consisting of one to five digits. When a statement is punched on a card, the label goes in the first five columns. Statements that are referred to elsewhere in the program must have labels. Therefore, format statements are always labeled while instruction labels are more or less optional.

The form GO TO *label* is used for unconditional control.

Each of the six relational symbols is represented by a pair of letters, preceded and followed by a period:

.EQ.	equal	.NE.	not equal
.GT.	greater	.LE.	less or equal (not greater)
.LT.	less	.GE.	greater or equal (not less)

The two expressions in an assertion must be of the same type, and the assertion must be enclosed in parentheses:

IF (*assertion*) GO TO *label*

Payroll:

```
      READ (5, 82) TH, BR
82    FORMAT (8 F 10.0)
      GP = BR * TH
      IF (TH .LE. 40.0) GO TO 160
C     OVERTIME BONUS CALCULATION
      GP = GP + 0.5 * BR * (TH - 40.0)
160   WRITE (6, 83) GP
83    FORMAT (F 15.2)
```

PL/1 Labels in PL/1 have the same form as variables. A label is separated by a colon from the instruction that it identifies.

The form go to *label* is used for the unconditional control instruction.

The six relations are available in PL/1, but the symbols ≠, ≤, and ≥ are represented by the following *symbol pairs*:

¬=	not equal
<=	less or equal (not greater)
>=	greater or equal (not less)

The conditional go to instruction is written as:

if *assertion* then go to *label*;

Payroll:

```
ONE: procedure options (main);
       declare (TOTALHRS, BASERATE,
          GROSSPAY) float;
       get list (TOTALHRS, BASERATE);
       GROSSPAY = BASERATE *
          TOTAL HRS;
       If TOTALHRS <= 40 then go to
          NOVT;
       /* Overtime Bonus calculation */
       GROSSPAY = GROSSPAY + 0.5 *
          BASERATE * (TOTALHRS - 40);
   NOVT: put skip list (GROSSPAY);
       end ONE;
```

Words to Remember

loop	go to instruction	assertion
control instruction	condition	relation
label	branching	

Exercises

1. Write a program that will either add or multiply a pair of fractions, keeping the numerators and denominators separate. Input should consist of values for A, B, C, D, and K. If K is 1, print the numerator and denominator of the *sum* of (A/B) and (C/D) as explained in Section 2.7. If K is 2, print the numerator and denominator of the *product* of (A/B) and (C/D). If K is zero, *stop*; otherwise, go back and read more data after printing the required results.

	first numerator	*first denominator*	*second numerator*	*second denominator*	*operation*
a.	1	2	2	3	add
b.	35	51	49	72	add
c.	2	3	4	1	add
d.	1	1	1	1	add
e.	2	3	4	1	multiply
f.	34	55	89	144	multiply
g.	1	1	1	1	multiply
h.					stop

2. Write a program to solve the quadratic equation $Ax^2 + Bx + C = 0$. Read the values of the coefficients A, B, and C. If the value of A is zero, the solution is the value of $-C \div B$ unless the value of B is zero. If the value of A is not zero, compute the *discriminant*

```
Disc ← B × B − 4 × A × C
```

and the two related values

```
P ← − B ÷ (2 × A)
Q ← √|Disc| ÷ (2 × A)
```

If the value of the discriminant is zero or positive, compute and print the two real roots:

```
R1 ← P + Q
R2 ← P − Q
```

If the value of the discriminant is negative, on the other hand, print the values of P and Q, which are the real and imaginary parts of the complex roots.

	value of A	*value of* B	*value of* C		*value of* A	*value of* B	*value of* C
a.	1	−2	−10	f.	1	4	−4
b.	3	9	1	g.	1	0	1
c.	6	3	2	h.	6	5	−4
d.	0	3	2	i.	1	−4	−3.9999999
e.	4	20	25	j.	1	−0.000001	1

3.5 *What Are Some Examples of Programs with Control Instructions?*

As examples of programs using control instructions, we return to the algorithms of Fibonacci and Euclid, introduced in Section 3.3.

The Fibonacci Algorithm As a first move in developing a more general program to generate the Fibonacci sequence, let us write out the first few steps in linear form.

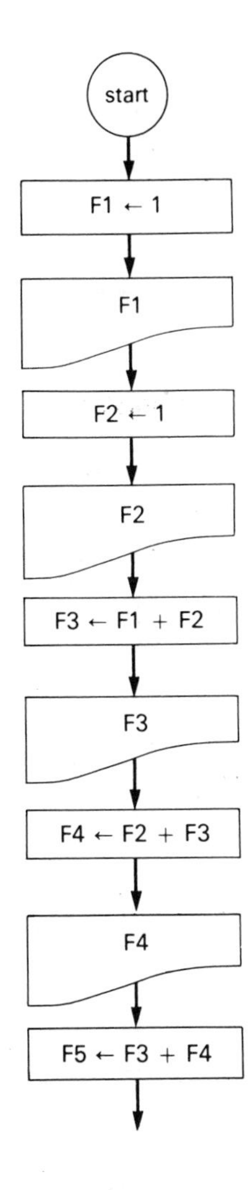

If we continue in this way much longer, we will need a great many variables. But we notice that F1 is never again used after F3 is calculated, so we could use the same cell for F3 as for F1. Similarly, F4 could use the same cell as F2. When we change all the odd-numbered variables to use the same cell as F1 and all the even-numbered variables to use the same cell as F2, we see that a short sequence of steps is repeated (after the first few steps, which are not repeated):

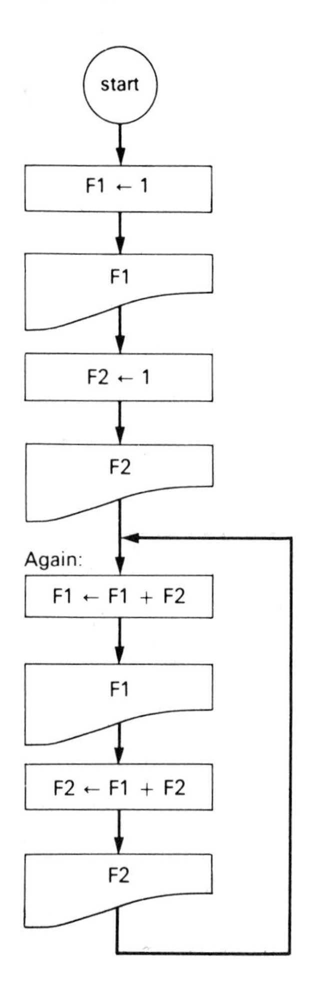

Of course we must add some criterion for terminating the algorithm. For instance, we could add (just below the last box of the flowchart):

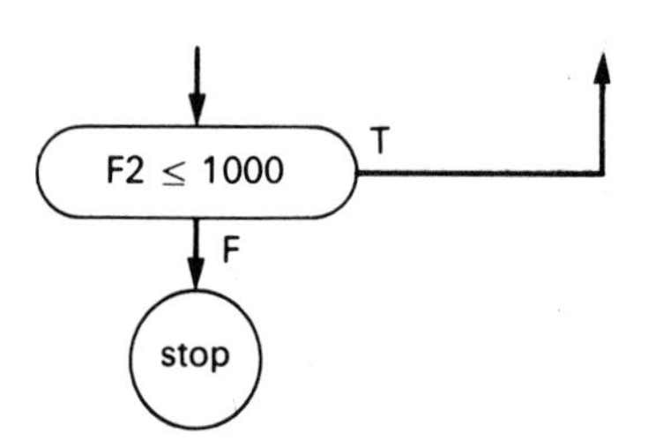

The Euclidean Algorithm We begin with the instructions written in Section 3.3:

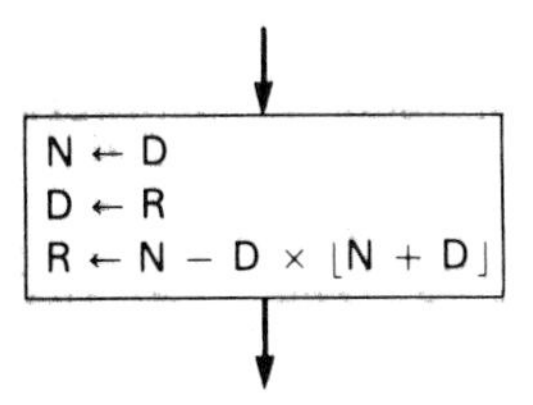

and add steps corresponding to the explanation given there— "As we repeat this process, the value of R becomes smaller until, eventually, we find zero as a remainder. When that happens, the number which is the current value of D is . . . the greatest common divisor of the two original numbers." Thus, we have the following simple program:

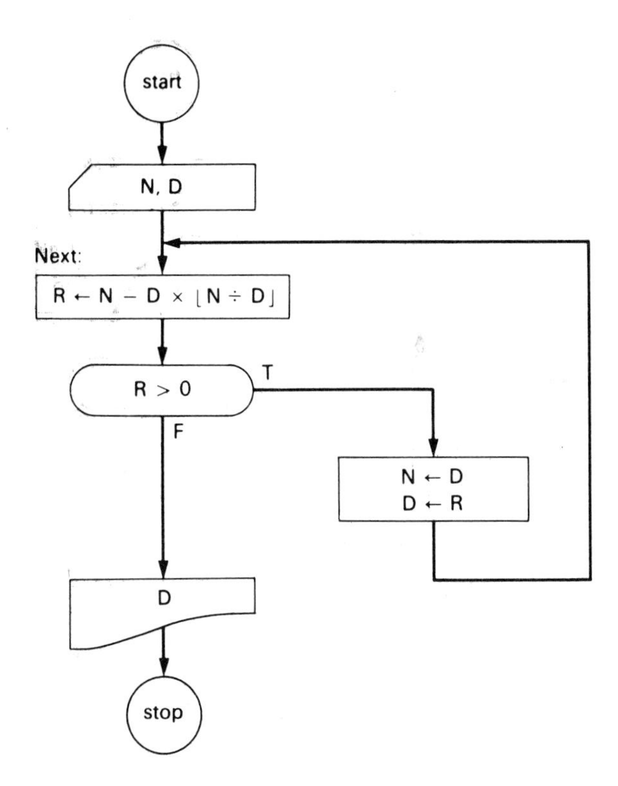

We have said that at the first step the value of N should be larger than the value of D, but if we arrange the steps as shown this is no longer important. If N has a smaller value, the value of $\lfloor N \div D \rfloor$ will be zero and R at the first step will have the same value as N. In this case, one extra execution of the loop is required, during which the values of N and D are interchanged.

Flowchart

Fibonacci algorithm:

Euclidean algorithm:

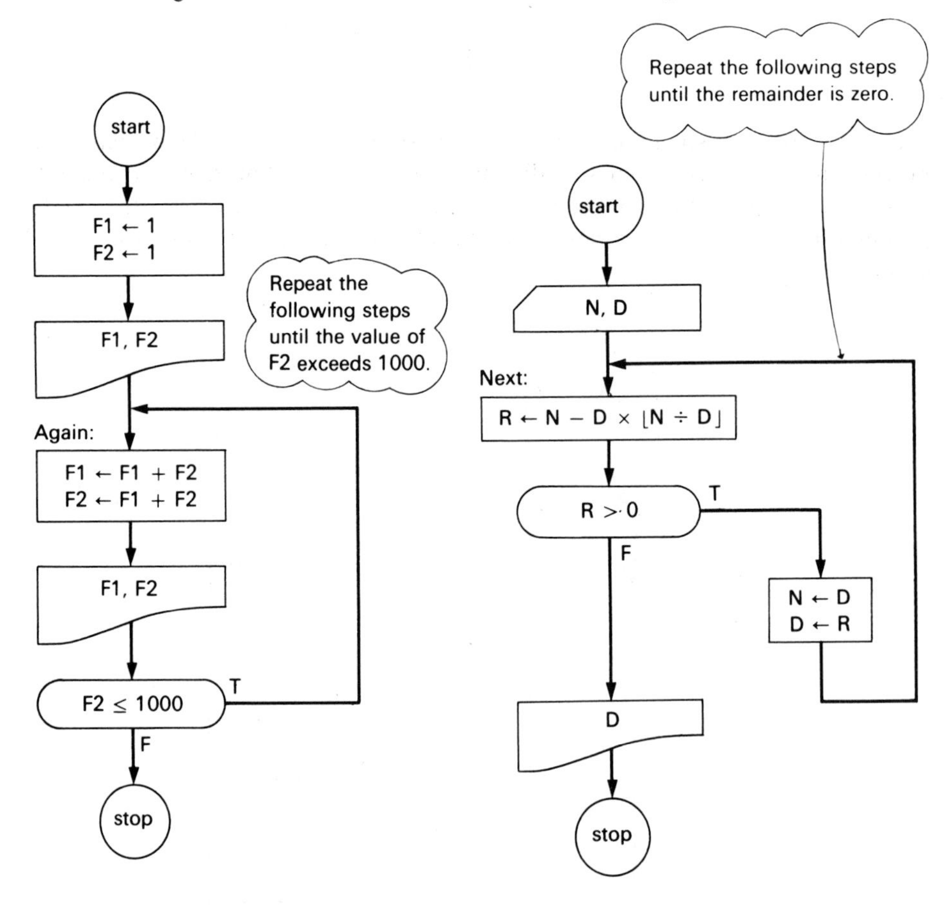

Basic

Fibonacci algorithm:

```
110 INPUT F1, F2
120 PRINT F1, F2
130 REM: REPEAT UNTIL F2 > 1000
140 LET F1 = F1 + F2
150 LET F2 = F1 + F2
160 PRINT F1, F2
170 IF F2 <= 1000 THEN 130
180 END
```

Euclidean Algorithm:

```
310 INPUT N, D
320 REM: REPEAT UNTIL R = 0
330 LET R = N - D  * INT (N / D)
340 IF R = 0 THEN 380
350 LET N = D
360 LET D = R
370 GO TO 320
380 PRINT D
390 END
```

Fortran Note the use of the STOP instruction in the following examples.

Fibonacci algorithm:

```
      READ (5, 81) IF1, IF2
81    FORMAT (8 I 10)
      WRITE (6, 81) IF1, IF2
C     REPEAT UNTIL IF2 EXCEEDS 1000
130   IF1 = IF1 + IF2
      IF2 = IF1 + IF2
      WRITE (6, 81) IF1, IF2
      IF (IF2 .LE. 1000) GO TO 130
      STOP
```

Euclidean algorithm:

```
      READ (5, 81) N, ID
81    FORMAT (8 I 10)
C     REPEAT UNTIL IR = 0
320   IR = MOD (N, ID)
      IF (IR .EQ. 0) GO TO 380
      N = ID
      ID = IR
      GO TO 320
380   WRITE (6, 81) ID
      STOP
```

PL/1 Note that the program name has been truncated to seven characters.

```
FIBONAC: procedure options (main);
      declare (F1, F2) float;
      F1 = 1; F2 = 1;
      put skip list (F1, F2);
AGAIN: F1 = F1 + F2;
      F2 = F1 + F2;
      put skip list (F1, F2);
      if F2 <= 1000 then go to AGAIN;
      end FIBONAC;

EUCLID: procedure options (main);
      declare (N, D, R) fixed;
      get list (N, D);
NEXT: R = MOD (N, D);
      if R = 0 then go to DONE;
      N = D; D = R;
      go to NEXT;
DONE: put skip list (D);
      end EUCLID;
```

Word to Remember

contingent instruction

Exercises

1. E. Lucas (1876) generalized the Fibonacci sequence, and obtained the Lucas sequence: 1, 3, 4, 7, 11, 18, 29, 47, ... Write a program to compute and print the first few numbers of this sequence.
2. During execution of the Euclidean algorithm, the values of the two original variables are destroyed. Write a modified program that will preserve the original values, so that they can be used (if desired) after their greatest common divisor has been computed.

Value of N	308	144	720	60
Value of D	1575	89	960	75

3. Combine the program for adding fractions (Section 2.7) with the Euclidean algorithm to produce a program to add two fractions, (A/B) and (C/D), with the

numerator and denominator expressed as separate integers, and to reduce the numerator and the denominator of the sum to lowest terms before printing them.

	first numerator	*first denominator*	*second numerator*	*second denominator*
a.	5	9	1	16
b.	15	30	8	32
c.	1	2	1	6

4. The *least common denominator* of two fractions (A/B) and (C/D) is given by the expression (B × D) ÷ G where G is the greatest common divisor of B and D. The numerator of the fraction which is the sum of (A/B) and (C/D), with the least common denominator, is given by the expression A × (D ÷ G) + (B ÷ G) × C. Will the fraction having this numerator and this denominator always be in lowest terms? Does this method of adding fractions, which is similar to the methods usually taught in elementary school, have any advantage over the method suggested in exercise 3? (If not, why would this method be taught in the schools?)

3.6 *How Can We Control the Termination of a Calculation?*

We now consider a number of ways of using control instructions to detect the conditions under which we wish to terminate a calculation. First we consider conditions involving the values of variables that are already available for other purposes in the calculation. Our first two examples are drawn from technical applications.

The following program computes the *cube root* of the value of X by Newton's method. Beginning with Q ← 1, find the "new" value of Q by adding together two-thirds of the value of Q and one-third of the value of X ÷ (Q × Q). (The latter expression would be equal to Q in value, if Q were the cube root of X.) Repeat until the difference between the two latest values of Q is less than some desired tolerance (say, 0.0001).

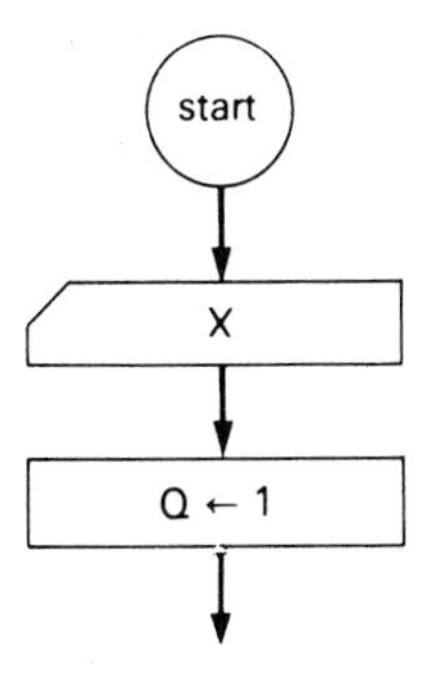

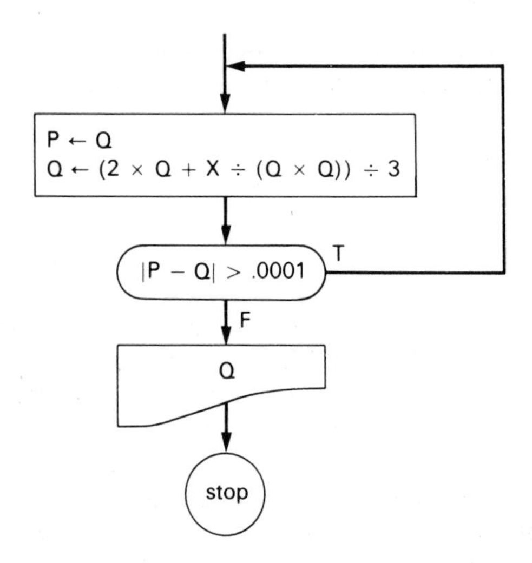

The negative exponential function, exp (−X), for small values of X, can be found by adding together a series of terms (the Taylor series). The first term is 1, and the second is − X. The third term is − X ÷ 2 times the second; the fourth is − X ÷ 3 times the third, and so on. The series is continued until the next term to be added is smaller than a given tolerance.

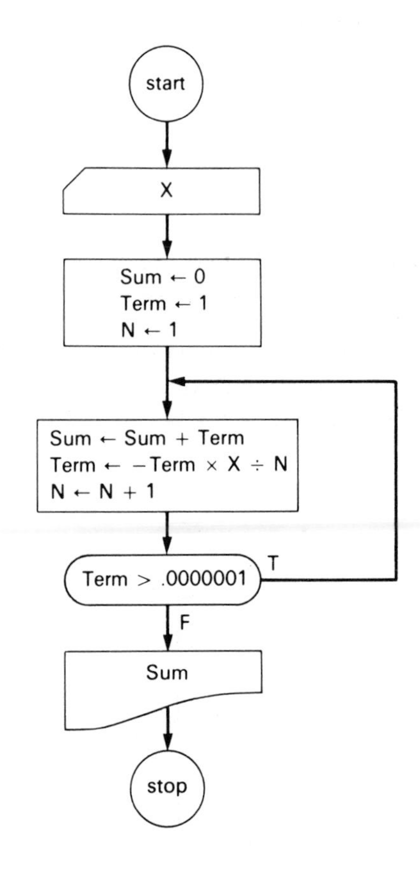

Control Based on Input Data Values The Blank Manufacturing Company would find it very wasteful to have a program that computes the gross pay for only one employee and then stops. To complete the payroll program, arrangements must be made to repeat all steps of the payroll calculation discussed in Section 3.4 for each employee. Let us also add a couple of steps to read and print the employee number for identification.

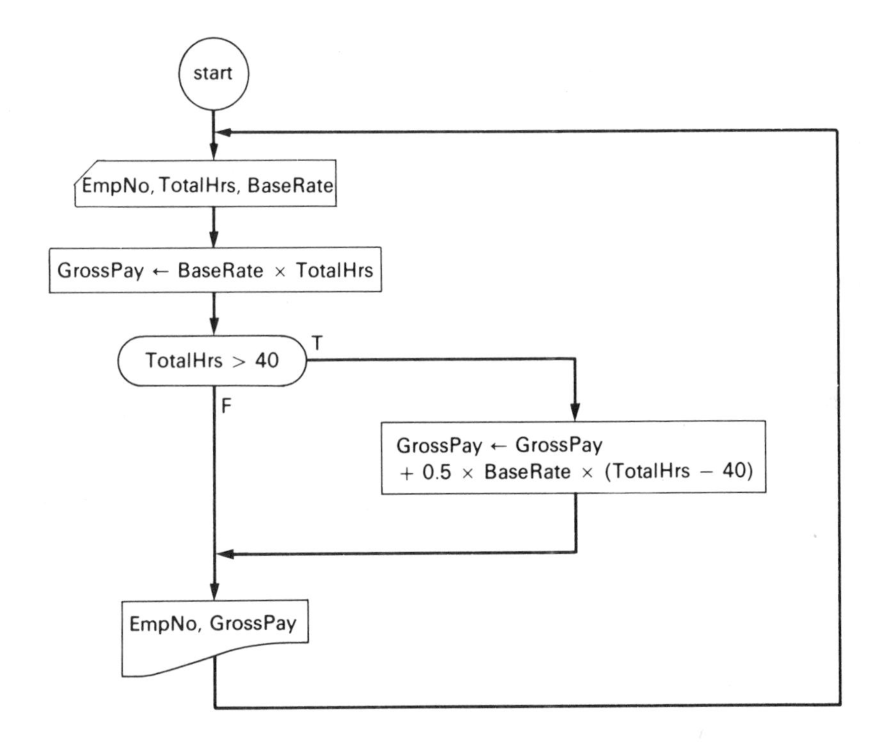

The repetition should stop after the payroll data for the last employee has been calculated. Suppose we know that the last data items in the reader queue pertain to Mr. X. Y. Zwettler, whose employee number is 81307. Then we arrange that, when the final instruction of the loop is reached, the loop will be repeated unless the employee number just processed is 81307.

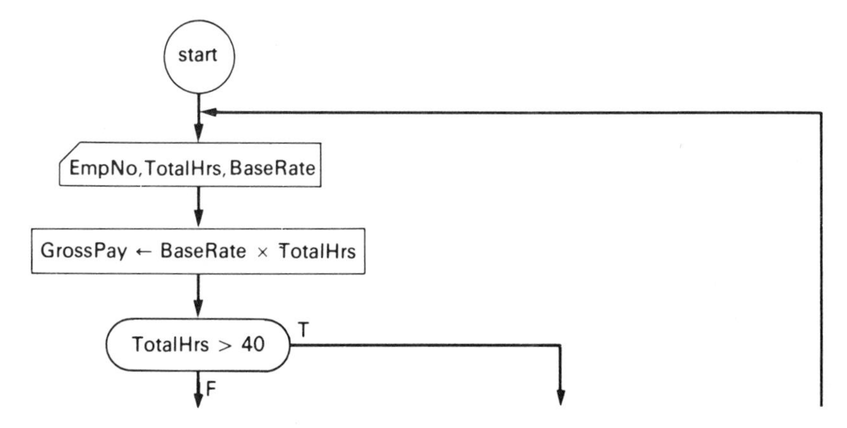

A more satisfactory procedure in most cases is to give an "end of data" indication to the program, by inserting in the reader a *sentinel* item with the same form as a regular data item but with a value outside the normal range for the corresponding variable. For example, suppose that the Blank Manufacturing Company uses employee numbers in the range from 10000 to 99998.

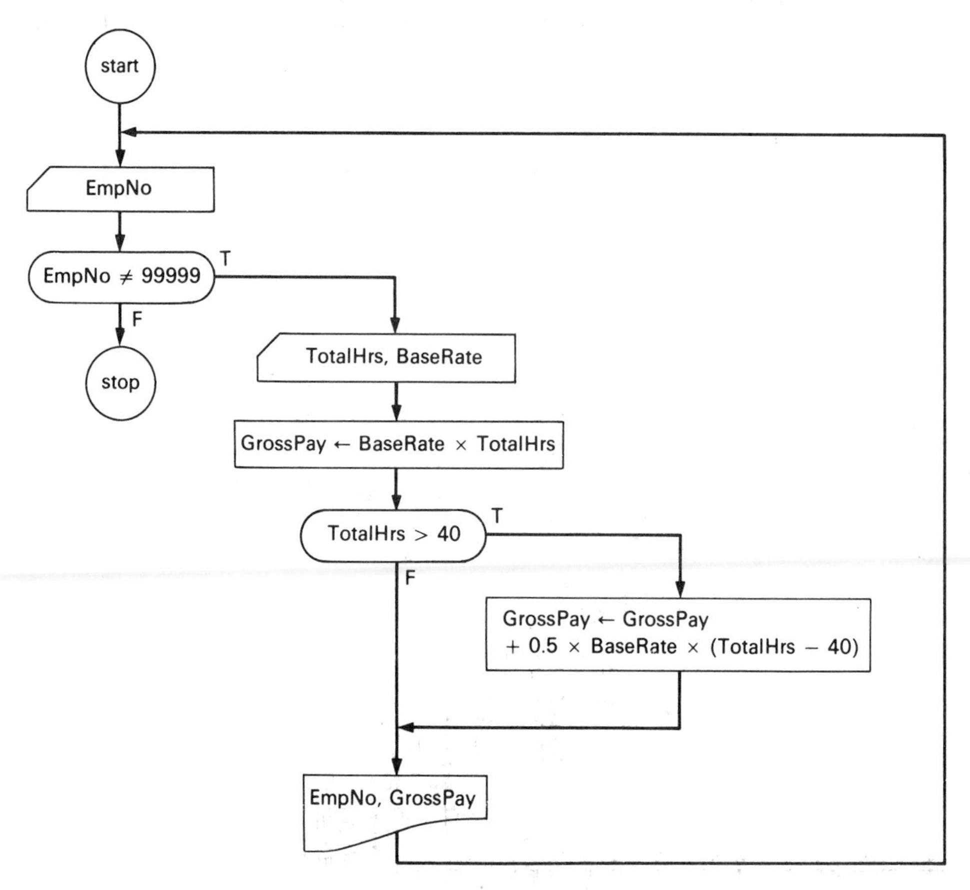

The person who prepares payroll data in the reader always adds, after the data referring to Mr. Zwettler, a sentinel data set for the fictitious employee number 99999. Alternatively, the condition for termination could be based on a fictitious employee number smaller than 10000: an acceptable sentinel value then would be 00000. Or the sentinel could be recognized by a negative value for TotalHrs or BaseRate, since valid values for these variables are never negative. In this case, the test for the sentinel would follow the box that reads the variable involved in the test.

Count-Controlled Branching In another important case, we control the sequence of instructions by *counting* the number of repetitions of a loop. As a counter, we use a variable whose value at any step is the number of times the loop has been repeated. An instruction in the loop increases the value of this counter by one each time the loop is executed. The repetition ends when the value of this counter reaches a predetermined number.

For example, the Blank Manufacturing Company has 763 employees. The payroll is set up as shown in the following flowchart. The variable N is the counter. This variable is *initialized* to the value 1 before the loop begins. Then it is increased by 1 each time the loop is executed. The loop is repeated until the value of N exceeds 763.

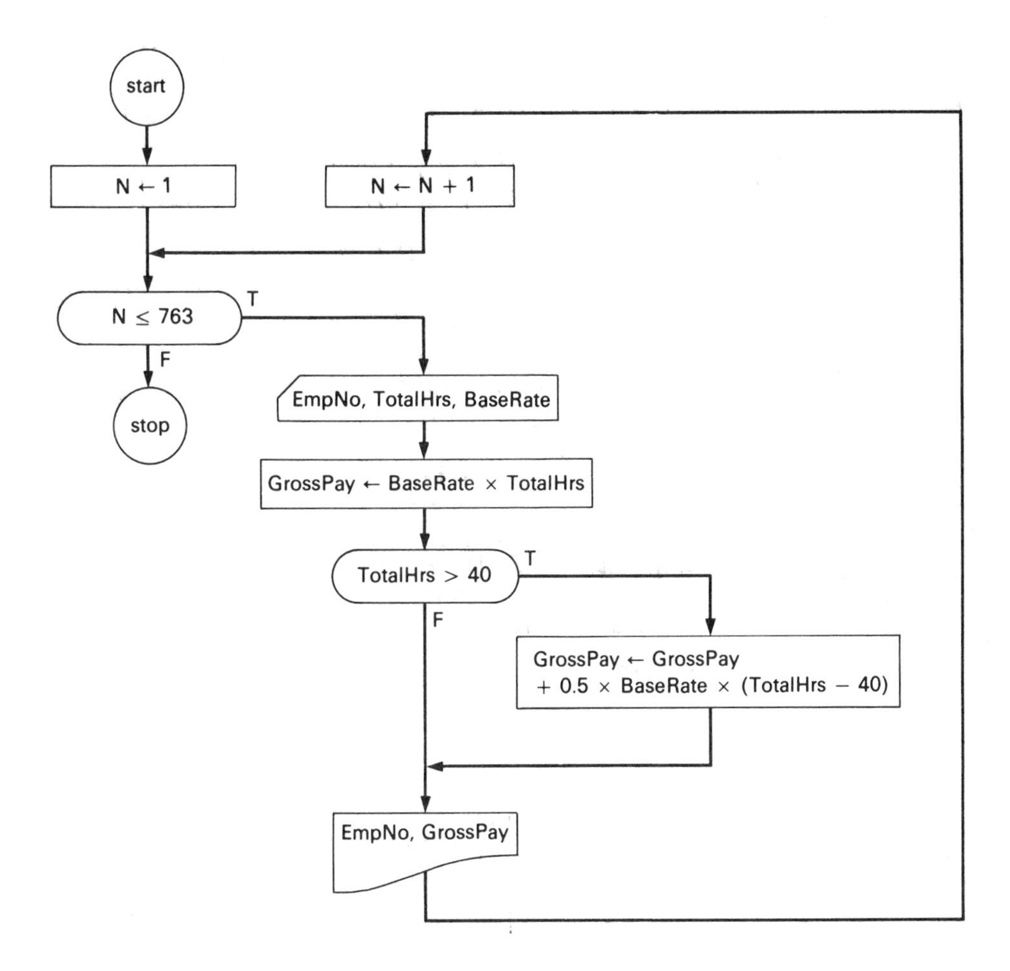

In the following somewhat more general version, a separate read instruction is executed at the beginning. The number of employees is included as a separate item of data, preceding the regular payroll data in the reader queue. Thus, the program need not be revised each time the number of employees changes.

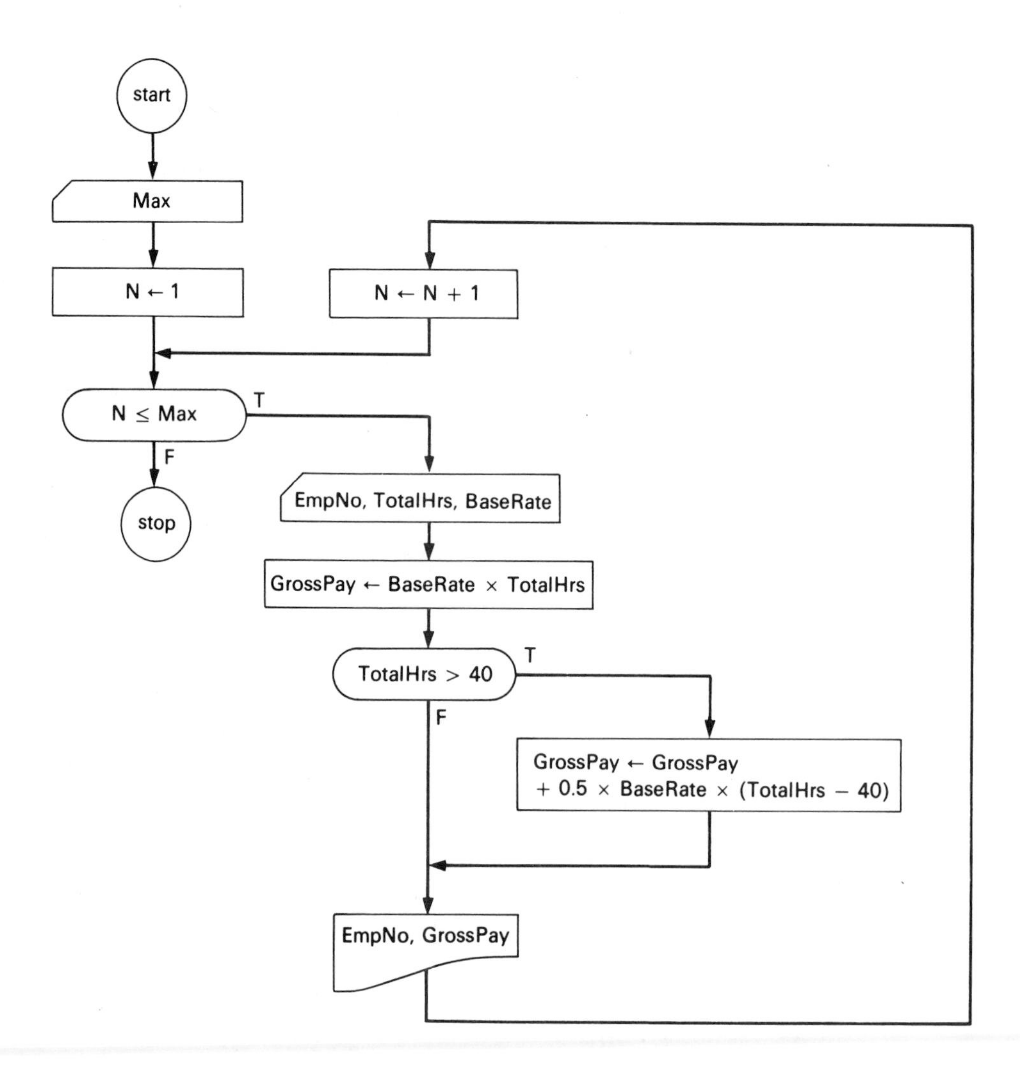

Flowchart

Payroll with count control:

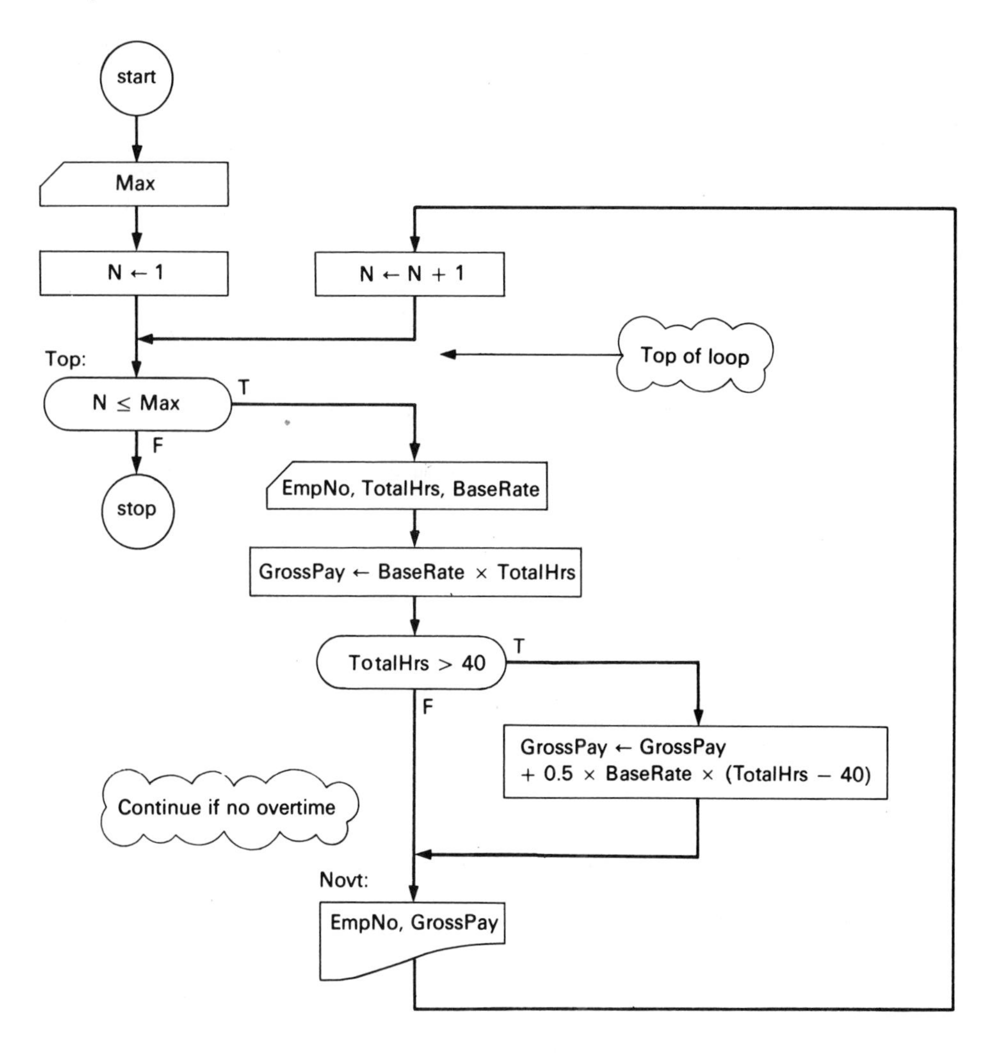

Basic

Payroll with count control:

```
600 INPUT M
605 LET N = 1
610 REM... TOP OF LOOP
615 IF N > MAX THEN 660
620 INPUT E, H, R
625 LET G = R * H
630 IF H <= 40 THEN 640
635 LET G = G + 0.5 * R * (H - 40)
640 REM... CONTINUE IF NO OVERTIME
645 PRINT E, G
650 LET N = N + 1
655 GO TO 610
660 END
```

Basic has no stop instruction. In many programs, the INPUT operation serves this purpose by returning control to the user. Otherwise, a conditional or unconditional control instruction referring to the line number of the END statement may be used.

A line number followed by REM may be used when a label is needed between two instructions in the program.

There is no provision in Basic for more than one instruction on a line, or for an instruction too long to fit on one line. A long instruction can usually be rewritten using two or more shorter instructions.

Fortran

Payroll with count control:

```
    READ (5, 81) MAX
81  FORMAT (8 I 10)
    N = 1
C   TOP OF LOOP
610 IF (N .GT. MAX) STOP
    READ (5, 82) EMPNO, TH, BR
82  FORMAT (8 F 10.0)
    GP = BR * TH
    IF (TH .LE. 40.0) GO TO 640
    GP = GP + 0.5 * BR * (TH - 40.0)
C   CONTINUE IF NO OVERTIME
640 WRITE (6, 84) EMPNO, GP
84  FORMAT (F 15.0, F 15.2)
    N = N + 1
    GO TO 610
```

Unconditional and conditional STOP instructions are written as follows:

```
STOP
IF (assertion) STOP
```

A program may include more than one STOP instruction.

In Fortran, a READ, WRITE, or assignment instruction may also be used as the contingent instruction in a conditional instruction. The instruction following the assertion is executed if the assertion is *true*, and is ignored if the assertion is *false*. It is important to realize that the truth of the assertion has *no other effect* on the program. For instance, what will be the final value of A after execution of the following pair of instructions, and how will this final value of A depend on the value of I?

```
IF (I .EQ. 1) A = 1.0
A = 2.0
```

The answer is that the final value of A will *always* be 2.0. The assignment A = 1.0 may or may not occur, depending on the truth of the assertion, but this has *no effect* on the execution of the second assignment A = 2.0.

A labeled CONTINUE statement may be used as a reference point in Fortran:

```
72  CONTINUE
```

There is no provision for more than one Fortran statement on a line. However, long statements can be continued on additional lines with a character other than "blank" or 0 in column 6.

PL/1

Payroll with count control:

```
THREE: procedure options (main);
     declare (EMPNO, TOTALHRS,
        BASERATE, GROSSPAY) float;
     declare (MAX, N) fixed;
     get list (MAX);
     N = 1;
TOP: /* Top of loop */
     if N > MAX then go to BOTTOM;
     get list (EMPNO, TOTALHRS,
        BASERATE);
     GROSSPAY = BASERATE *
        TOTALHRS;
     if TOTALHRS < = 40 then go to NOVT;
     GROSSPAY = GROSSPAY + 0.5 *
        BASERATE * (TOTALHRS - 40);
NOVT: /* Continue if no overtime */
     put skip list (EMPNO, GROSSPAY);
     N = N + 1;
     go to TOP;
BOTTOM: end THREE;
```

The last statement of a PL/1 program has the form end *label*;, where the label following end is the name of the program. This end statement may itself have a label. Execution of the program is terminated when control reaches this statement.

In PL/1, the contingent instruction that follows if *assertion* then need not be a go to instruction. It may also be a get list, put skip list, or assignment instruction:

if *assertion* then *contingent instruction*;

The contingent instruction is executed if the assertion is *true* and is ignored if the assertion is *false*. It is important to realize that the truth of the assertion has *no other effect* on the program. For instance, what will be the final value of A after execution of the following pair of instructions, and how will this final value of A depend on the value of I?

```
if I = 1 then A = 1.0;
A = 2.0;
```

The answer is that the final value of A will *always* be 2.0. The assignment A = 1.0 may or may not occur, depending on the truth of the assertion; but this has *no effect* on the execution of the second assignment A = 2.0.

In PL/1, a semicolon indicates the end of an instruction. Therefore, a semicolon with nothing preceding it is a null instruction. Such an instruction may have a label:

```
HERE:;
```

Several PL/1 instructions may be combined on one line, or one instruction may extend over several lines—the end of a line has no special significance in PL/1 instructions, since they are delimited by semicolons.

Words to Remember

sentinel   count control   initialize

Exercises

1. Modify the payroll problem, assuming that a negative value of TotalHrs will be used as the sentinel. Tell what advantage the use of a sentinel might have as compared to a termination criterion based on the last actual employee number.
2. Write a program to *read* the value of K, multiply together all of the positive integers ($1 \times 2 \times 3 \times \ldots$) that are not larger than K and *print* the result. (This product is called the *factorial* of K.) Use the following values for K: (a) 4; (b) 6; (c) 20; (d) 1.
3. Write a program to *read* values of A and K, then multiply A by itself K times (to compute the *K*th power of A), and *print* the result. Values are:

A	19	2744	18	2	1
K	3	1	7	-23	0

4. Write a program to compute compound interest. An amount of money, Prin, becomes

Prin + (Rate ÷ 4) × Prin

after three months. *Read* Prin, Rate (the annual rate), and the length of time in years. (Hint: Find the length of time in quarter years, and use a new variable for the quarterly rate, Rate ÷ 4.)

	Principal	*Rate*	*Time (Years)*	*Times Compounded per Year*
a.	$100.00	6%	0.25	4
b.	$1500.00	$7\frac{1}{4}$%	4	4
c.	$700.00	20%	2	4
d.	$1000.00	6%	12	4
e.	$1000.00	6%	10	4
f.	$1000.00	6%	10	12
g.	$1000.00	6%	10	365

5. Modify the program of exercise 4, to read in the number of times per year that the interest rate is to be compounded. This number will replace the constant 4 in the preceding exercise. (See the data above.)
6. To find the cube root of a number whose cube root is less than 0.0001, it would not seem sensible to stop when two successive approximations agree to within 0.0001 in absolute value. Modify the cube root program to base the stopping criterion on *relative* error, that is, on the difference between two successive approximations *relative* to the size of the cube root itself.

3.7 *How Can We Avoid Mistakes in Loop Structures?*

Many programs include a structure of one or more instructions to be repeated a specified number of times. For example, we saw in Section 3.6 how to control a payroll calculation when the number of employees is given as the first item of data to be read.

Here is another example—a program reads a value for K and multiplies together all of the positive integers (1 × 2 × 3 × ...) that are not larger than K in value. The resulting product is called the factorial of K.

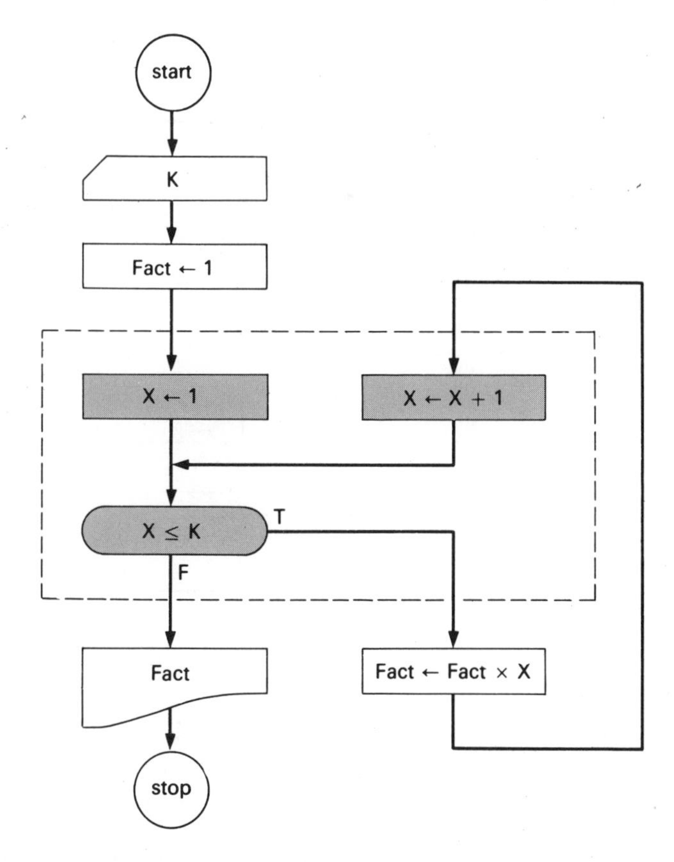

This program demonstrates a typical count-controlled loop pattern. The central purpose of the loop in this case is to repeatedly execute (*iterate*) the single instruction

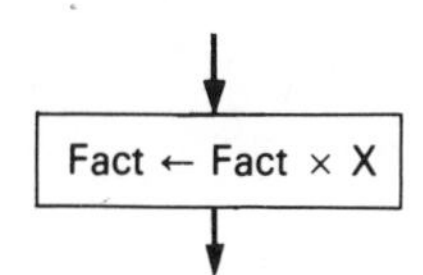

The three instructions inside the dashed outline control the loop. Loop control involves an *index* variable, in this case X. This variable is *initialized* (initially set) to the value 1. Its value is *incremented* (increased by 1) each time the loop is repeated. The iteration continues until the value of the index variable exceeds the *upper limit*, in this case the value of K. When the upper limit is exceeded, the iteration terminates and the next instruction beyond the loop is executed.

In some cases the index may be used in the central portion of the loop. But this is not mandatory; below is a program to multiply the value of A by itself K times (to compute the *K*th power of A):

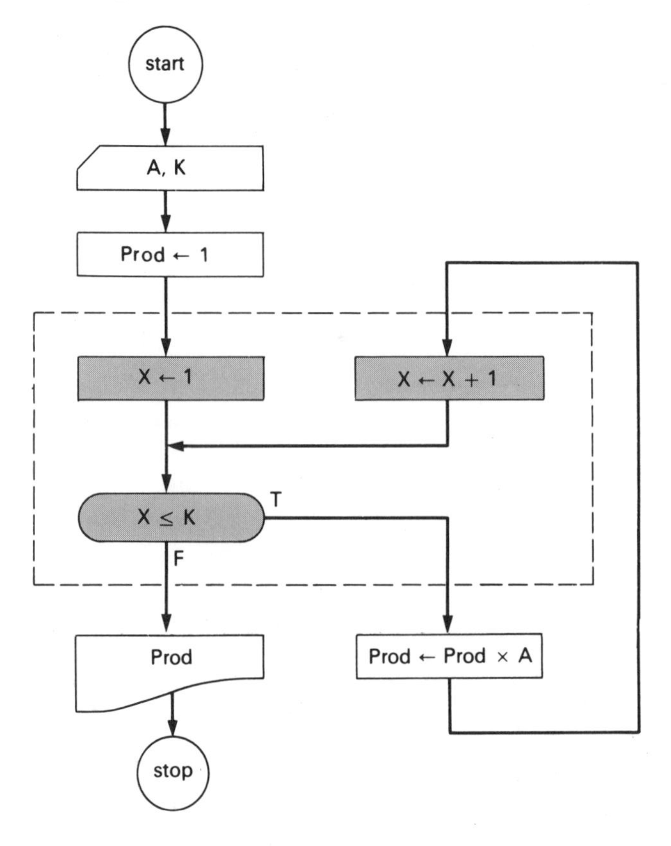

The Iteration Instruction Because the count-controlled loop pattern occurs so frequently, programming languages provide a special automatic "shorthand" for it to reduce the likelihood of programming errors. In flowcharts we shall replace the three boxes inside the dashed outline by a single composite box, corresponding to the iteration instruction in conventional programming languages. This box has the form

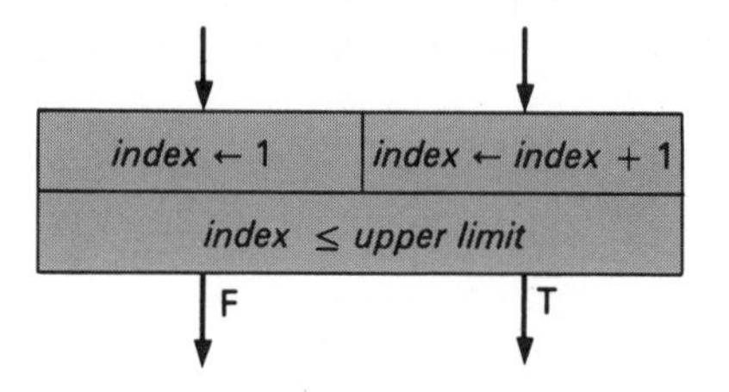

Note that there are always exactly two lines entering the top of this iteration box, and two lines leaving the bottom. The instructions being controlled are connected to the outgoing line labeled *T*. The *F* exit leads to the action that is to be taken when the loop is completed. For example, we may rewrite the two previous examples with iteration boxes, as follows:

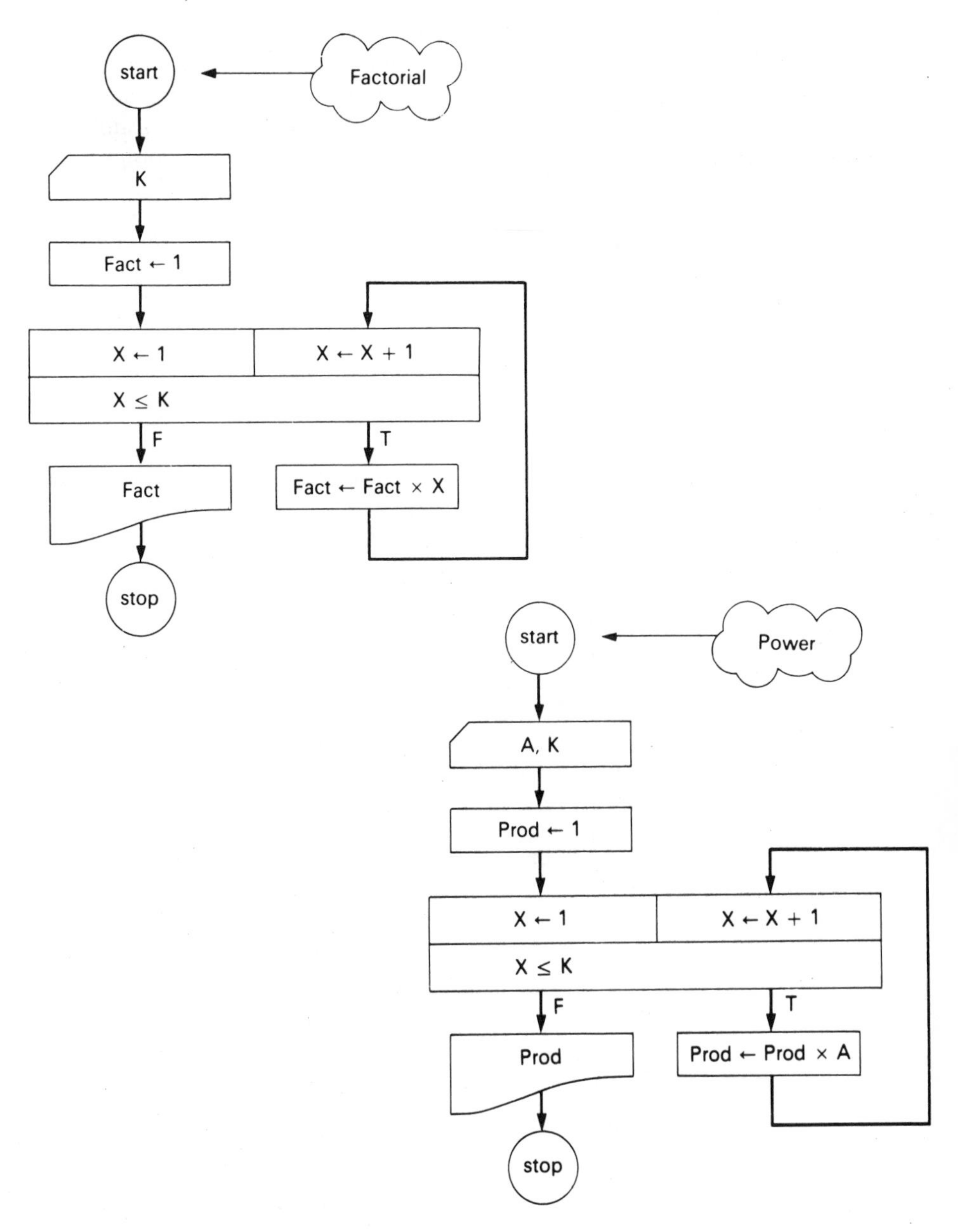

This compressed form clearly separates the loop control instructions from the instructions being controlled. We no longer have to consider each time we write a loop, for example, whether to use $<$ or $\leq$ when we compare the index with the upper limit.

Note that the index in an iteration must be a variable. The limit may be any expression. The integer *nearest* the value of this expression is the number of times the loop will be repeated (that is, the limit expression will be rounded if necessary). If this integer is less than 1, the loop is repeated 0 times; that is, it is skipped over, and execution continues at the next instruction beyond the loop.

Instruction Blocks In the two previous examples, the central portion of the loop consisted of a single instruction. More commonly, however, the calculation repeated is a group of instructions, or *block*. We rewrite the payroll program, using the iteration instruction, to illustrate the repetition of a block of instructions.

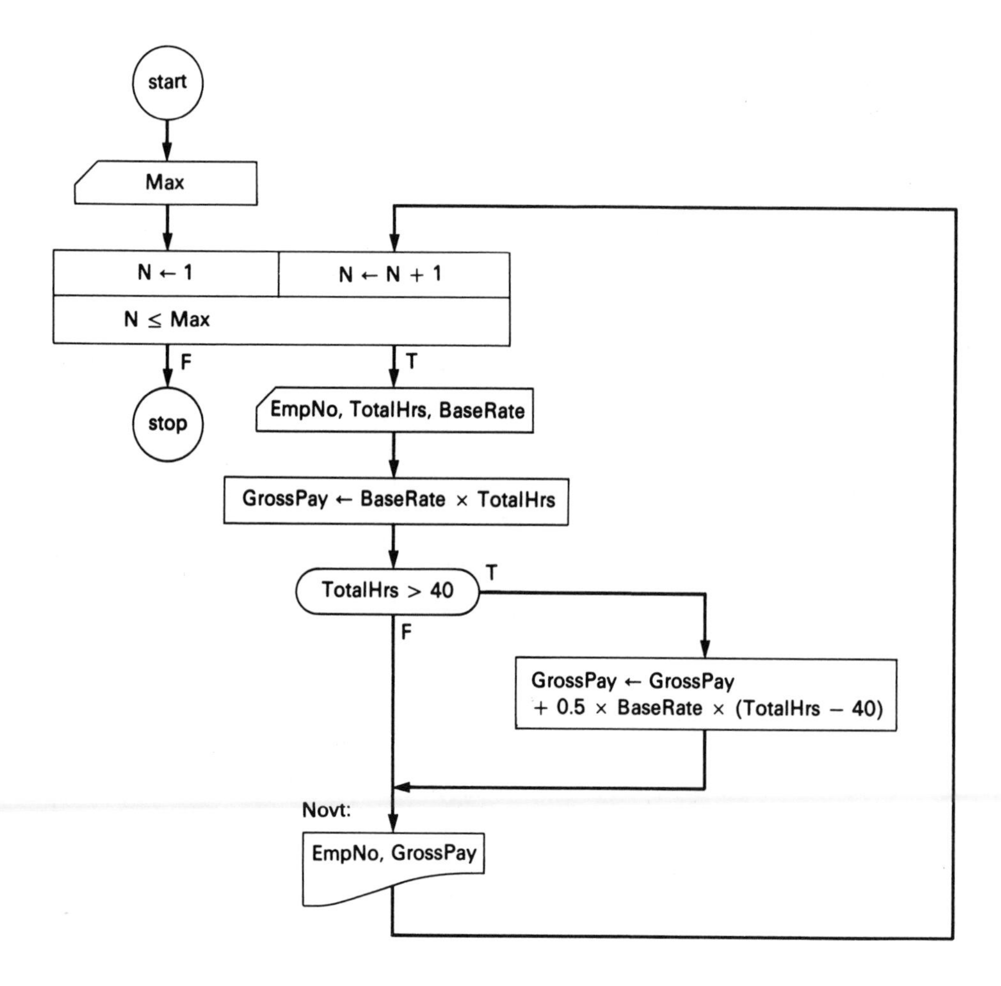

Nested Iterations Suppose we want to determine all the divisors of a given large integer (such as 72). One way to do this is to divide the given integer by each integer less than or equal to it, and look at the remainder.

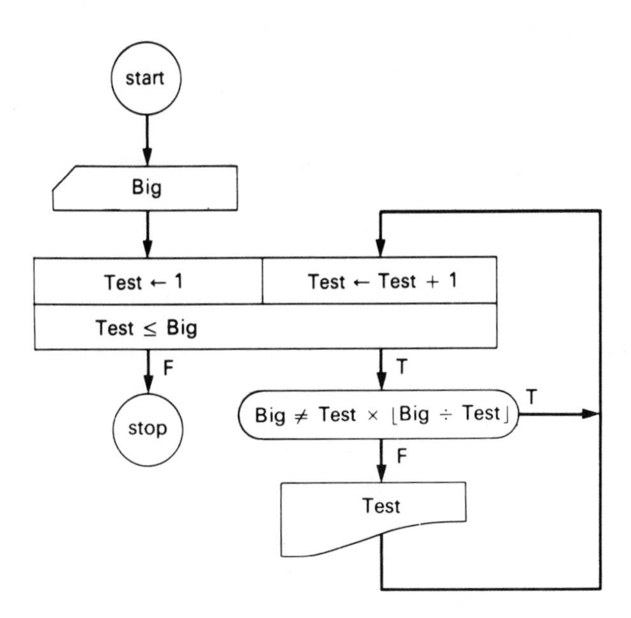

If we want to do this for all integers between 1 and 100, we make this entire program (except for the read and stop instructions) into a block, which in turn is controlled by another iteration instruction.

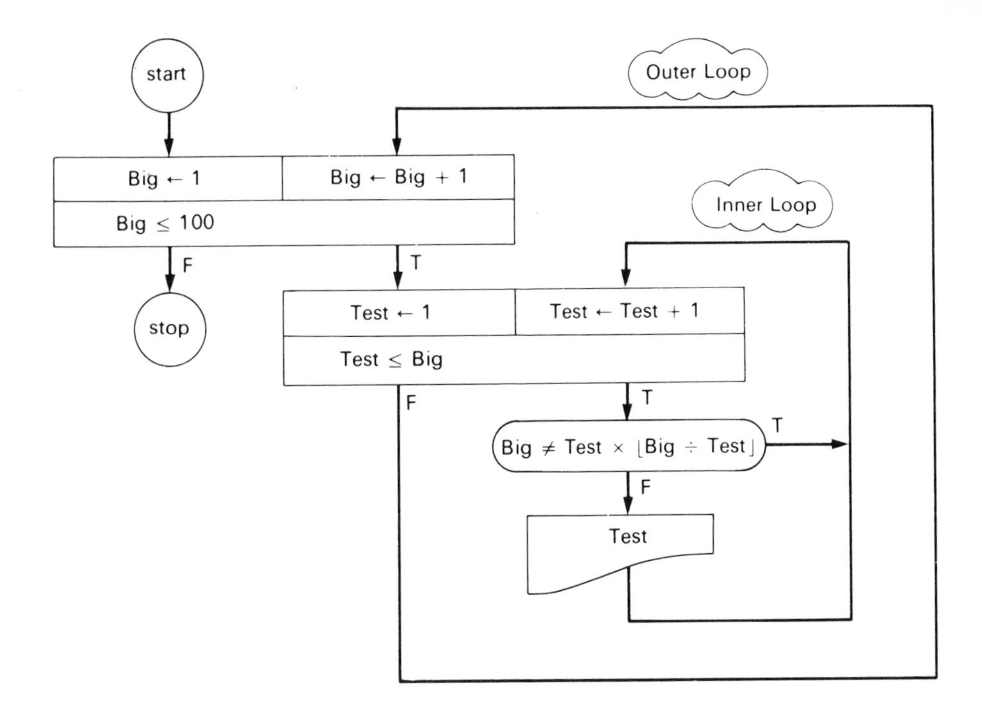

Tracing this flowchart as it will be executed, we see that Big is first given the value 1, and the "inner loop" is executed exactly as in the previous example. When the value of Test in this inner loop reaches its upper limit, the inner loop is complete. The value of Big is then incremented and the process is repeated. Thus the entire inner loop is executed once for each value of Big from 1 to 100.

A Warning The whole purpose of the iteration shorthand is to make use of count-controlled loops nearly automatic, so that the user can write the instructions correctly with almost no thought. Still, a little care is needed to avoid some common blunders.

For instance, it is dangerous to assign a new value to the index variable by means of instructions within the block controlled by the index. Assigning a small value (less than the limit) to the index, each time the loop is executed, would prevent the loop from terminating. Although an assignment to this variable might be valid in some cases, the experienced programmer develops a natural inhibition that causes him to question each such occurrence until he is certain of its validity.

In some programming languages, the expression used as the upper limit of the iteration is evaluated once, before the block is executed, and any assignment within the block to the variables occurring in this expression will have no effect. In other languages this is not the case, however, and strange effects may result if the iteration parameters are modified during execution of the loop.

Exit from a Loop When a block of instructions is used in a count-controlled loop, we may wish to discontinue execution of the loop prematurely, before the index has reached its upper limit. We accomplish this by incorporating within the block a control instruction referring to a point outside the loop.

For example, we may want to find the smallest divisor (other than 1) of a given positive integer greater than 1. To find the smallest divisor, there is no need to find all the divisors and choose the smallest. A better procedure is to test the trial divisors in increasing order, so that the first divisor found must be the smallest. We test 2 to see if it divides the given number; if so, 2 is the smallest divisor. Otherwise, we try 3, 4, and so on. As soon as we have found a divisor, the loop should be discontinued so that the computer will not waste time looking for other divisors.

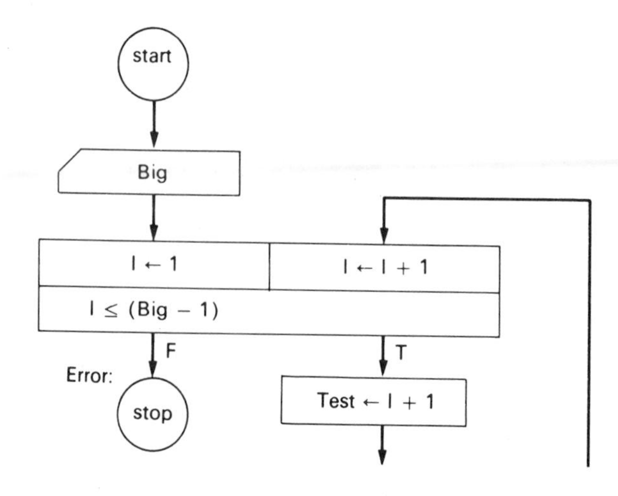

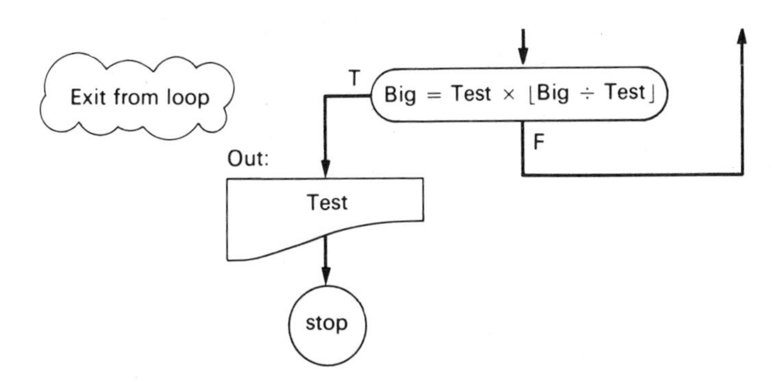

Notice the stop instruction with the label Error. Analysis of this example shows that the loop should never terminate in the regular way (by the value of the counter I exceeding that of the upper limit (Big − 1)) unless Big has the value 1. This is because a positive integer always has *itself* as a divisor and, if no smaller divisor has been found first, this divisor will be found when the loop is executed with the value of I equal to that of (Big − 1). However, when using the iteration box the programmer is compelled to consider what course of action he would take if the loop were to terminate "normally," even if he never expects this to occur. In this example, it probably indicates an error such as a negative value in the input data.

If the upper limit will never be reached, why is a count-controlled loop used at all? The answer is that a counter (Test) is needed in any case in the evaluation of the expression that appears in the test for the divisor. The user may consider the iteration pattern a convenience, and to use it he needs to supply only one additional parameter (the upper limit), but this parameter has some value as a safety feature as well.

Flowchart

Factorial:

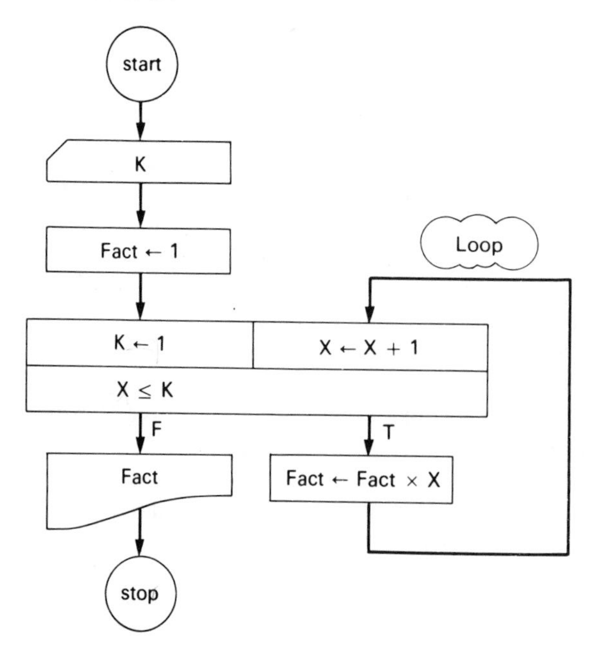

Power:

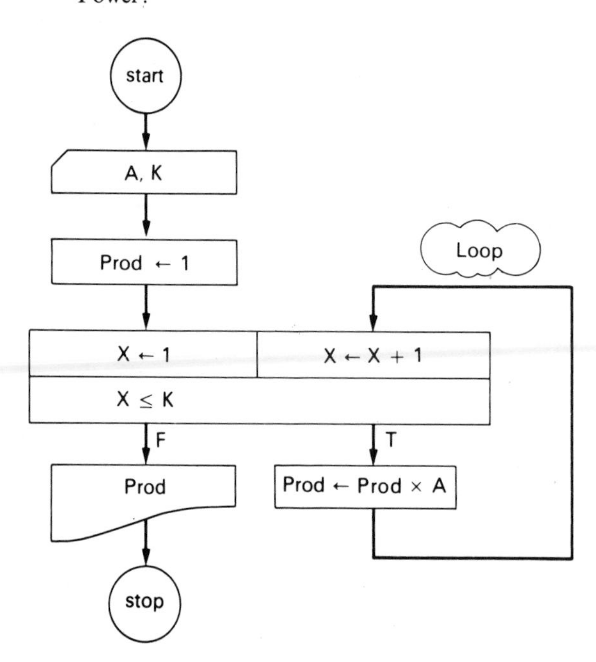

Payroll with iteration instruction:

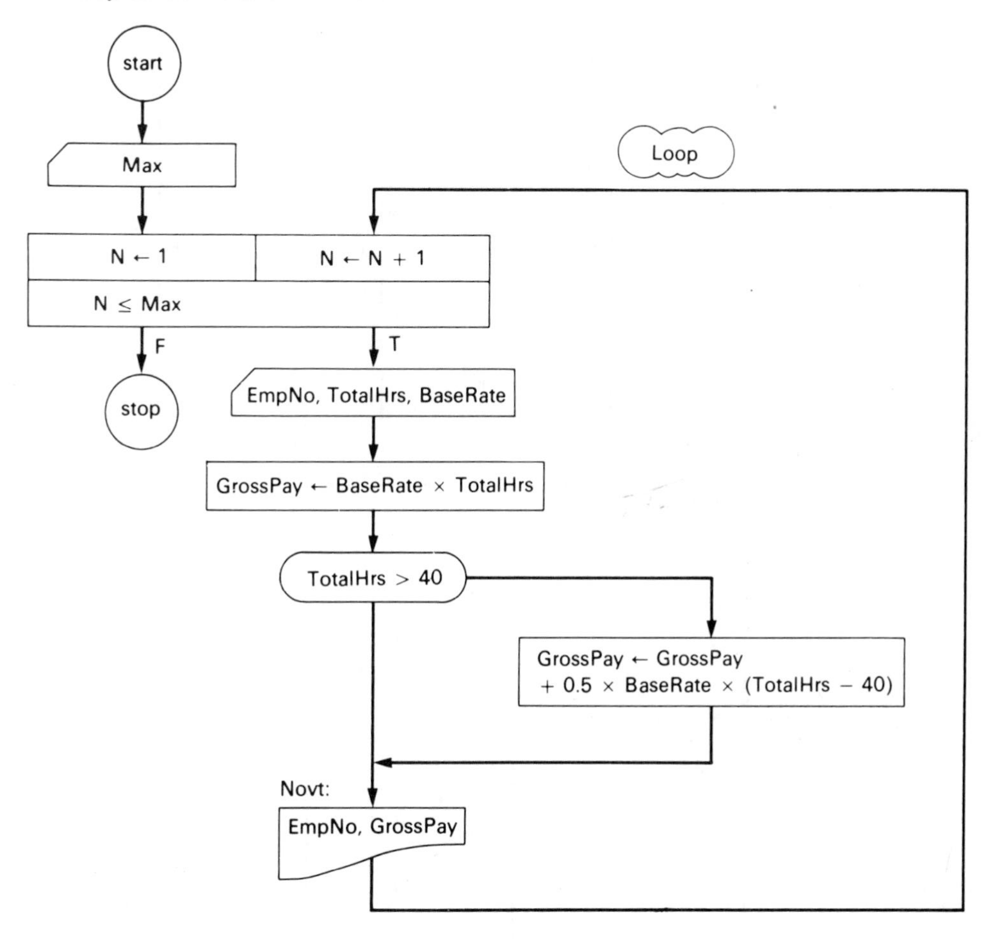

All divisors of integers up to 100:

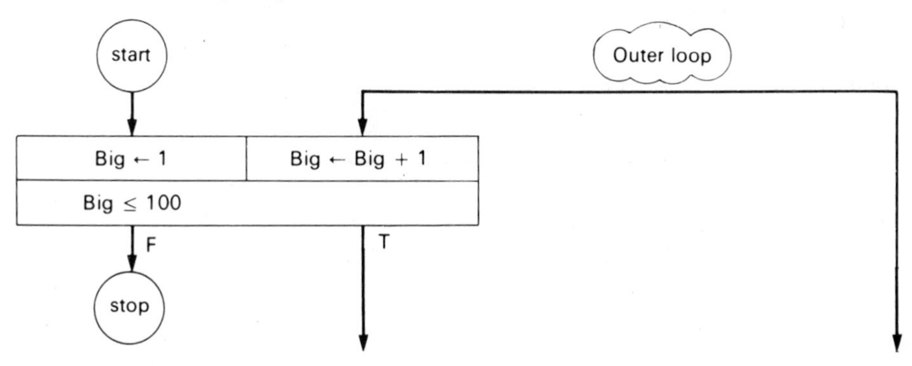

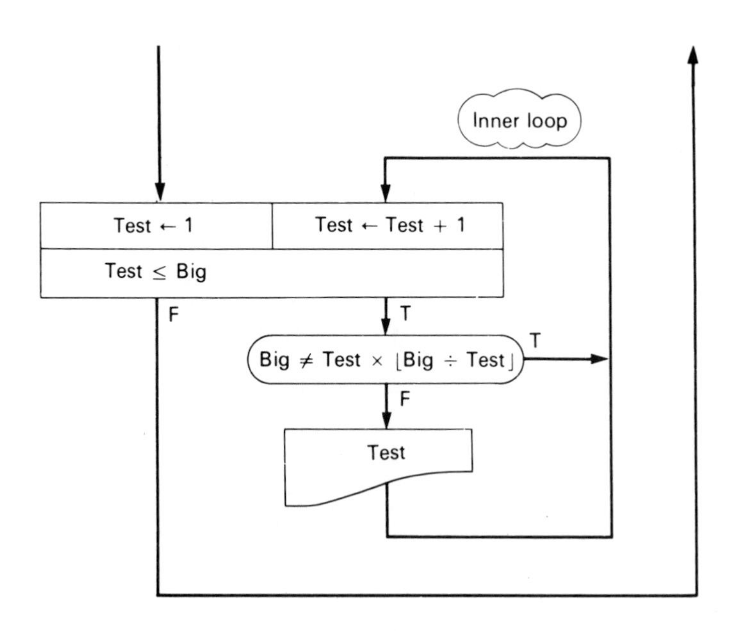

Smallest divisor of an integer:

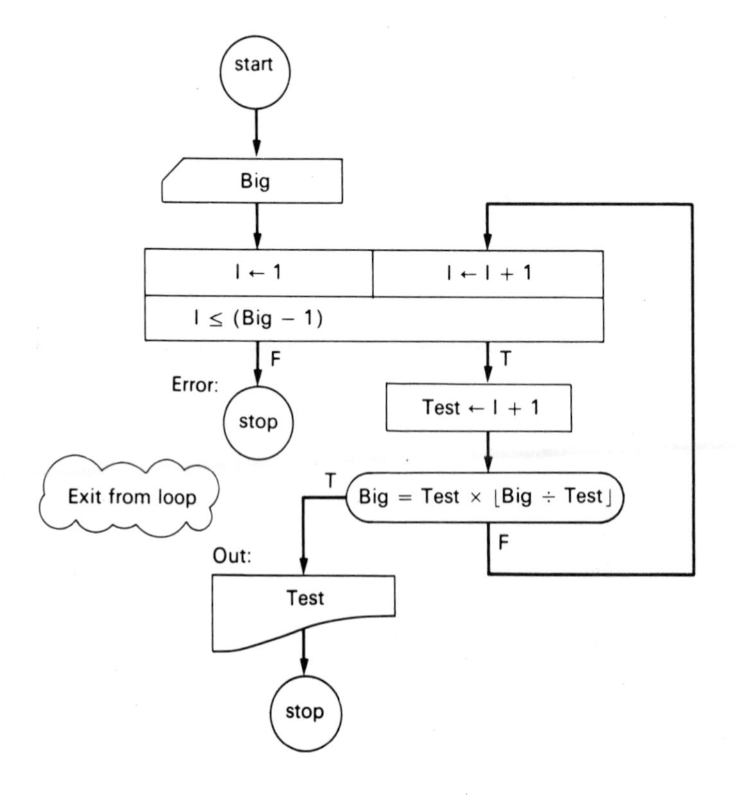

Basic The iteration instruction in Basic has the form

FOR *index* = 1 TO *upper limit*

and the block of instructions to be repeated is terminated by the statement

NEXT *index*

A lower limit other than 1 may also be used.

Factorial:

```
100 INPUT K
110 LET F = 1
120 FOR X = 1 TO K
130    LET F = F * X
140    NEXT X
150 PRINT F
160 END
```

Power:

```
200 INPUT A, K
210 LET P = 1
220 FOR X = 1 TO K
230    LET P = P * A
240    NEXT X
250 PRINT P
260 END
```

Payroll with iteration instruction:

```
300 INPUT M
310 FOR N = 1 TO M
320    INPUT E, H, R
330    LET G = R * H
340    IF H <= 40 THEN 360
350    LET G = G + 0.5 * R * (H - 40)
360    PRINT E, G
370    NEXT N
380 END
```

All divisors of integers up to 100:

```
410 FOR B = 1 TO 100
420    FOR T = 1 TO B
430       IF B <> T * INT (B/T) THEN 460
440       PRINT T
450       NEXT T
460    NEXT B
470 END
```

Smallest divisor of an integer:

```
500 INPUT B
510 FOR T = 2 TO B
520    REM: POSSIBLE EXIT FROM LOOP
530    IF B = T * INT (B/T) THEN 560
540    NEXT T
550 GO TO 570
560 PRINT T
570 END
```

Fortran The iteration instruction in Fortran has the form

DO *label index* = 1, *upper limit*

and the block of instructions to be repeated is terminated by the statement

label CONTINUE

A lower limit other than 1 may also be used. In Fortran, the block of instructions is executed *once* (rather than 0 times) if the upper limit is smaller than the lower limit. Therefore, we may find it necessary to insert extra instructions to skip the entire loop in this case. When the lower limit is 1 and the upper limit has a positive integer value, no such extra instructions are needed.

Factorial:

```
      READ (5, 81) K
81    FORMAT (8 I 10)
      IFACT = 1
      IF (K .LT. 1) GO TO 141
      DO 140 IX = 1, K
         IFACT = IFACT * IX
140      CONTINUE
141   CONTINUE
      WRITE (6, 81) IFACT
      STOP
```

Note that the index and the limit of a DO instruction must be of integer type.

Power:

```
      READ (5, 82) A
82    FORMAT (8 F 10.0)
      READ (5, 81) K
81    FORMAT (8 I 10)
      PROD = 1.0
      DO 240 IX = 1, K
         PROD = PROD * A
240      CONTINUE
      WRITE (6, 83) PROD
83    FORMAT (8 F 15.3)
      STOP
```

Payroll with iteration instruction:

```
      READ (5, 81) MAX
81    FORMAT (8 I 10)
      IF (MAX .LT. 1) STOP
      DO 370 N = 1, MAX
         READ (5, 82) EMPNO, TH, BR
         GP = TH * BR
         IF (TH .LE. 40.0) GO TO 360
         GP = GP + 0.5 * BR * (TH - 40.0)
360      WRITE (6, 84) EMPNO, GP
84       FORMAT (F 15.0, F 15.2)
370      CONTINUE
      STOP
```

All divisors of integers up to 100:

```
      DO 460 IB = 1, 100
         DO 450 IT = 1, IB
            IF (MOD (IB, IT) .NE. 0)
     1         GO TO 460
            WRITE (6, 81) IT
81          FORMAT (8 I 10)
450         CONTINUE
460      CONTINUE
      STOP
```

Smallest divisor of an integer:

```
      READ (5, 81) IB
81    FORMAT (8 I 10)
      IF (IB .LT. 2) STOP
      DO 540 IT = 2, IB
C        POSSIBLE EXIT FROM LOOP
         IF (MOD (IB, IT) .EQ. 0) GO TO 560
540      CONTINUE
      STOP
560   WRITE (6, 81) IT
      STOP
```

PL/1 The iteration instruction in PL/1 has the form

```
do index = 1 to upper limit;
```

and the block of instructions to be repeated is terminated by the statement

```
end;
```

A lower limit other than 1 may also be used. The index and the upper limit should be declared to be of fixed type.

Factorial:

```
FACTORIAL: procedure options (main);
    declare (K, X, FACT) fixed;
    get list (K);
    FACT = 1;
    do X = 1 to K;
       FACT = FACT * X;
       end;
    put skip list (FACT);
    end FACTORIAL;
```

Power:

```
POWER: procedure options (main);
    declare (A, PROD) float;
    declare (K, X) fixed;
    get list (A, K);
    PROD = 1;
    do X = 1 to K;
       PROD = PROD * A;
       end;
    put skip list (PROD);
    end POWER;
```

Payroll with iteration instruction:

```
PAYROLL: procedure options (main);
    declare (EMPNO, TOTALHRS,
       BASERATE, GROSSPAY) float;
    declare (MAX, N) fixed;
    get list (MAX);
    do N = 1 to MAX;
       GET LIST (EMPNO, TOTALHRS,
          BASERATE);
       GROSSPAY = BASERATE *
          TOTALHRS;
       if TOTALHRS < = 40 then go to
          NOVT;
       GROSSPAY = GROSSPAY + 0.5 *
          BASERATE * (TOTALHRS - 40);
    NOVT: put skip list (EMPNO,
          GROSSPAY);
       end;
    end PAYROLL;
```

All divisors of integer up to 100:

```
FACTORS1: procedure options (main);
    declare (BIG, TEST) fixed;
    do BIG = 1 to 100;
       do TEST = 1 to BIG;
          if MOD (BIG, TEST) = 0 then go
             to C;
          put skip list (TEST);
       C: end;
       end;
    end FACTORS1;
```

Smallest divisor of an integer:

```
FACTORS2: procedure options (main);
    declare (BIG, TEST) fixed;
    get list (BIG);
    do TEST = 2 to BIG;
       /* Possible exit from loop */
       if MOD (BIG, TEST) ¬ 0 then go to
          OUT;
       end;
ERROR: go to LAST;
OUT: put skip list (TEST);
LAST: end FACTORS2;
```

Words to Remember

index	upper limit	nested iteration
increment	iteration instruction	

Exercises

1. The Fibonacci algorithm given in Section 3.5 does not use a count-controlled loop. Modify the algorithm so that it can use the iteration notation. The block of instructions to be repeated should consist of the two assignments

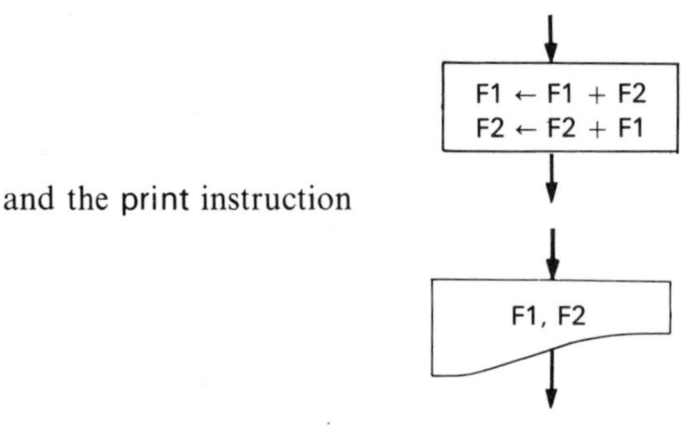

and the print instruction

and should be repeated 10 times.

2. It can easily be proved (Knuth, 1968, p. 80) that the greatest common divisor of two consecutive Fibonacci numbers is 1. Modify the program of exercise 1 to verify this fact by computing and printing the greatest common divisor of F1 and F2 after repeating the loop 10 times.
3. Note that if Test is a divisor of Big, then Big ÷ Test is another divisor. Thus the divisors of Big can be found two at a time. So, to determine all the divisors of a given integer, we can replace the program given on p. 119 with the following:

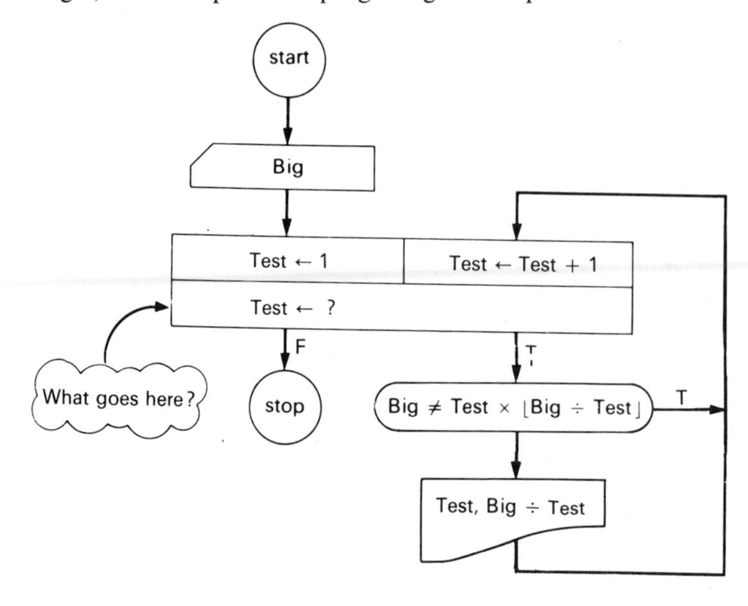

What value should be used as the upper limit of the iteration for each of the following values of Big: (a) 720, (b) 960, (c) 144, (d) 75, (e) 3, (f) 89, (g) 1?

4. Write a program to find the *largest* divisor of Big, other than Big itself. Use the values of Big given in exercise 3.
5. Write a program to do some ordinary statistical calculations. Suppose that the first number in the reader queue is the value of N, the number of cases to be analyzed. This is followed by N pairs, each consisting of an X value and a Y value. The program should compute the average of the X values (called Xavg) and the average of the Y values (called Yavg), also the X and Y variances and standard deviations (Xvar, Yvar, Xdev, Ydev), the XY covariance (Cov), and the correlation (Corr). The formulas are:

$$Xavg = \frac{\Sigma X}{N} \qquad Yavg = \frac{\Sigma Y}{N}$$

$$Xvar = \frac{\Sigma X^2}{N} - Xavg^2 \qquad Yvar = \frac{\Sigma Y^2}{N} - Yavg^2$$

$$Xdev = Xvar \qquad Ydev = Yvar$$

$$Cov = \frac{\Sigma XY}{N} - (Xavg \times Yavg) \qquad Corr = \frac{Cov}{(Xdev \times Ydev)}$$

Use the following values of X and Y:

X	480	500	520	540	560	493	503	600
Y	56	61	78	71	82	59	67	94

6. The factorial calculation can be used to find the numbers in Pascal's triangle (see Section 2.4, exercise 3; Section 2.7, exercise 9). The binomial coefficient $C[n, r]$ is given by the formula

$$\frac{n!}{r!(n-r)!}$$

In this notation, $n!$ denotes the factorial of n. Notice that, in the program we have given for calculating factorials, the value 1 emerges when we try to find the factorial of 0. This agrees with the convention used in mathematics for the definition of factorials; further, it produces the correct result for entries along the edges of Pascal's triangle. Use the formula involving factorials to compute and print the first eight lines of Pascal's triangle. (Note: This is by no means the most efficient way to compute these values. See Kemeny and Kurtz (1971, p. 17) for a better way to calculate the individual numbers. If the entire triangle is to be computed, the method of Section 2.7, exercise 9, is much better.)

7. Write a program that will read a sequence of numbers, until a sentinel value of 0 is encountered, and will then print the largest number and its position in the sequence. (Hint: Keep track of the "largest number found so far.")

a. (19, 13, 5, 27, 1, 26, 31, 16, 2, 9, 11, 21)
b. (2, 1, 4, 3, 6, 5, 8, 7, 10, 9)
c. (19, 13, 8, 27, 3, 14, 7, 16, 17, 6, 24)

3.8 What Are Some Numerical Computations That Have Many Applications?

At this point we have surveyed all the essential techniques of programming for problems with ordinary numeric variables and constants. Let us now look at a few examples of numerical computations which, while rather simple, appear in many different computer applications.

Dynamics Often, one or several variables will represent quantities that change in time according to some simple law. Examples of such variables are competing biological populations, the swinging pendulum or the electronic oscillator in physics and electronics, radioactive decay in nuclear chemistry, demographic models of the population explosion, and even compound interest. In all these applications, the *rate* of change of some quantity is proportional to the *amount* present.

In the computational model, we invent a variable for each dynamic quantity. After establishing the initial values of the variables, we begin executing a loop. Each iteration represents a time lapse—a fraction of a second for a problem in radioactivity or electronics, some months for compound interest, years for demography.

At each iteration, we change each of the variables according to the simple law based on the values of the variables. In some cases, such as radioactive decay or compound interest, there is only one dynamic quantity, whose changes over each time interval is directly proportional to the value of the quantity itself. In the formula $P \leftarrow P + (K \times T) \times P$, P is the amount (of the isotope or the principal), T is the time interval, and K is the proportionality constant (isotope decay rate or interest rate per time interval).

For problems with competing populations or with oscillations, we may have two variables, with the change in each depending on the amount of the other. For instance, some foxes and some rabbits live in a certain valley; the fox population will increase faster if there are more rabbits, and the rabbit population will decrease faster if there are more foxes. In problems of this kind, there may be an equilibrium level at which both populations can remain in balance.

```
F ← F + (K1 × T) × (R − Ro)
R ← R − (K2 × T) × (F − Fo)
```

In an oscillation problem such as a pendulum, the two variables are position and velocity, each measured positively in one direction and negatively in the other. When the velocity is large and toward the right, the displacement increases rapidly in the same direction. When the displacement toward the right becomes large enough, there is a large acceleration toward the left, causing the velocity (measured toward the right) to decrease:

```
D ← D + (K1 × T) × V
V ← V − (K2 × T) × D
```

Some programs for these examples are included as exercises at the end of this section.

Random Processes One of the most important uses of a computer is for simulation. In some cases, the process to be simulated is completely deterministic—that is, the effects of all the variables on the others are completely known. We have made this sort of assumption in the examples in dynamics. But in other processes we may need to simulate the effects of chance. Many programming languages provide a *random number generator*, which can be used as a source of randomly chosen numerical values. Where such a feature is not available, a simple program can be constructed to generate random numbers.[3]

We assume that there is a special name, rnd, that can be used in any expression where we need to obtain random numbers. In effect, the name acts like the name of a cell in which nothing can be stored, but which keeps changing its value so that each time we use the name in an expression we get a different value. The values vary uniformly between 0 and 1; that is, if we divide the interval from 0 to 1 into equal subintervals, over the long run the values will tend to fall equally often into each subinterval.

To simulate a process that occurs with probability P, where P is a value between 0 and 1, we fetch a value of rnd, and test whether its value is less than P. For instance, in the long run the value obtained from rnd will be less than 0.25 exactly one-fourth of the time.

An Example: Simulating the Game of Table Tennis ("Ping-Pong") One way to simulate a game of table tennis is to assume that the "effectiveness" of each player as a server is known. That is, there is a known probability, P1, that Player 1 will make the point when it is his turn to serve, and a probability, P2, that Player 2 will make the point when it is his serve. (In this game, one player or the other wins a point each time the ball is served.)

For each serve, we fetch a value of rnd, and compare it with P1 or P2 (depending on which player is serving). The server wins a point if the random number is less than his "effectiveness" value; otherwise, his opponent wins the point.

Let us assume that Player 1 always serves first. The serve changes after every five serves. The game ends when one player has 21 points, if the other player has 19 or less. In case the score 20–20 is reached, the players alternate service from then on until one player has two more points than the other.

---

[3] A program to find the next number in the Fibonacci sequence, then to take the remainder after dividing this number by some reasonably large constant, is adequate for some purposes. Much better sequences can be constructed with little difficulty, but the details depend on the characteristics of particular computer number representations (see Knuth, 1969).

Another Example: Arithmetic Drill A student is having trouble learning part of his multiplication table, and we want to try to use the computer to help him learn. We propose to use an on-line computer system and to present the problems to the student at the console. The computer will keep track of his progress by evaluating his responses.

Let us suppose that the student has progressed fairly well and needs further drill on only four multiplication facts:

(a) $6 \times 8 = 48$ (b) $7 \times 8 = 56$
(c) $6 \times 9 = 54$ (d) $7 \times 9 = 63$

We want to present these four problems to the student in random sequence, but in such a way that he is drilled more often on the questions that he misses, and less often on those he answers correctly.

We give each question a *weight*, or relative frequency, which at the beginning is the same for each question. Whenever he misses a question we double its weight, and whenever he answers it correctly we halve its weight. For example, we start with the four weights (1, 1, 1, 1). The first question presented happens to be 6×9. The student answers incorrectly, so we change the weights to (1, 1, 2, 1), doubling the weight of the question that was missed. The next question happens to be 7×9, which he answers correctly. The weights are adjusted to $(1, 1, 2, \frac{1}{2})$. Next he answers 6×8 correctly, and the weights become $(\frac{1}{2}, 1, 2, \frac{1}{2})$.

Dividing each weight by the sum of all the weights, which is now 4, we obtain the desired *probabilities* (0.125, 0.25, 0.5, 0.125). Thus, if the weights did not change, we would want to ask question (c) half the time, question (b) one-fourth of the time, and questions (a) and (d) each one-eighth of the time.

How can we arrange to ask each question with the proper relative frequency? The trick is to compute a list of *cumulative* probabilities, adding to each of the given probabilities all those values that precede it. Thus we obtain

$$(0.125, 0.375, 0.875, 1.000)$$

As a check, we note that the last entry in this list, which must be the sum of all the entries in the given probability list, is 1.000.

We now generate a random number (uniformly distributed between zero and one), and compare it with each entry in the cumulative probability list, until we find a list entry that is larger than our random number. The first entry, 0.125, will be larger than the random number about one-eighth of the time. About one-fourth of the time, the first entry will be smaller and the second entry will be larger than the random number. About half the time, the third entry will be the first to exceed the random number, and the last number will remain to be chosen about one-eighth of the time. Thus we will ask the third question whenever the random number is between 0.375 and 0.875; this will happen about half the time, which is just what we want (Figure 3.8a).

The program is simple. We present a question and evaluate the response. Then we adjust the weight of the question just asked, so that it will be asked more frequently if the response was incorrect and less frequently if the response was correct. We immediately calculate the sum of the new weights, and divide this into each of the weights to find the new desired probabilities; also we compute the new cumulative probabilities. We then generate a new random number, and compare it with the

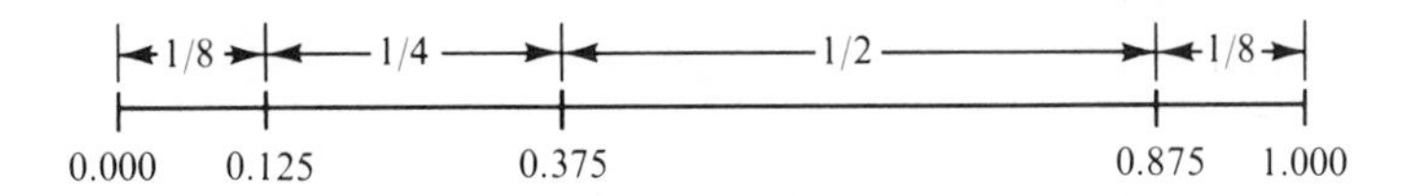

3.8a. *Cumulative probabilities are used to define the endpoints of a set of partitions of the interval from 0 to 1. A "uniform" random number will fall into any one of the partitions with a frequency depending on the width of the partition.*

entries in the cumulative probability list; finally we choose the next question to present, corresponding to the first entry on the cumulative list that exceeds the new random number.

Basic Basic includes the function RND for obtaining random numbers. The function name RND may be used in any expression; each time it appears, a different value (between 0 and 1) is generated.

Fortran Standard Fortran does not include a function for obtaining random numbers, but such a function (under some name) is supplied at almost every installation. Obtain details from your computer center.

PL/1 There is no standard PL/1 function for obtaining random numbers. However, such a function may be supplied at your computer center. (Otherwise, see Knuth, 1969.)

Exercises

1. The rate of radioactive decay is usually expressed in terms of the half-life, H (the time required for the isotope to decay to one-half its original mass). The proportionality constant K is approximately 0.7 ÷ H. For an isotope whose half-life is 100 seconds, compute the amount remaining after each second, from an initial time at which 40 grams were present. (The amount after 100 seconds should be about 20 grams.)
2. For compound interest, K is just the annual interest rate. Use the data suggested for exercises 4 and 5, Section 3.6, to compute compound interest.
3. Compute the problem of competing populations over a period of three years, adjusting the population figures once each month.
   Equilibrium level: 100 rabbits, 40 foxes.

Rates: K1 = 0.04; K2 = 0.25; T = 1 month.
Initial levels: 160 rabbits, 22 foxes.

4. For a swinging pendulum, K1 is 1, and K2 is ($g \div L$), where g is the gravitational acceleration constant (9.8 meters per second^2) and L is the length of the pendulum in meters. Compute the displacement and velocity of a pendulum, over a period of 10 seconds, if the length is 5 meters. Begin with an initial displacement of 0.1 meter, and an initial velocity of 0. Repeat the calculation with a time interval of 0.1 second, 0.025 second, and 0.01 second. Do the results agree? Explain, if you can.
5. Using an on-line system, write a program for an arithmetic drill on four problems. Include instructions to stop when the sum of the weights is less than 0.1.
6. Write a program to simulate a sequence of three complete table tennis games. For game 1, use P1 = 0.46, P2 = 0.43; for game 2, P1 = 0.39, P2 = 0.47; for game 3, P1 = 0.63, P2 = 0.71. For each serve, print an indication of the server and the cumulative score.

References

Alfred, U., "Exploring Fibonacci Numbers." *Fibonacci Quarterly* 1: 57–63 (1963).

American National Standards Institute, *American National Standard Vocabulary for Information Processing*. New York: American National Standards Institute, 1966.

Ball, W. W. R., *Mathematical Recreations and Essays*, rev. ed. New York: Macmillan, 1960.

Banerji, R. B., *Theory of Problem Solving; An Approach to Artificial Intelligence*. New York: American Elsevier, 1969.

Bellman, R., K. Cooke, and J. Lockett, *Algorithms, Graphs, and Computers*. New York: Academic Press, 1970.

Birkhoff, G., and T. C. Bartree, *Modern Applied Algebra*. New York: McGraw-Hill, 1970.

Cole, R. W., *Introduction to Computing*. New York: McGraw-Hill, 1969.

Encyclopaedia Britannica. Chicago: Encyclopaedia Britannica, Inc., 1970.

Forsythe, A. I., et al., *Computer Science: A Primer*. New York: Wiley, 1969.

Gies, J., and F. Gies, *Leonard of Pisa and the New Mathematics of the Middle Ages*. New York: Crowell, 1969.

Hoggatt, V. E., *Fibonacci and Lucas Numbers*. Boston: Houghton Mifflin, 1969.

Hull, T. E., and D. D. F. Day, *Computers and Problem Solving*. Reading, Mass.: Addison-Wesley, 1968.

Kemeny, J. G., and T. E. Kurtz, *Basic Programming*, second ed. New York: Wiley, 1971.

Knuth, D. E., *The Art of Computer Programming*: Vol. 1, *Fundamental Algorithms*. Reading, Mass.: Addison-Wesley, 1968.

Knuth, D. E., *The Art of Computer Programming*: Vol. 2, *Seminumerical Algorithms*. Reading, Mass.: Addison-Wesley, 1969.

Land, F., *The Language of Mathematics*. Garden City, N.Y.: Doubleday, 1963.

Maisel, H., *Introduction to Electronic Digital Computers*. New York: McGraw-Hill, 1969.

Markov, A. A., *Theory of Algorithms*. Translated by J. J. Schorr-Kon et al. Jerusalem: Israel Program for Scientific Translations, 1961.

McCoy, N. H., *Introduction to Modern Algebra*. Boston: Allyn and Bacon, 1960.

Minsky, M., *Computation, Finite and Infinite Machines*. Englewood Cliffs, N.J.: Prentice-Hall, 1967.

Newman, J. R., ed., *The World of Mathematics*, 4 vols. New York: Simon and Schuster, 1956.

Polya, G., *How to Solve It*. Princeton, N.J.: Princeton University Press, 1945.

Rice, J. K., and J. R. Rice, *Introduction to Computer Science*. New York: Holt, Reinhart and Winston, 1969.

School Mathematics Study Group, *Algorithms, Computation, and Mathematics*. Stanford, Calif.: Stanford University Press, 1965.

Simon, H. A., "The General Problem Solver," in M. Greenberber, ed., *Computers and the World of the Future*. Cambridge, Mass.: M.I.T. Press, 1962. Pp. 97–114.

Sterling, T. D., and S. V. Pollack, *Computing and Computer Science*. New York: Macmillan, 1970.

Trakhtenbrot, B. A., *Algorithms and Automatic Computing Machines*. Lexington, Mass.: Heath, 1963.

Turnbull, H. W., *The Great Mathematicians*. London: Methuen, 1951.

Walker, T. M., and W. W. Cotterman, *An Introduction to Computer Science and Algorithmic Processes*. Boston: Allyn and Bacon, 1.970.

Weland, K., "Some Rabbit Production Results Involving Fibonacci Numbers." *Fibonacci Quarterly* 5: 195–200 (1967).

4 Representation of Information

4.1 How Much Information Does It Take to Express an Idea?

As we have noted, the purpose of information is to communicate an idea by means of a change in the physical universe, so that it can be used to re-create a similar idea at another place or time. We are now ready to consider *quantity* of information. Suppose we have a message or piece of information and a machine for transmitting or storing information. We want to know how much *time* will be required to transmit the message, or how much *space* will be needed to store it.

Of course, an idea can be expressed in any of several ways. If the idea concerns a physical phenomenon, such as an object, we sometimes express the idea by simulation—that is, by a physical phenomenon similar to the one we wish to represent. For the idea "dog," we might use a picture of a dog or a barking noise. Even if only a few features of the true object are included in the image, it may still more or less faithfully reproduce the original idea. The light waves or sound waves from the simulated object affect the observer's senses in somewhat the same way as the original object would. This simulation works because the brain is very effective at supplying missing parts of a perceived scene. A dog standing behind a fence so that only the head and tail are visible, the silhouette or shadow of a dog, the sound of a dog barking—all have closely related psychological results. Much the same principle—the representation of an entire concept by an image possessing only a few properties—is perhaps behind the use of tallies for counting (such as 5 sticks or a pile of 5 stones to represent 5 sheep or even 5 days).

At the next higher level of abstraction, an idea is "verbalized," replaced by a group of one or more words. These words may be spoken at first, but later some permanent written marks (logograms, ideograms, or hieroglyphics) come to be used. The shapes of these marks are often derived from conventionalized pictorial images. For example, the letter *A* may have originated as a rudimentary picture of an ox. Later, instead of trying to draw a real ox, people began simulating previous drawings, roughly copying the figures they had seen used previously for the same idea. Gradually, a particular written shape came to be associated with a specific spoken sound, thus acquiring a phonetic value independent of the meaning of the sign.

These logograms, syllabograms, and phonemes developed into an alphabet (*Encyclopaedia Britannica*, 1970, Vol. 23, pp. 817–819).

Written alphabets have proved to be very flexible and are easily made permanent in form. So they are quite efficient for storage of information (for example, in libraries and filing cabinets) so long as space requirements are not too stringent, and for transmission (for example, by mail) if sufficient time can be allowed. Although modern media have been developed recently for particular purposes, it is quite clear that books and other written marks on paper will continue to be widely and heavily used for both storage and transmission of information.

Patterns in Black and White The Jacquard "automatic" loom is often mentioned (Bernstein, 1964; Wilkes, 1956; Cole, 1969) as a predecessor to machines such as Hollerith's punched-card tabulators (which were used to analyze data for the U.S. census in 1890) and Babbage's proposed "analytical engine" (1833). In the history of numerical computation, the Jacquard loom (Fox, 1911; Watson, 1946) probably has little further significance, but in terms of the storage and transmission of information it has considerable importance.

In a Jacquard fabric, all the threads are the same color; the pattern is formed by the arrangement of warp (lengthwise) and weft (crosswise) threads. A typical pattern for a fabric woven on a Jacquard loom uses about 500 rows of threads, with about 500 threads in each row. The loom is controlled by a set of punched cards, one card for each row. Each time the shuttle carries the weft across the loom, a new card is brought into position. The card has 500 positions in which a hole can be punched. For each position that is actually punched, one warp thread is lifted so that the weft can pass underneath it. For each position that is left unpunched, the corresponding warp thread is not lifted and the weft passes above it. The 500 cards are strung into an endless chain and are used repeatedly as the fabric is woven. Furthermore, the pattern may be repeated several times across the fabric, with each hole in the card simultaneously controlling a warp thread in each repetition of the pattern. Larger looms and more complicated patterns may use as many as 2000 cards, each controlling up to 2000 warp threads per pattern repetition (Watson, 1946, pp. 211–219).

The automatic loom was certainly not the first device to use stored information. Music boxes, and looms using smaller amounts of stored information, had been in use for many years. First, a cam or tappet and, later, a chain of small cards or drilled wooden blocks (Dobby loom) controlled a sequence of heddles, each attached to a group of warp threads. Tappet looms and Dobby looms are still widely used for weaving ordinary fabrics. The Jacquard loom differs only in that it uses a larger card, with one hole position for each warp thread in the pattern. The flexibility made possible by this larger *quantity* of information produced a real *qualitative* change in the way a loom is used. The patterns that can be produced on a simple loom, with 10 or 20 cams or hole positions, are suitable for entirely different purposes than the patterns that can be woven on a Jacquard loom with 500 or 2000 hole positions.

For instance, we are told (Wilkes, 1956) that a portrait of Joseph Marie Jacquard himself was woven on a Jacquard loom. Charles Babbage saw this tapestry, learned how it was produced, and was inspired to invent a card-sequenced computer. Babbage might have succeeded except for two difficulties—first, the mechanical technology his machine would have required was beyond the state of the art in his day; second, Babbage was a bit too much of a dreamer and could not wait to finish one machine before going on to invent the next one. Practical use of punched cards for computing began with the U.S. census of 1890. The Census Bureau finished tabulating the 1880 census data in about 1887 and predicted that the 1890 tabulations would not be completed by 1900. The Bureau commissioned Hermann Hollerith to solve the problem, and he constructed and successfully used machines that are the direct predecessors of the IBM card tabulating machines used today.

In the cards that control a Jacquard loom, each hole corresponds to a crossing of the warp over the weft, and each unpunched hole position corresponds to a crossing of the weft over the warp. The simplest kind of information seems to be this form, expressed as the existence or nonexistence of a hole or some other elementary mark. A unit of information represented in this way is called a *bit*. Thus, the quantity of information is measured in bits; for example, a Jacquard pattern described by 500 cards, each having 500 potential hole positions, would contain 250,000 bits. There are about the same number of usable positions in a standard black-and-white television picture.

Figure 4.1a shows a computer-generated pattern, 12 bits wide by 16 bits high. If you stand far away from this picture or squint, or move it about as you hold it, you may be able to recognize it as a portrait of a famous American. Figure 4.1b shows the

4.1a. *Portrait of a famous American, represented in about 200 bits. If you stand far away from this picture, or squint or move about as you hold it, you may be able to recognize him.*

same portrait, represented in various numbers of bits. The 6 by 8 bit version is obviously unrecognizable. With 2,700 bits (45 by 60) almost anyone can recognize the portrait. Somewhere between 200 and 2,000 bits is the threshold for recognition of a picture of this kind. Harmon (1973) describes an experiment with 14 fellow employees, who achieved a 50 percent chance of success in identifying 400-bit pictures of one another.

The quantity of information needed for recognition is also important to designers of devices for displaying mechanically generated numbers, letters, and other symbols. A scheme that is widely used for simple numeric displays has seven line segments, which can be illuminated in any combination (Figure 4.1c). One bit indicates whether to illuminate a segment; thus, the display uses seven bits to represent a digit. A display that includes alphabetical characters needs more than seven bits. A commonly used 35-bit scheme is illustrated in Figure 4.1d. This scheme (used on IBM key punches) includes capital letters, numerical digits, and more than 20 special symbols. I once worked out a set of characters for displaying computer-processed text that includes upper and lower case letters, Greek letters, numerical digits, and many different mathematical symbols. The display uses a 9 by 12 basic grid plus some additional information to specify vertical position—about 110 bits per symbol. Some of these characters are illustrated in Figure 4.1e.

Words to Remember

quantity of information   bit

Exercise

The following paragraph describes Abraham Lincoln at the time of his election to Congress in 1847. Estimate the quantity of information in this description. (This will depend on your assumptions about how the characters are represented.) Compare this with the amount of information that might be needed to convey a similar description in pictorial form.

> In the small clique of Springfield Whigs who had come to wield party controls, the opposition dubbed Lincoln the "Goliath of the Junto." On streets, in crowds or gatherings, Lincoln's tall frame stood out. He was noticed, pointed out, questions asked about him. He couldn't slide into any group of standing people without all eyes finding he was there. His head surmounting a group was gaunt and strange, onlookers remembering the high cheekbones, deep eye sockets, the coarse black hair bushy and tangled, the nose large and well shaped, the wide full-lipped mouth of many subtle changes from straight face to wide beaming smile. He was loose-jointed and comic with appeals in street-corner slang and dialect from the public square hitching posts; yet at moments he was as strange and far-off as the last dark sands of a red sunset, solemn as naked facts of death and hunger. He was a seeker. Among others and deep in his own inner self, he was a seeker. (Sandburg, 1954, p. 92.)

4.1b. *The same famous American's portrait, represented in various numbers of bits.*

6 by 8 (48 bits)
12 by 16 (192 bits)
24 by 32 (768 bits)
45 by 60 (2,700 bits)
90 by 120 (10,800 bits)
180 by 240 (43,200 bits)
360 by 480 (172,800 bits)

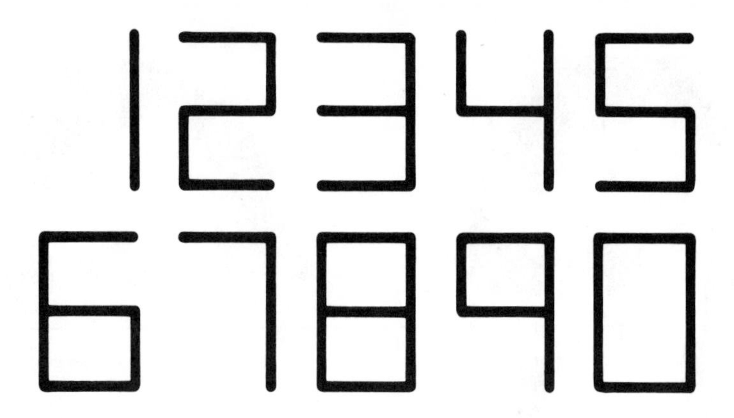

4.1c. Seven-bit representation of decimal digits.

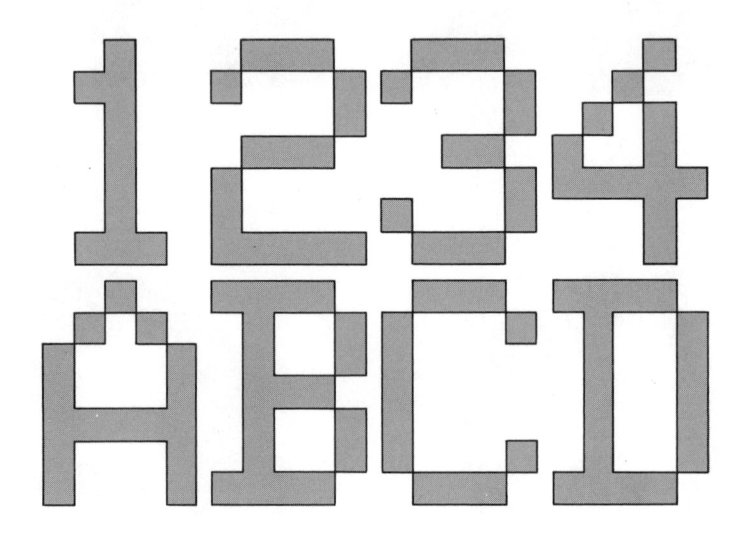

4.1d. 35-bit representation of digits and capital letters.

4.1e. 110-bit representation of upper- and lower-case letters.

4.2 *To What Extent Can Information Be Compressed?*

If 2,700 bits are undeniably adequate to permit us to recognize a portrait, what do we gain by using a quarter million bits? If we examine several different portraits of a person, made at different times, we can detect a number of characteristics (such as age and emotional mood) that would surely be lost in a reduction to 45 by 60 bits. The extra 247,300 bits convey secondary information which, while not absolutely necessary to insure bare recognition of the individual, are useful for other purposes. We can consider most of the information in a picture *redundant* (or extraneous) for the simple problem of recognizing the individual. However, the concept of redundancy is relative to the particular purpose to which information is applied.

Alphabetical, numerical, and symbolic characters need more bits for generating a recognizable pictorial *display* than for storing or transmitting a *coded* form from which the display can be reconstructed. In 1829, Louis Braille invented an alphabet for the blind, which uses only six bits for each capital letter. Early versions of the Teletypewriter used a five-channel paper tape, with each letter coded as a row of five bits across the tape. This code is known as Baudot code (Harper, 1971).

One bit can express two possibilities: * or ○, black or white, yes or no, on or off, presence or absence of a hole, 1 or 0. Two bits can express four possibilities (Figure 4.2a). If a third bit is added (which can be either * or ○), a total of eight different configurations are possible. Similarly, four bits can be arranged in 16 different ways. Each time a bit is added, the number of configurations doubles (see Table 4.2a).

Table 4.2a *Configurations of Bits*

Number of bits $= n$	*Number of Configurations* $= 2^n$
0	1
1	2
2	4
3	8
4	16
5	32
6	64
7	128
8	256
9	512
10	1024
...	...

In the game Twenty Questions, a variation of this idea is used (Raisbeck, 1964). An object is secretly chosen, and the player tries to identify the object by asking questions that can be answered "yes" or "no." The answer to each question provides one bit of information about the identity of the chosen object. If we know that the chosen object must be one of two known objects, we can surely identify it with a single question. If we know that it is one of four, we can divide the objects into two

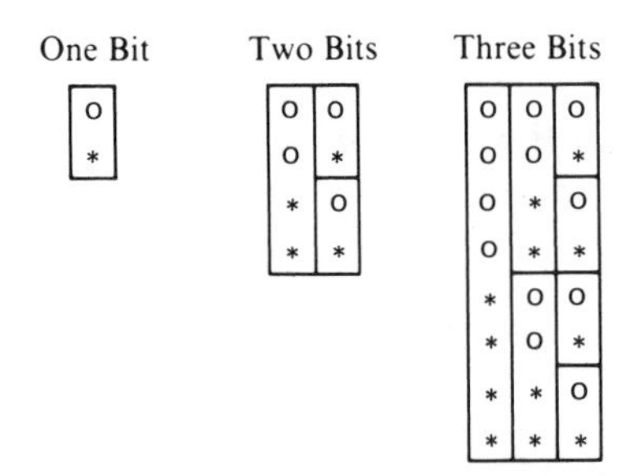

4.2a. Possible combinations of one, two, or three bits.

pairs, and use the first question to find out which pair the object is in; the second question would then complete the identification of the object since we then would know it to be one of the objects in a certain pair. Similarly, with three questions and eight objects, the first question locates the object within one of two groups of four, the second question locates it within a certain pair of objects, and the third question completes the identification.

Five bits can designate one of 32 objects. In the five-bit Baudot (paper tape) code, 26 of the 32 combinations are assigned to letters of the alphabet. Since more than 32 characters are needed, including numeric digits and punctuation, an elaborate shifting procedure is provided for the additional characters.

Six-bit codes are much more commonly used for representing alphabetical and other characters. Of the 64 possible combinations, 26 are assigned for the upper-case alphabet and ten for the digits, leaving 28 for basic punctuation and arithmetic signs and for special purposes. A six-bit code is almost adequate for upper- and lower-case letters, but the 52 letters and ten digits would leave only two combinations of the six bits for punctuation of all kinds. Until recently, the high-speed line printers on most computers have not provided for lower-case letters and have featured six-bit character codes.

The American Standard Code for Information Interchange (ASCII), adopted in 1967, uses seven bits and provides 96 different symbols (52 upper- and lower-case letters, ten numeric digits, and 34 other characters). Thirty-two "control codes" complete the total of 128 different seven-bit combinations (see Table 5.1a).

The eight-bit EBCDIC (Extended Binary-Coded Decimal Interchange Code) representation is used on IBM 360 and 370 series computers. The 256 possible combinations in an eight-bit code can be chosen to include enough mathematical and other symbols to handle almost all the characters in any one technical article or book (see Table 5.1b).

Four bits are sufficient to designate one of the numeric digits of the decimal number system, which require perhaps seven bits to display. Six bits can encode upper-case letters, digits, and a few symbols, any of which can be displayed pictorially with about 35 bits. Eight bits will designate any of 256 characters, while 110 bits seem to be required for an adequate graphical representation of this many different characters. In each case, the extra bits required for the pictures are redundant from the standpoint of information storage and transmission. Especially in applications requiring large quantities of information, we want to eliminate this redundant information and use the coded form for bulk handling of the characters. This assumes that we can interpret the codes and reconstruct the pictorial displays when we need them.

As an example, suppose that we are developing a computer-controlled instructional dialog to be used with an on-line computer system (see Section 2.9). A student will sit at a console, and the computer will print questions that the student answers by using the typewriter-like console keyboard. It has been estimated that 100,000 characters of text and control information will have to be stored for each hour of instruction. Storing all this information will cost perhaps one cent per hundred bits. If we foolishly use 35 bits per character, storage for one hour's lesson costs $350. Coding each character in five bits reduces this cost to $50. Of course, we would also have to store the decoding information, including perhaps one copy of each different character in 35-bit or 110-bit form. But even 200 different characters in 110-bit form, to be used for final graphic display, require only 22,000 bits and would cost $2.20 to store.

Can we reduce the $50 cost of storing 100,000 characters even further? We recall that in 1838 Samuel F. B. Morse had the idea of using shorter codes for letters that occur more frequently. This code was intended for representation as pulse durations rather than as bits, so any comparison must (unfairly) assume some specific bit coding. Let us use ○ for dot, ○* for dash, and ** to end a character (see Table 4.2b). Based on a letter-frequency table for the English language (Pratt, 1939, pp. 252–278), this code requires 26,800 bits to represent a text of 5,500 characters (including spaces between words), or about 4.87 bits per letter.

Table 4.2b *Morse Code Represented as Bits*

Letter	*Morse Code*	*Bit Code*	*Number of Bits*	*Frequency*
Blank		**	2	995
E	·	○**	3	591
T	–	○***	4	473
A	·–	○○***	5	368
O	–––	○*○*○***	8	360
N	–·	○*○**	5	320
R	·–·	○○*○**	6	308
I	··	○○**	4	286
S	···	○○○**	5	275
H	····	○○○○**	6	237
D	–··	○*○○**	6	171
L	·–··	○○*○○**	7	153
F	··–·	○○○*○**	7	132
C	·–·–	○○*○○***	8	124
M	––	○*○***	6	114
U	··–	○○○***	6	111
G	––·	○*○*○**	7	90
Y	–·––	○*○○*○***	9	89
P	·––·	○○*○*○**	8	89
W	·––	○○*○***	7	68
B	–···	○*○○○**	7	65
V	···–	○○○○***	7	41
K	–·–	○*○○***	7	19
X	–··–	○*○○○***	8	7
J	·–––	○○*○*○***	9	6
Q	––·–	○*○*○○***	9	5
Z	––··	○*○*○○**	8	3

We can do slightly better if we abandon the dots and dashes but retain Morse's general idea of using fewer bits for the more frequent characters. Huffman (1952; see also Pierce, 1961, p. 94; Knuth, 1968, pp. 402–405) has shown how to construct a more efficient code requiring 4.08 bits per character (Table 4.2c). The principles of information theory show that, if each character in a text occurs with known *independent* probability, it is impossible to devise a code that requires fewer total bits than an "ideal" code. In this ideal code, the number of bits used for the ith character would be $\log_2(1/p_i)$, where p_i is the relative frequency of the ith character. The overall average number of bits per character for the ideal is $\Sigma p_i \log_2 (1/p_i)$. This theoretical minimum value can be calculated for any given frequency table. Based on Table 4.2c, we calculate the value 4.06 for this sum, which the Huffman code approaches remarkably well.

Table 4.2c *Huffman "Optimum" Code Based on Letter Frequency*

Letter	*Frequency*	*Code*	*Number of Bits*	*Ideal*
Blank	995	***	3	2.467
E	591	**O	3	3.218
T	473	*O**	4	3.540
A	368	**O*	4	3.902
O	360	O*O*	4	3.933
N	320	*OO*	4	4.103
R	308	*O*O	4	4.159
I	286	OO*O	4	4.265
S	275	**OO	4	4.322
H	237	*OO**	5	4.537
D	171	*OOO*	5	5.007
L	153	*O*OO	5	5.168
F	132	O**OO	5	5.381
C	124	**OOO	5	5.471
M	114	O*OOO	5	5.592
U	111	*OOOO	5	5.631
G	90	*OOO**	6	5.933
Y	89	OOOO**	6	5.950
P	89	*OOOO*	6	5.950
W	68	*OOOOO	6	6.338
B	65	OOOOOO	6	6.403
V	41	*OOOOO*	7	7.068
K	19	*OOOOOO*	8	8.177
X	7	**OOOOOOO*	10	9.618
J	6	O*OOOOOOO*	10	9.840
Q	5	*OOOOOOOO*	10	10.10
Z	3	OOOOOOOOO*	10	10.84

However, in text from any natural language such as English, clearly the frequencies of the individual characters are *not* independent. For instance, the letter *T* has a definite tendency to be followed by *H*, blanks are very likely to be preceded by *E* or *S*, and *Q* is almost invariably followed by *U*. Taking advantage of just these four cases, we code the four digraphs (letter pairs) *TH*, *E*-blank, *S*-blank, and *QU* as single characters, and obtain a Huffman code (Table 4.2d) requiring only 3.97 bits per character. With Huffman coding for all 729 digraphs, the number of bits per character is reduced to about 3.57 (Goldman, 1953).

Table 4.2d *Optimum Code Based on Frequencies of Letters and Letter Combinations*

Letter	*Frequency*	*Code*	*Number of Bits*	*Ideal*
Blank	635	*○*	3	2.968
E	368	*○**	4	3.755
A	368	****	4	3.755
O	360	○***	4	3.768
N	320	*○○*	4	3.956
R	308	**○○	4	4.011
T	305	○*○○	4	4.026
I	286	*○○○	4	4.118
E-blank	223	***○	4	4.477
D	171	*○○○*	5	4.860
TH	168	○○○○*	5	4.886
L	153	*○○○○	5	5.021
S	138	**○*○	5	5.170
S-blank	137	○*○*○	5	5.180
F	132	*○○*○	5	5.234
C	124	*○**○	5	5.324
M	114	○○**○	5	5.445
U	106	**○○**	6	5.550
G	90	○*○○**	6	5.786
Y	89	*○○○**	6	5.803
P	89	○○○○**	6	5.803
H	69	*○○○○○	6	6.170
W	68	*○○○*○	6	6.191
B	65	○○○○*○	6	6.256
V	41	*○○○○○○	7	6.921
K	19	*○○○○○○○	8	8.030
X	7	**○○○○○○○○	10	9.471
J	6	○*○○○○○○○○	10	9.693
QU	5	*○○○○○○○○○	10	9.956
Z	3	○○○○○○○○○○	10	10.69

It seems entirely practical, therefore, to store large amounts of text in a code averaging about four bits per character, including a small amount of extra space for the program steps and the tables needed in decoding. Thus the text for an hour of dialog could reasonably be stored at a cost of $40, if our estimate of 100,000 characters at one cent per hundred bits is correct.

Of course the frequency of letters, and hence the code, will depend on the language. The high frequency of occurrence in English of the letter *E* does not hold in German, for example, and even in English it varies somewhat between technical writing and fiction, among various authors, and so on. But the use of a code designed for one set of frequencies on a different sort of text does not result in a drastic loss, so long as the mismatch is not too great.

The general principle is that efficiency can be gained when we reduce the cost of frequently occurring patterns. With care we can maintain this improvement even if the less frequent cases are increased in cost at the same time. Taking this principle to the extreme, we realize that very little would be lost if we left out the most common characters entirely. In cryptograms the space between words is almost always omitted, and some languages omit vowels, with very little loss in legibility. In fact, we may consider such systems as early intuitive attempts to compress information, along the lines later adopted by Morse and justified formally by information theory.

The coding of miscellaneous-sized groups of letters as a single character is also possible. In level two Braille, for instance, 185 contractions are used, allowing a group of as many as 10 letters to be written in 6 or 12 bits. Gregg and other shorthand systems have similar contractions or short forms, whereby common phrases are represented in much the same manner as a single word.

Another suggestion (Pierce, 1961) is to use a dictionary of about 16,000 of the most common words. Each word in the dictionary could be specified with 14 bits, with enough codes left over so that less common words could be spelled out letter by letter. In this way, text could be coded with about $2\frac{1}{2}$ bits per character. Perhaps even fewer bits would be required if Huffman coding were applied on the basis of word frequency.

This principle has been extended even further in commercial codes. In the heyday of submarine cable telegraphy, the high cost of communication ($5 to $10 a word) made it essential to compress messages to the greatest possible extent. To this end, a company would provide its overseas branches with a long list of likely messages, giving for each message a short code word to be transmitted by cable. Later, several standard compilations were provided, containing many useful messages on a variety of commercial subjects, and leaving some code words with unspecified messages to be inserted by a company using the code. In an appendix to one such compilation (International Cable Directory, 1903) it is stated that code words must contain 10 characters or less and must be actual words from one of seven European languages. (For artificial code words, higher rates were applied—presumably to compensate for the higher error rate and more likely need to retransmit.) This compilation, a thick book of 800 large pages, lists 157,418 different words and gives an associated message for most of them. Here are some coded messages listed in the directory, along with their meanings:

BODENLOCH, [JONES], ESTORSAO, ELCOMA: Reported that [Jones] has made an assignment. Investigate thoroughly and report by telegraph. Has recently purchased a large stock of goods and is looked on with suspicion.

JURADURIAS, EPATICHE, DOVELAR, HOERROHRES: When does case come up for trial? Give us the information at the earliest possible moment. A member of the firm will be in readiness to come immediately on being advised. It is important that we should be officially represented.

TUFFOLONE: We advise marketing the early thin mackerel as soon as ready for shipment.

TRAFRODERO: The following cars on track loaded and have been ready for several days. Railroad has made no effort to take them. Have them moved at once, otherwise our tracks will be blocked and will be compelled to close down.

Of course, this compression was achieved at the cost of restricting the number of different messages that could be sent by the compressed code. Also, the entire code book had to be transmitted at some time, although presumably by slower and less expensive means than submarine cable.

Shannon and Weaver (1949; see also Pierce, 1961, pp. 129–130; Raisbeck, 1964, pp. 19–20) describe several ways of estimating the "ultimate" information content of a single alphabetical character in a sentence or paragraph of text. They conclude that each character actually carries about one bit of "message," and additional bits are therefore redundant. Elaborate coding schemes to approach this ideal are theoretically possible. A system similar to the submarine cable directory might contain 250,000 messages, averaging 18 characters in length and requiring 18 bits to specify a message. This scheme would approach the ideal.

However, any scheme that reduces the average storage requirement much below four bits per character seems to require large amounts of extra storage or considerable extra processing time and to decrease flexibility as well; thus, the practical value of those codes for such applications as instructional dialog is doubtful.

Error Detection Practitioners of information theory, particularly as it applies to transmission of coded messages over "noisy" channels (that is, transmission lines subject to a significant error rate, such as those between a ground station and a space satellite), are deeply concerned with the problem of detecting errors. This problem is also important in some of the less reliable portions of a computer's information system. Fairly complete treatments of the question, with a minimum of technical detail, are found in textbooks in this field (Pierce, 1961; Raisbeck, 1964; Goldman, 1953).

A simple method of detecting errors is to transmit every message twice; then, unless a pair of compensating errors occurs (which is far less likely than a single error), a difference between the two transmitted messages would indicate an error in one or the other. Unfortunately, the user is left in the same quandary as the ship captain with two clocks—he has two different answers but doesn't know which is correct. (This is why, before the days of long-range radio, a ship always carried three clocks.) The user resolves his dilemma by requesting that a portion of the message be transmitted a third time.

A more economical way of detecting errors is to use a *parity* scheme. The principle here has long been used by accountants in the form of "casting out nines." The modern ASCII telegraph code with its extended character set uses seven regular bits per character. With each character, an eighth bit is transmitted, the extra bit being chosen to make an odd number of "holes" or "on" bits among the eight. For example, the seven-bit ASCII code for the digit 1 is ○**○○○*, while the code for 9 is ○***○○*. Thus an error in the transmission of the fourth bit could change a 9 into a 1, or vice versa. If single-bit transmission erors are independent and occur with a frequency of .001, say, an error would occur every 1000 bits or approximately every 143 characters. Now let us add a *parity bit* at the left, so that 1 becomes ○○**○○○* and 9 is *○***○○*. The parity bit is chosen as ○ in the first case because the number of * bits is already odd; * is chosen in the second case to change the number of * bits from 4 (an even number) to 5 (an odd number).

Now suppose that the fifth bit is transmitted incorrectly, and the 1 character is received as ○○***○○*. This no longer looks like a 9, because the parity bit does not agree. In fact, the number of * bits is now even, which gives a clear warning that a transmission error may have occurred, and a request for retransmission can be made either manually or automatically. Any single transmission error will change a * to a ○ or a ○ to a * and thus will change the number of * bits from an odd number to an even number, resulting in a detectable error. It is possible that *two* bits in the same character will be changed, but the expected frequency of this event is once per 35,000 characters.[1] Thus the frequency of *undetected* errors is reduced from one every two lines to one every 500 lines, at an extra transmission cost of 15 percent. More sophisticated error correcting codes are also used to give even better performance at slightly greater cost. (See Pierce, 1961, pp. 159–163; Goldman, 1953.)

The information added for the error detecting system may be considered redundant. For duplicate transmission, the redundancy is 50 percent of the total transmitted, but the eighth odd parity bit in ASCII produces only 12.5 percent redundancy. Using sufficiently sophisticated coding, the general principle is that more redundancy makes it easier to detect and to correct errors. Put another way, a noisier channel requires more redundancy in coding to achieve a given standard of reliability.

Ordinary verbal communication incorporates considerable redundancy. Because of this, a long-distance telephone circuit with considerable background noise can still be used and, as we have noted, *MSSGS FRM WHCH LL TH VWLS HV BN MTTD CN STLL B RD.*

Words to Remember

redundancy	Huffman code	noise
coded information	efficiency	error detection
Morse code	compression	parity

Exercises

1. You are playing a game similar to Twenty Questions, except that the number of questions is much more restricted. Tell how many questions (to be answered "yes" or "no") you need and what strategy you will use to identify each of the following:
   a. One of the decimal digits (0, 1, 2, 3, 4, 5, 6, 7, 8, 9).
   b. A letter of the alphabet.
   c. One of the characters (including upper- and lower-case alphabets, numerical digits, and other symbols) that can be typed on a standard typewriter.
   d. The amount of money I have in my coin purse, if I tell you in advance that it is less than $10.00.

---

[1] If the probability of an error in a single bit is e, then the probability of an error in a seven-bit character is $7e$ (if e is small enough, and if errors occur independently), and the probability of two errors in an eight-bit character is $28e^2$. We are using $e = .001$ in this example; 143 is $1/7e$ and 35,000 is $1/28e^2$.

2. Seven-bit ASCII codes for the digits from 0 to 9 are listed below. Construct the corresponding eight-bit codes, by adding a parity bit at the left. Choose the parity bit so that the total number of * bits in the eight-bit code will be odd.

0	○**○○○○
1	○**○○○*
2	○**○○*○
3	○**○○**
4	○**○*○○
5	○**○*○*
6	○**○**○
7	○**○***
8	○***○○○
9	○***○○*

3. For each seven-bit code listed above, which other codes on the list could result if a single bit were changed from ○ to * or vice versa? Verify that all such errors would be detected in the eight-bit coding scheme.
4. Given any integer, we may cast out nines by finding the remainder when the value is divided by nine. The result will be an integer value between 0 and 8. To "complement" this value, subtract it from eight. Now suppose that we have a nine-digit integer, such as a Social Security number. If we append the complement of the remainder at the left, we obtain a ten-digit integer. This ten-digit integer has the property that the remainder, when it is divided by nine, is always 8. However, if a digit is transcribed incorrectly, some other remainder will result. Hence the complement is used as a "check digit" to guard against errors in transcription, and is the decimal analogue of a parity bit. A simple way to cast out nines manually without dividing is to add the digits, then add the digits of the sum until a single digit remains. At any stage, any occurrence of the digit 9, or any combination of digits adding to nine, such as (6, 3) or (4, 2, 3) can be "cast out" and omitted when calculating the sum. For instance, from the number

   564-98-2374

   we can cast out the 9 and the combinations (5, 4), (6, 3), and (2, 7), leaving an 8 and a 4. We add these to obtain 12, and add the two digits of this result to obtain 3, the required result. (Checking, we see that 564,982,374 $\div$ 9 = 62,775,819, remainder 3.) The complement of this remainder is (8 $-$ 3), or 5; thus we form the "guarded" Social Security number

   5-564-98-2374

   Casting out nines (or dividing by nine) again, we obtain 8 as predicted. However, any one-digit error such as 5-564-98-2574 will produce a remainder other than 8. Write a program to read a nine-digit integer, determine the check digit, and print the guarded ten-digit integer. (Multiply the check digit by 1,000,000,000 and add it to the nine-digit number.) Manually check the result to see that 8 is obtained from it. Use the following data: (a) 564-98-2374, (b) 347-42-8927, (c) 736-31-4327, (d) 136-59-0875.

4.3 What Is in a Cell?

We have seen (Table 4.2a) that one bit can uniquely represent two different cases, two bits give four cases, three bits give eight, four bits give 16, and so on. The number of cases doubles each time we add a bit. The number of cases for n bits is 2^n (the nth power of 2).

For numerical computations, we use bits to represent the consecutive integers, starting at 0. Thus with one bit we have (0, 1), with two bits (0, 1, 2, 3), with three bits (0, 1, 2, 3, 4, 5, 6, 7), and with four bits we can go from 0 to 15. The 2^n combinations of n bits represent the integers from 0 to $2^n - 1$. An individual bit is written as either 0 or 1. Strings of these 1 and 0 digits are *binary numerals*. The meaning of numerals in the binary (or base 2) number system can be understood by analogy to ordinary decimal (or base 10) numerals. For instance, the 7 in the *decimal* numeral 74 has *ten* times the value of the 7 in 47; similarly, the 1 in the *binary* numeral 10 has *two* times the value of the 1 in 01. In the decimal system the values of the digits are multiplied by one, ten, one hundred, as we proceed from right to left; in the binary system the multipliers are one, two, four, eight, sixteen. Thus 11 in binary is two plus one, or three; 101 is four plus one, or five; 110 is four plus two, or six. The integers that can be expressed with five bits are listed in Table 4.3a. As in the decimal system, zeros on the left are not significant and are not usually written down.

Table 4.3a *Correspondence between Binary and Decimal Integers*

Binary	*Decimal*	*Binary*	*Decimal*	*Binary*	*Decimal*
0	0	1011	11	10110	22
1	1	1100	12	10111	23
10	2	1101	13	11000	24
11	3	1110	14	11001	25
100	4	1111	15	11010	26
101	5	10000	16	11011	27
110	6	10001	17	11100	28
111	7	10010	18	11101	29
1000	8	10011	19	11110	30
1001	9	10100	20	11111	31
1010	10	10101	21		

The meaning of fractional binary numerals can also be determined by analogy to ordinary decimal fractions. Each binary digit to the right of the "binary point" decreases in value by 1/2. Thus, 0.1 is equivalent to the decimal $\frac{1}{2}$, 0.01 is $\frac{1}{4}$, 0.11 is $\frac{1}{2} + \frac{1}{4}$ or $\frac{3}{4}$. Table 4.3b shows the 16 binary fractions that have four bits to the right of the binary point.

Table 4.3b Decimal Equivalents of Some Binary Fractions

Binary Fraction	Decimal Equivalent	Binary Fraction	Decimal Equivalent
0.0001	1/16	0.1001	9/16
0.0010	2/16 or 1/8	0.1010	10/16 or 5/8
0.0011	3/16	0.1011	11/16
0.0100	4/16 or 1/4	0.1100	12/16 or 3/4
0.0101	5/16	0.1101	13/16
0.0110	6/16 or 3/8	0.1110	14/16 or 7/8
0.0111	7/16	0.1111	15/16
0.1000	8/16 or 1/2		

Although it is important to understand the principle of construction of these tables, it is not necessary to be proficient at binary arithmetic to use computers. Although practically all computers use binary numerals internally, they generously permit their human users to furnish data in decimal form, and they return the results in decimal form as well. The process of translation between decimal and binary numerals is just one of the jobs that is distasteful to people but is efficiently accomplished by a computer. Therefore, we can get along very well for a while without knowing or caring about the way the computer represents numbers in its internal storage cells.

We have now reached the point where we need to understand more clearly what is in a cell, however, so that we can see why different types of data representation are needed for various purposes. Present-day computers generally have cells consisting of 32 to 60 bits.[2] Each bit is represented by a physical device such as a magnetic core or a solid-state chip (Scientific American, 1970). Such a device can be switched into either of two states, by the application of an electronic write signal. Once switched into a state, it remains set in that state until it receives another write signal. Meanwhile, read signals can be applied to sense the state of the device without changing it.

The two possible states of the device are associated more or less arbitrarily with the values 0 and 1 of a bit in the binary number system. Thirty-two or more of these devices are grouped together to form a cell. In *storing* a number, electronic write signals are sent to switch each device into a state corresponding to a bit in the binary representation of the number to be stored in the cell. To fetch a number from storage, read signals are applied to all the devices that compose the cell.

In a 32-bit cell, we reserve one bit for the sign. The remaining 31 bits can represent the integers from 0 to $2^{31} - 1$ or 2,147,483,647. If all our work were done with integers, we could work throughout the range from $-2{,}147{,}483{,}647$ to $+2{,}147{,}483{,}647$. For practical purposes, however, we need to interpret the bits in a different way, so that we can represent both fractions and the approximate values of much larger numbers.

[2] Some computers have "words" or addressable "bytes" of fewer than 32 bits. However, when a problem-oriented language is implemented on one of these computers, the usual practice is to use a group of adjacent words or bytes to store the value of a single variable or constant. Hence we consider a cell in such systems to be the group of words or bytes used to store a number. It has been found that 32 bits is about the smallest practical cell size for general numeric work.

Numbers with Scale Factors When we want to write the number of seconds in a year, or the number of miles to the nearest star, or the number of tons of silt deposited by the Mississippi at its delta in a year, we are dealing with rather large numbers. Usually, however, when a number is very large we are not interested in its exact value, so quite a lot of the digits we write are insignificant. Sometimes we replace the insignificant digits by zeros, writing the result as a "round number" such as 175,000,000. Instead of writing zeros for the insignificant digits, we can convey the essential information in a more compact form such as 175 (6), indicating that there are six insignificant digits following, which have been omitted for the sake of compactness. The usual notation recognizes that 10^6 is 1,000,000 and that 175,000,000 is 175 × 1,000,000 or 175 × 10^6; this is sometimes written $175_{10}6$. By the same reasoning, we could also write in the same number as $17.5_{10}7$ or $1.75_{10}8$ or $0.175_{10}9$. The last of these is in a standard or *normalized* form; the *significant part* is at least 0.1 but less than 1.0, and the *scale factor* in this case is $_{10}9$ or 1,000,000,000. The number 9 is called the exponent part of the scale factor. When a number is written with a scale factor, the exponent tells how many places we should move the decimal point to the right to recover the original number.

At the other extreme, 0.000 000 000 321 is 0.321 × 0.000 000 001; the decimal point in 0.321 must be moved nine places to the left to recover the original number, so we write it with a scale factor having a negative exponent: $0.321_{10} - 9$. This "scientific" or floating decimal notation greatly extends the range of values that can be represented by the bits in a cell.

Since we can express a decimal digit with 4 bits, we can express a five-digit significant part in 20 bits, and a two-digit exponent in 8 bits. Using two additional bits, for the sign of the significant part and the sign of the exponent, a total of 30 bits is required. Thus $0.175_{10}9$ can be represented as

$$+0.17500_{10}+09 \qquad \text{or} \qquad +17500+09$$

since, in the cell, the decimal point and the zero to the left, as well as the $_{10}$ portion of the scale factor, do not have to be stored explicitly. In bits, we use 0 for + and 1 for −, and represent the decimal digits as shown in the first ten rows of Table 4.3a:

0	0001	0111	0101	0000	0000	0	0000	1001
+	1	7	5	0	0	+	0	9

Similarly, $0.321_{10}-9$ becomes

0	0011	0010	0001	0000	0000	1	0000	1001
+	3	2	1	0	0	−	0	9

The numbers that can be represented in this way consist of 90,000 fractions (significant parts from 0.10000 to 0.99999) in each of 200 different ranges, corresponding to the exponents from −99 to +99. The smallest range includes values between $0.1_{10} - 99$ and $1.0_{10} - 99$; these values if written without scale factors would have 99 zeros after the decimal point. The largest range goes from $0.1_{10}99$ to $1.0_{10}99$; these values are rather large numbers. Because the significant part holds only five digits, these large numerals would have 94 nonsignificant zeros to the left of the decimal point.

Exponent values from +1 to +5 provide exact representations for the integers from 1 to 99,999. The integers in floating decimal form have to be written with a significant part that is 0.1 or greater but is less than 1. Thus 1 has to be written as $0.1_{10}1$. Integers between 10 and 99 are normalized with 10^2 as scale factor; those between 100 and 999 have scale factor 10^3, and so on. One minor inconvenience of floating decimal form is this awkward representation of simple integers. The integer 99,999 becomes $(99{,}999/100{,}000) \times 10^5$ since 10^5 is the same as 100,000. Integers larger than 99,999 do not necessarily have exact representations—the fraction 100,001/100,000, for instance, cannot be expressed exactly within the five-digit allowance for the significant part.

Examples

3.1416								
0	0011	0001	0100	0001	0110	0	0000	0001
+	3	1	4	1	6	+	0	1
.0008761								
0	1000	0111	0110	0001	0000	1	0000	0011
+	8	7	6	1	0	−	0	3
.33333								
0	0011	0011	0011	0011	0011	0	0000	0000
+	3	3	3	3	3	+	0	0
15,874,000,000								
0	0001	0101	1000	0111	0100	0	0001	0001
+	1	5	8	7	4	+	1	1
602,300,000,000,000,000,000,000								
0	0110	0000	0010	0011	0000	0	0010	0100
+	6	0	2	3	0	+	2	4
−6670								
1	0110	0110	0111	0000	0000	0	0000	0100
−	6	6	7	0	0	+	0	4
−522.4								
1	0101	0010	0010	0100	0000	0	0000	0011
−	5	2	2	4	0	+	0	3

Floating Binary Numerals This notation can easily be adapted to binary numerals as well, using a scale factor based on powers of 2 instead of 10. In practically all computers this is the form actually used. Floating binary numerals are normalized so that the significant part is at least $\frac{1}{2}$ but is less than 1. The scale factor indicates the power of two needed to recover the original number from the normalized form. If the original number was larger than 1, the binary exponent will be positive; if it was smaller than $\frac{1}{2}$ the binary exponent will be negative.

Various present-day computers use a significant part of 23 to 48 bits, and a scale factor containing 7 to 10 bits. Two additional bits are required, for the sign of the significant part and the sign of the exponent. We shall give a number of examples based on a "typical" computer with a 32-bit cell size, using 23 bits for the significant part (besides the sign), and 7 bits for the exponent (besides the sign). The exponent represents a scale factor between 2^{-127} and 2^{+127} or approximately 10^{-38} to 10^{+38}.

The significant part includes 23 bits besides the sign; normalization guarantees that the leftmost of these 23 bits will not be a zero. This means that there are 2^{22} or 4,194,304 different possible bit combinations for the significant part. These represent all of the different fractions between $\frac{1}{2}$ and 1 in steps of 2^{-23} (or approximately 0.000 000 12).

Thus, instead of using the 32 bits of a cell for some 4 billion different possible *integer* values (half of which are negative), we use the bits instead for about 8 million positive and negative fractions in each of about 250 different ranges (the different ranges corresponding to the exponents from -127 to $+127$). The smallest range includes values between $\frac{1}{2} \times 2^{-127}$ and 1×2^{-127}; this range is near 10^{-38} and in decimal notation would begin with about 38 zeros after the decimal point. The largest range goes from $\frac{1}{2} \times 2^{+127}$ to 2^{+127}; these values are near 10^{38} and are thus rather large numbers.

Exponent values from $+1$ to $+23$ provide exact representations for the integers from one to $2^{23} - 1$ or 8,388,607. The integers in floating binary form have to be written with a significant part that is $\frac{1}{2}$ or greater but is less than 1. Thus 1 has to be written as $\frac{1}{2} \times 2^1$; 2 and 3 are normalized with 2^2 as scale factor, as $\frac{1}{2} \times 2^2$ and $\frac{3}{4} \times 2^2$; 4, 5, 6, and 7 have scale factor 2^3 and fractional parts $\frac{4}{8}$, $\frac{5}{8}$, $\frac{6}{8}$, and $\frac{7}{8}$. The integer 8,388,607 becomes $(8{,}388{,}607/8{,}388{,}608) \times 2^{23}$, since 2^{23} has the same value as the decimal numeral 8,388,608. Integers larger than 2^{23} do not necessarily have exact representations—the fraction 8,388,609/8,388,608, for instance, cannot be expressed exactly within the 23-bit allowance for the significant part.

In the following examples, the first bit is the sign of the number, the next 23 are the significant part expressed as a binary fraction, the next is the sign of the exponent, and the last seven are the value of the exponent. For example, the integer 15, in normalized form, becomes $\frac{15}{16} \times 2^4$. The binary fraction representing the significant part, $\frac{15}{16}$, is 0.1111; this is written (without the binary point or the zero on the left) as the 23-bit pattern 111 1000 0000 0000 0000 0000. The entire 32-bit floating binary representation of the integer 15, then, is

```
  0   111   1000   0000   0000   0000   0000   0   000   0100
  +   (                  15/16                 )  +   (   4   )
```

meaning $+(15/16) \times 2^{(+4)}$. Other examples:

```
13
  0   110   1000   0000   0000   0000   0000   0   000   0100
  +   (                  13/16                 )  +   (   4   )
2
  0   100   0000   0000   0000   0000   0000   0   000   0010
  +   (                   1/2                  )  +   (   2   )
8,388,607
  0   111   1111   1111   1111   1111   1111   0   001   0111
  +   (         8,388,607 / 8,388,608)            +   (  23   )
1/2 × 2^127
  0   100   0000   0000   0000   0000   0000   0   111   1111
  +   (                   1/2                  )  +   (  127  )
(1/128) = 1/2 × 2^-6
  0   1000  0000   0000   0000   0000   0000   1   000   0110
  +   (                   1/2                  )  -   (   6   )
```

Data Types and Declarations Using the 32-bit floating binary representation, a cell may be used to store a floating-point number between approximately 10^{-38} and 10^{38}, with a precision of one part in 8,388,608. Or the same 32-bit cell can be used to store an integer between −2,147,483,647 and 2,147,483,647, or a portion of a row of a Jacquard loom pattern. Or 18 bits can be used to identify one of the 157,418 messages from the International Cable Directory. When an instruction to be executed has the form

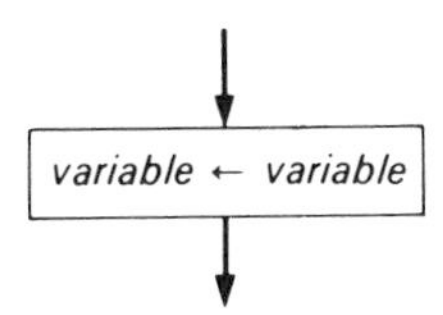

the pattern of bits stored in the cell named on the right is moved intact to the cell named on the left, without any manipulation. So long as the bits remain in storage or are moved in groups from one cell to another, no "meaning" is attached to them, and questions of interpretation never arise. On the other hand, when an arithmetic expression is evaluated, the bit patterns found in the referenced cells are interpreted as numbers—even if they were supposed to be Jacquard loom patterns or Huffman codes when they were stored in the cells. The arithmetic unit is omnivorous—it "digests" whatever bits are given to it.

Most computers have arithmetic units capable of performing either integer arithmetic or floating binary arithmetic on demand. (Some computers can do decimal arithmetic as well.) When told to do integer arithmetic, the arithmetic unit interprets the pattern of bits given to it as a binary integer: that is, it associates the value 1 with the rightmost bit, the value 2 with the next bit to the left, and so on, interpreting the last bit on the left as the sign. On the other hand, if a floating binary arithmetic operation is requested, the arithmetic unit will interpret the last eight bits on the right as an exponent (with its sign), and the other 24 bits as a fractional significant part (including sign).

The machine code described in Section 2.9 must be augmented to include separate numeric operation codes for integer arithmetic and for floating binary arithmetic. For instance, the numeric operation code 2 would be used for *integer add*, but another code such as 7 would indicate *floating add*. (The *fetch* and *store* operations do not need separate forms for integer and floating binary bit patterns.)

The compiler must generate the proper numeric code to be used in the evaluation of each expression. It does this by associating a *data type* with each variable or constant in the calculation, to indicate whether the bit pattern in the corresponding cell is to be interpreted as an integer or as a floating binary numeral. A data type indicator is included in the symbol table, along with the address of the cell allocated to each variable or constant.

How does the compiler know which type is intended for a particular variable or constant? In most programming languages, *constants* are identified as to type by the way they are written in the source program. (In Fortran, for instance, constants in the source program that are to be stored in floating binary form are distinguished by the fact that they include a decimal point; in PL/1, they are written with an e.) For *variables*, a type declaration is included in the source program. However, most languages permit some of the type declarations to be omitted by providing a system of "default" assumptions for variables in the source program that are not included in a type declaration. In our flowcharts, for instance, we assume that all variables are of floating type unless otherwise specified.

When it generates the machine instructions corresponding to a source-language expression, the compiler incorporates a floating arithmetic operation code if the cells involved contain floating binary patterns, or an integer operation code if they contain integer bit patterns.

The type of a variable or expression must also be taken into account by the compiler when it generates the machine code for read and print instructions. As we have noted, the computer converts data and results between the binary internal form and the decimal external form. This conversion will be somewhat different depending on whether the binary internal form is in an integer or a floating binary pattern.

Octal and Hexadecimal Notations When computers with binary arithmetic first came into widespread use about 1950, people soon found it awkward to write long strings of 1 and 0 digits to represent numbers. It was then realized that individual binary digits could be grouped, and each group of binary digits could be written with a single symbol. For instance, taking binary digits in groups of three, we can use the octal representation:

Binary Digit Group	*Octal Digit*
000	0
001	1
010	2
011	3
100	4
101	5
110	6
111	7

Thus a long binary numeral such as

11 110 100 001 001 000 000

becomes much more manageable when written in octal as

3 6 4 1 1 0 0

Or, instead of using groups of three binary digits, we can use groups of four, which give a hexadecimal representation:

Binary Digit Group	*Hexadecimal Digit*	*Binary Digit Group*	*Hexadecimal Digit*
0000	0	1000	8
0001	1	1001	9
0010	2	1010	A (ten)
0011	3	1011	B (eleven)
0100	4	1100	C (twelve)
0101	5	1101	D (thirteen)
1110	6	1110	E (fourteen)
0111	7	1111	F (fifteen)

The binary numeral 11110100001001000000, regrouped into sets of four binary digits, becomes

1111 0100 0010 0100 0000

and in hexadecimal notation

F 4 2 4 0

Of course it takes a while to get used to the idea of letters being used as numerical digits. (Any six symbols would do just as well for the extra digits, and some different schemes have been proposed, but these have been abandoned mostly for typographical reasons.)

Binary fractions can also be written in octal or hexadecimal. For example, 1/10 in binary is

0.000110011001100110011001100110011001 . . .

which in octal becomes

0.063146314631 . . .

and in hexadecimal

0.199999999 . . .

Octal and hexadecimal notations can, in fact, be used for any string of bits. In representing the bit pattern of a cell, we might wish to use octal if the number of bits is a multiple of 3, or hexadecimal if it is a multiple of 4. The 30-bit floating decimal pattern used in the foregoing illustrations has a 21-bit significant part (including sign), and a nine-bit exponent (including sign). Each of these makes a nice octal group. Thus +17500 + 9 has the bit pattern 000 010 111 010 100 000 000 000 001 001 or in octal 0272400 011. On the other hand, the 32-bit floating binary representation is much more convenient in hexadecimal:

15

0111	1000	0000	0000	0000	0000	0000	0100
7	8	0	0	0	0	0	4

8,388,607

0111	1111	1111	1111	1111	1111	0001	0111
7	F	F	F	F	F	1	7

$1/2 \times 2^{127}$

0100	0000	0000	0000	0000	0000	0111	1111
4	0	0	0	0	0	7	F

Rather than thinking of octal and hexadecimal numbers as binary numbers written in a shorter form, we can consider them as numbers in base 8 or 16, just as decimal numbers are in base 10 and binary numbers are in base 2. The IBM 360 and 370 series computers, in fact, use a floating hexadecimal internal representation rather than floating binary; the significant part is normalized between 1/16 and 1, and the scale factor indicates a power of 16. (Exponents from −64 to +63 are allowed, giving scale factors from 16^{-64} to 16^{+63}, or the approximate equivalent of decimal scale factors from 10^{-77} to 10^{+76}.)

Serious computer users may find it necessary to know something about either the octal or hexadecimal numeral system. Some bugs in large or complicated programs are difficult to analyze in human-oriented programming languages. In these cases, it may be necessary to resort to a *dump*, which is a printed list of the internal bit patterns of the cells inside the computer. On a dump, bit patterns will be represented in either octal or hexadecimal, depending on the computer.

Flowchart In our flowcharts, the data type of numerical variables and constants is assumed to be *floating*, and the floating binary internal representation will be assumed. We assume that constants in the source program, as well as data values prepared for input, may have any of three external forms:

Integer external form:

```
1      2      356      123456789
```

Fixed decimal external form:

```
14.37  .00547  0.33333  0.000000000321
```

Floating decimal external form:

```
0.175e9      0.321e–9     0.31416e1
0.8761e–3    0.33333e0    0.15874e11
0.6023e24    –0.667e4     –0.5224e3
```

Note that the letter e (for *exponent*) and not the subscript $_{10}$ is used, for typographical reasons. The only characters permitted in a constant or in a data value are digits, a decimal point, a minus sign, and e.

We assume that numbers will be printed in fixed decimal external form whenever possible, but otherwise in floating decimal external form.

Variables may be declared to be of *integer* type. The form of the declaration is

```
integer list
```

where the list consists of one or more variables, as in the following example:

```
integer A, B, C
```

Basic All numerical variables and constants are of *floating* type, and a floating binary internal representation is used. Constants in the source program, however, as well as data values entered by READ or INPUT instructions, may have any of three external forms:

Integer external form:

```
1      2      356      123456789
```

Fixed decimal external form:

```
14.37  .00547  0.33333  0.000000000321
```

Floating decimal external form:

```
0.175E9      0.321E–9     0.31416E1
0.8761E–3    0.33333E0    0.15874E11
0.6023E24    –0.667E4     –0.5224E3
```

Note that the letter E (for *exponent*) and not the subscript $_{10}$ is used, for typographical reasons. The only characters permitted in a constant or in a data value are digits, a decimal point, a minus sign, and E.

Numbers will be printed in fixed decimal external form whenever possible, but otherwise in floating decimal external form.

Fortran Variables and constants of *integer* type in Fortran (on most computers) use a binary integer internal representation. Variables and constants of *real* type use a floating binary internal representation. Integer constants in the source program, and data prepared as input values for integer variables, are in integer external form:

```
1        2        356        123456789
```

Real constants, and data prepared as input values for real variables, use fixed or floating decimal external form:

```
14.37               .00547       0.33333
0.000000000321      0.175E9      0.321E-9
0.31416E1           0.8761E-3    0.33333E0
0.15874E11          0.6023E24    -0.667E4
-0.5224E3
```

Note that the letter E (for *exponent*) and not the subscript $_{10}$ is used, for typographical reasons. The only characters permitted in constants and input data values are digits, a decimal point, a minus sign, and E.

REAL and INTEGER type declarations may be used to override the "implicit" type determination based on the initial letter of a variable. These declarations have the forms

REAL *list*
INTEGER *list*

where the list consists of one or more variables, as in the following examples:

```
REAL IN, MI, JA, L, ISAX, MAX
INTEGER X, R3, XY, XDEV, A, T
```

If a variable is not listed in a type declaration, its type is determined by its initial letter.

The format specification for data to be read and for results to be printed must take the variable *types* into account. The input or output format specification for integer variables has the form

I*w*

where *w* is the field width, that is, the number of columns on the input card or on the printer line that are allocated for this value.

For input data of *real* type, the format

F*w*.0

may be used, regardless of whether the data values are represented in fixed or floating decimal external form. The F in this format specification merely indicates that the data is to be converted to floating binary internal form during input, and *w* specifies the field width.

Data of real type may be printed using either of two external forms. An F format specification is used when real data are to be printed with a decimal point but no scale factor. The specification

F*w*.*d*

causes data to be printed in field *w* columns wide, with *d* columns to the right of the decimal point. An E format is used when a scale factor is needed. The specification

E*w*.*d*

provides a total field width of *w* columns, with *d* significant digits to the right of the decimal point. To provide room for the scale factor, sign, and space between fields, *w* should be at least 7 or 8 more than *d*. Also a G specification may be used, which will cause a number to print in either E or F form depending on its magnitude.

PL/1 Numbers in PL/1 may have any of three internal forms—fixed decimal, fixed binary, or floating binary. Variables may be declared to be of *fixed* or *float* scale, and of *decimal* or *binary* base. In the absence of explicit declarations of some or all of these attributes, default attributes are assigned. Therefore, we find it sufficient to list all of the variables in our examples in one of the following declarations[3]:

```
declare (list) fixed;
declare (list) fixed binary;
declare (list) float;
```

A constant written as a string of digits, with or without a decimal point, is represented internally in fixed decimal form. Thus, the internal fixed decimal form corresponds to either the integer external form or the fixed decimal external form.

Integer external form:

```
1        2        356        123456789
```

Fixed decimal external form:

```
14.37   .00547   0.33333   0.000000000321
```

Constants written in floating decimal external form are represented internally in floating binary form.

Floating decimal external form:

```
0.175e9        0.321e-9        0.31416e1
0.8761e-3      0.33333e0       0.15874e11
0.6023e24      -0.667e4        -0.5224e3
```

Note that the letter e, for *exponent*, and not the subscript $_{10}$, is used for typographical reasons.

It is one of the anomalies of PL/1 that there is no external form signifying that a constant is to be represented internally in fixed binary form. In our examples we write such constants in integer external form, and we rely on the fact that PL/1 provides automatic conversion when such constants are used in an expression along with variables of fixed binary type.

When a get list instruction is used, input data values for variables of any given type should be prepared in the same form as constants of that type. When results are printed by a put skip list instruction, the external form is determined by the variable attributes.

[3] See Fike, 1970, pp. 8–13. When a variable is simply declared fixed or float, the default base is *decimal*. However, numbers of *float decimal* type are represented internally in floating binary form.

Words to Remember

binary numeral
scale factor
normalize
significant part
exponent
floating decimal
floating binary
type declaration

Exercises

1. Write each of the following as a floating decimal numeral, normalized so that the significant part is a fraction, less than 1 but at least 1/10 in value.
   a. 0.000031416 b. 6385000000000000 c. −0.00000000000003
   d. 0.0000268 e. 909300 f. 3628800
2. What numerical value cannot be written in normalized form as either a floating decimal numeral or a floating binary numeral? Suggest a way of representing this numerical value. Try to find out how this value is represented on the computer you are using.
3. Explain why two signs are needed for floating binary numerals. Why is the second bit from the left in the 32-bit floating binary representation (that is, the most significant bit of the significant part) always a 1?
4. Write each of the following decimal numbers in floating binary form: 1, 5, 11, 3.25, 7.5, 1.75, 6.5, 3.75, 3.5, 13.0, 1.875, 7.0.
5. Rewrite the machine-language program of Section 2.9, using the following floating binary operation codes:

floating add	7
floating multiply	8
floating divide	9

   Assume that the variable values are stored in cells 4221 through 4228, as before, but that they are now stored in floating binary form.
6. Try to find out how many bits are used for the significant part and for the exponent on the computer you are using. For what purposes is your computer's representation better or worse than the 32-bit system explained here?
7. Convert the following bit patterns to octal or hexadecimal form. (These are taken from a pictorial representation of the capital letter *R*.)

   100000000001
   111111111111
   100000100001
   100000100000
   100000100000
   100000110000
   010001001100
   011011000011
   000100000001

4.4 *How Does Internal Number Representation Affect the Results of a Computation?*

The most important implication of the internal number representation, for numerical calculations, is that some numerical values cannot be represented exactly with 32 (or even 60) bits. In fact, only a few billion, of the infinity of possible numbers, can be represented exactly with 32 bits, because there are only a few billion possible bit patterns. Of course, the integer and floating binary representations are deliberately chosen so that many of the numerical values most likely to be encountered (the first eight million or so positive and negative integers, many common fractions, and so on) are among the several billion values that can be represented.

For any system of numerals, however, there will inevitably be some quite simple numbers that cannot be represented exactly. In the decimal number system, $\frac{1}{3}$ cannot be represented exactly with five or even 25 digits, but must be truncated. One minor inconvenience of the binary number system is that $\frac{1}{10}$ and most other simple decimal fractions cannot be represented exactly.

One reasonably effective way to solve this part of the problem is to represent any *rational number* (any fraction or "mixed number" having an integer part and a fractional part) as a *pair* of integers. Any rational number can be converted to a numerator and a denominator. (For mixed numbers or improper fractions, the numerator may be larger than the denominator, but this will cause no difficulty.) As we saw in Section 2.7 (see also exercise 3, Section 3.5), it is fairly easy to combine two such pairs of integers and form the numerator and denominator of the sum or product.

Even with integers, however, some problems arise. If all of our calculations were of the simple variety found in elementary school arithmetic textbooks, perhaps a few million or billion integers would be enough. In most basic accounting work (excluding a very large corporation or bank) few problems would arise. In statistical computations, on the other hand, even if we start with integers of reasonable size, a sequence of operations such as a sum of squares will soon produce much larger numerical values.

A more serious problem arises with operations such as square roots, because the square root of a rational number (or even of an integer)[4] will hardly ever be expressible as a rational number. Hence, even a system that interprets a pair of integers as the numerator and denominator of a fraction will be unable to represent square roots and other irrational numbers exactly.

Let us write a program to print all perfect square integers between 1 and 1000 —these are the numbers whose square root is an integer.

[4] The discovery that square roots of rational numbers can be irrational is attributed to Pythagoras (sixth century B.C., about 200 years before Euclid). This discovery is said to have led Greek mathematicians away from algebra in favor of geometry.

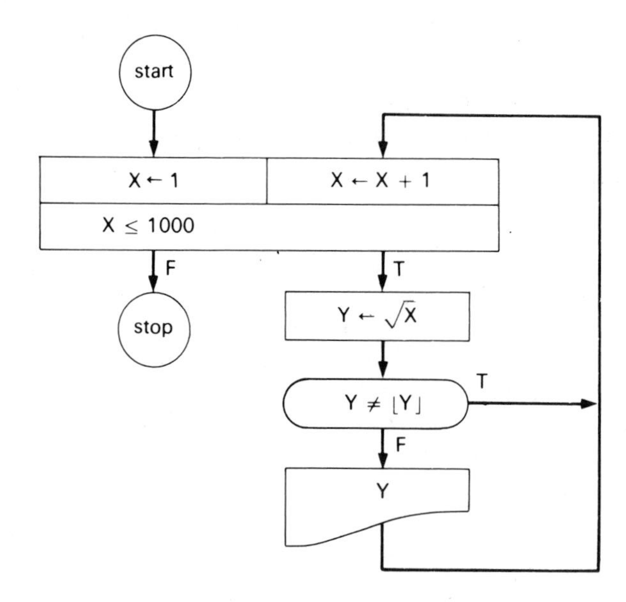

We expect this program to print the sequence of values

1 4 9 16 25 36 49 64 81 100 121 144 169 196 225 256
289 324 361 400 441 484 529 576 625 676 729 784 841
900 961

This perfectly reasonable simple program will fail, however (by omitting some of the correct values), on any computer on which the result of the square root function deviates from the exact value in the slightest degree.

Again, let us attempt to compute the value of some expression containing the variable T, for values of T between 0 and 1, in steps of 0.01, without using an iteration instruction.

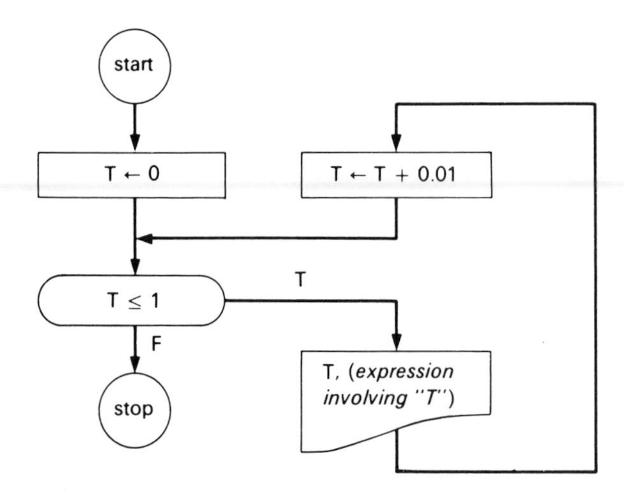

What value of T will be the last one printed? 1.00? On a binary computer, *no*. The binary representation of 0.01 is a little less than the true value, so after one hundred steps the value of T will still be a tiny bit less than 1 (although the printed value at this step may appear as 1.00), and the loop will be executed one more time, with 1.01 as the value of T.

Similarly, in many cases where two expressions are to be compared, their values may be calculated in slightly different ways and may thus be different when we expect them to be the same. Instead of using a condition of the form

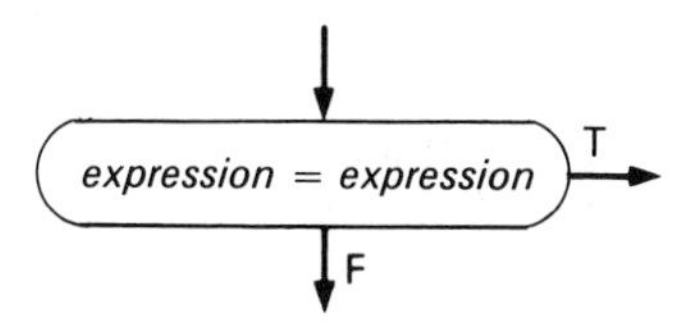

we must ask whether the two values are equal to within some small tolerance. A way to express this condition is to ask whether the absolute value of the difference between the expressions is less than the acceptable tolerance. The tolerance must be chosen carefully, to avoid admitting a pair of values as equal when they are "really" different.

A different sort of difficulty can arise, either when the numerical values in a calculation are *very different* or when they are *very nearly equal* in size (relative to the amount of significance provided by the number representation being used). Operations such as multiplication or subtraction sometimes tend to exaggerate the relative sizes of numbers. For instance, if we calculate the value of $X^2 - Y^2$ when X has the value 1000 and Y is 1, using a computer with floating decimal representation having five significant digits, we obtain the following computed values:

X	$0.10000_{10}4$
Y	$0.10000_{10}1$
X × X	$0.10000_{10}7$
Y × Y	$0.10000_{10}1$
(X × X) − (Y × Y)	$0.10000_{10}7$

Everything is perfectly correct until the last step. The value of X^2 is more than 10^5 times as large as the value of Y^2, even though the values of X and Y are not so widely separated, and so the subtraction of Y^2 has no significant effect compared to X^2 on a five-digit computer.

If we use this result in a further computation of

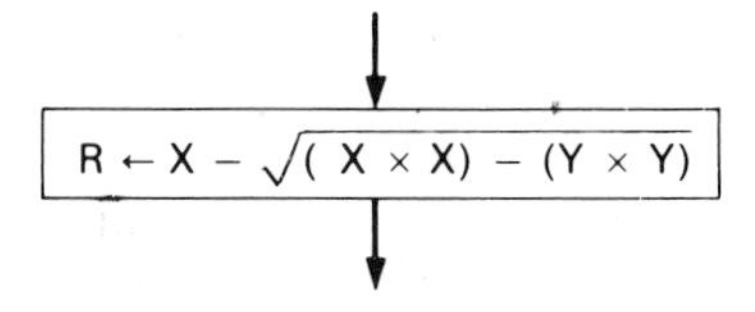

the square root of the *calculated* value will be equal to X, so the value of R will be zero, whereas the correct answer is very close to 0.0005 or $0.50000_{10} - 3$ in floating decimal notation.

When the values of X and Y are very close together, for example 1000 and 1001, we get the value 0.10020 $_{10}$ 7 as the square of 1001, truncated after the fifth significant decimal digit. The computed difference between the squares is 2000 or 0.20000 $_{10}$ 4, whereas the correct difference would be 2001 or 0.20010 $_{10}$ 4 in floating decimal.

With a computer that provides more significant digits (or the equivalent bits) in its floating decimal number representation, this potential problem must still be kept in mind. The sort of problem illustrated here with squares of very similar or very different numbers will still occur with higher powers, even on a computer that carries more significance. Forsythe (1970) gives a number of simple examples, such as solving quadratic equations or two simultaneous equations, where incorrect answers are obtained when a straightforward algebraic formula is simply translated into an algorithmic procedure. With a little care and attention to problems of significance, however, better results can be obtained.

Words to Remember

rational number   tolerance

Exercises

1. Rewrite the suggested program for finding all perfect square integers between 1 and 1000 in the language you are using, and run it on your computer. Are all of the correct results printed?
2. According to the Pythagorean theorem, one angle of a triangle is a right (90°) angle if the square of the length of the longest side equals the sum of the squares of the other two sides. In particular, any triangle whose sides are in the ratio 3:4:5 is a right triangle. Therefore, if n is a positive integer, a triangle with sides of lengths $3/n$, $4/n$, and $5/n$ is a right triangle. Write a program which, for all values of n from 1 to 20, will test whether A² + B² equals C², where A is $3/n$, B is $4/n$, and C is $5/n$. For each value of n, print an appropriate remark if the test passes. If the test fails, print the difference between the two expressions that should be equal, in floating decimal format. Can you explain what you find?

---

*Projects**

1. Write a program to solve the quadratic equation $Ax^2 + Bx + C = 0$ by the quadratic formula, as given in exercise 2, Section 3.4.
   (a) Apply this program to the case A = 1, B = 2, C = 1 − E. Use small values of E, starting with $\frac{1}{2}$ and dividing by 2 repeatedly until your computer cannot distinguish between 1 + E and 1. Your program might include a loop with the instructions

---

* Adapted from Forsythe (1970).

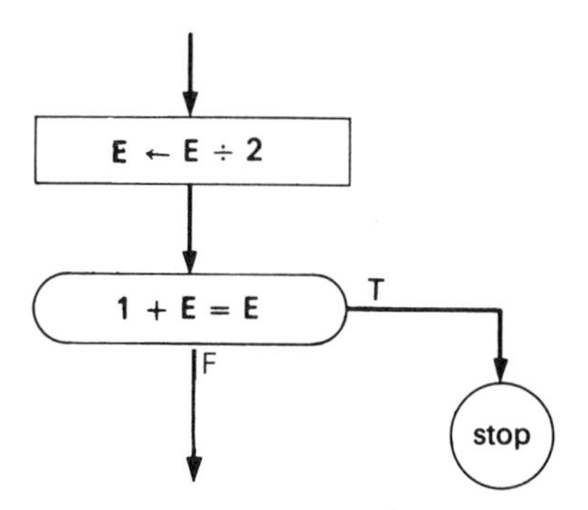

For each value of E as the loop is repeated, solve the equation. Note that the correct roots are $(-1 + \sqrt{E}, -1 - \sqrt{E})$. If the value of C is 0.99999999, for example ($E = 10^{-8}$), the roots should be 0.99990000 and 1.00010000 but your computer may give two equal roots, each having the value 1 to eight decimal places. This illustrates the fact that a small change in a coefficient may cause a larger change in a root. (In general, the roots of a polynomial equation tend to be *poorly determined* by the coefficients.)
(b) As another example, try A = 1, B = 2, C = E, using the same sequence of values of "E" as before. In this case, the roots are nearly $(-2 - \frac{1}{2}E, -\frac{1}{2}E)$. As the value of E becomes smaller, the computer will eventually produce the result $(-2, 0)$. The first root has a small relative error, but the relative error in the second root is 100 percent. Compare the results obtained by the following algorithm:

```
DD ← 1 − 4 × A × C ÷ (B × B)
P ← − B ÷ (2 × A)
QQ ← √|DD|
```

and, if DD is positive,

```
R1 ← P × (1 + QQ)
R2 ← C ÷ (R1 × A)
```

while, if DD is negative,

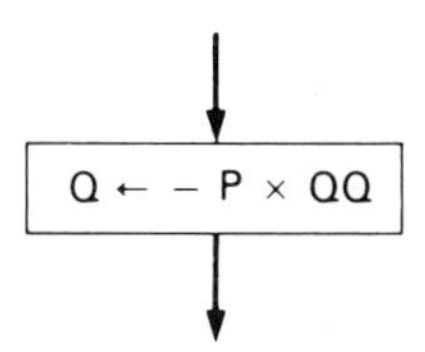

This algorithm should give better answers for small values of E than the straightforward application of the quadratic formula. Note, however, that this algorithm fails if the value of B is zero.

2. Compute the solution of a system of three simultaneous linear algebraic equations

$$\begin{aligned} A11 \times X1 + A12 \times X2 + A13 \times X3 &= B1 \\ A21 \times X1 + A22 \times X2 + A23 \times X3 &= B2 \\ A31 \times X1 + A32 \times X2 + A33 \times X3 &= B3 \end{aligned}$$

according to the following formulas:

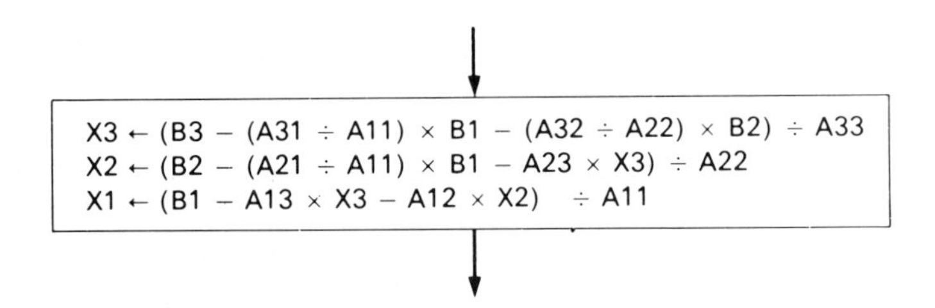

X3 ← (B3 − (A31 ÷ A11) × B1 − (A32 ÷ A22) × B2) ÷ A33
X2 ← (B2 − (A21 ÷ A11) × B1 − A23 × X3) ÷ A22
X1 ← (B1 − A13 × X3 − A12 × X2) ÷ A11

Sets of data to use are:

	A11	A12	A13	B1	A21	A22	A23	B2	A31	A32	A33	B3
a.	1	2	3	1	−2	−1	−2	0	3	1	4	0
b.	2	−1	3	1	3	−2	1	0	4	−3	2	0
c.	−1	2	−3	1	2	−1	4	1	3	−2	1	1

Difficulties may occur in the following cases:

d.	2	1	3	1	−1	4	7	1	4	2	6	1
e.	R1	R2	R3	1	R2	R3	R4	1	R3	R4	R5	1

where R1 = 1 ÷ N, R2 = 1 ÷ (N + 1), R3 = 1 ÷ (N + 2), R4 = 1 ÷ (N + 3), R5 = 1 ÷ (N + 4). Use large values of N, such as 1000 or more. Things will tend to get worse as N gets larger.

f.	e	1	0	1	1	1	0	2	0	0	1	1

Try various small values of e, as in Project 1, starting with $\frac{1}{2}$ and dividing by 2 repeatedly until your computer cannot distinguish 1 + e from 1.

g.	0	0	0	1	0	1	0	1	0	0	1	1

References

Bernstein, J., *The Analytical Engine: Computers—Past, Present, and Future*. New York: Random House, 1964.

Cole, R. W., *Introduction to Computing*. New York: McGraw-Hill, 1969.

Encyclopaedia Britannica. Chicago: Encyclopaedia Britannica, 1970.

Fike, C. T., *PL/1 for Scientific Programmers*. Englewood Cliffs, N.J.: Prentice-Hall, 1970.

Forsythe, G. E., "Pitfalls in Computation, or Why a Math Book Isn't Enough." *American Mathematical Monthly* 77: 931–955 (Nov 1970).

Fox, T. W., *The Mechanism of Weaving*. New York: Macmillan, 1911.

Goldman, S., *Information Theory*. Englewood Cliffs, N.J.: Prentice-Hall, 1953.

Harmon, L. D., "The Recognition of Faces." *Scientific American* 229: 71–82 (Nov 1973).

Harper, W. L., "The Remote World of Digital Switching." *Datamation* 17: 22–31 (Mar 15, 1971).

Huffmann, D. A., "A Method for the Construction of Minimum Redundancy Codes." *Proceedings of the Institute of Radio Engineers* 40: 1098–1011 (Sep 1952).

International Cable Directory Company, *Western Union Telegraphic Code*. 1903.

Knuth, D. E., *The Art of Computer Programming*: Vol. 1, *Fundamental Algorithms*. Reading, Mass.: Addison-Wesley, 1968.

Pierce, J. R., *Symbols, Signals, and Noise: The Nature and Process of Communication*. New York: Harper & Row, 1961.

Pratt, F., *Secret and Urgent: The Story of Codes and Ciphers*. Indianapolis: Bobbs-Merrill, 1939.

Raisbeck, G., *Information Theory: An Introduction for Scientists and Engineers*. Cambridge, Mass.: M.I.T. Press, 1964.

Sandburg, C., *Abraham Lincoln: The Prairie Years and the War Years*. New York: Harcourt Brace Jovanovich, 1954.

Scientific American eds., *Information*. San Francisco: W. H. Freeman, 1970.

Shannon, C. E., and W. Weaver, *The Mathematical Theory of Communication*. Urbana: University of Illinois Press, 1949.

Watson, W., *Textile Design and Colour*, rev. ed. London: Longmans Green, 1946.

Wilkes, M. V., *Automatic Digital Computers*. New York: Wiley, 1956.

5 *Information Structures*

5.1 *How Are Alphabetical and Symbolic Characters Processed by a Computer?*

All the programming examples in chapters 2 and 3 deal with *numerical* computations. We saw how to write instructions for reading and printing numbers, for evaluating arithmetic expressions, and for performing mathematical operations designated by functions such as absolute value, square root, and integer part.

In these computations, we assumed that the contents of a cell is a number.

In Chapter 4, we found out that the contents of a cell is actually coded as a pattern of bits. With equal ease, these bits can be interpreted as a letter, a digit, a punctuation mark, or any other symbolic character. The "meaning" of a pattern of bits is not inherent in the pattern—it depends entirely on the way the bits are to be used.

Character Representations: ASCII and EBCDIC The 64 possible combinations of six bits are sufficient to uniquely represent the 26 capital letters, 10 decimal digits, and 28 symbols (including blank and punctuation). However, most modern computers use seven or eight bits for each character, and include the lower-case alphabet as well as a variety of mathematical and other symbolic characters. Two different codes are widely used (ANSI, 1968): one of these is the American Standard Code for Information Interchange (ASCII) based on seven bits (or eight bits including parity). This code is shown in Table 5.1a. The ASCII code is used almost universally except on the IBM 360 and 370 series computers, which use the Extended Binary-Coded Decimal Interchange Code (EBCDIC), an eight-bit code. (See Table 5.1b.)

The operations we might wish to perform on characters include

storing
fetching
reading
printing
comparing

and normally do *not* include arithmetical operations or mathematical functions. Usually, it makes very little difference what bit patterns are used for the various letters, digits, and symbols, so long as the coding is consistent. We have seen that the operations of storing and fetching do not involve any *interpretation* of bit patterns. Reading and printing operations, which convert from internal to external form, require only that the internal code, whatever it is, be taken into account during the conversion process. Comparisons of *equality* can also be made without the particular internal coding being known. So long as the same character is consistently represented by the same bit pattern, the question of whether two bit patterns are identical can be settled without knowing the interpretation of either pattern.

Table 5.1a *Standard Seven-Bit ASCII Code*

<table>
<tr><td>32</td><td>space</td><td>64</td><td>@</td><td>96</td><td>`</td></tr>
<tr><td>33</td><td>!</td><td>65</td><td>A</td><td>97</td><td>a</td></tr>
<tr><td>34</td><td>"</td><td>66</td><td>B</td><td>98</td><td>b</td></tr>
<tr><td>35</td><td>#</td><td>67</td><td>C</td><td>99</td><td>c</td></tr>
<tr><td>36</td><td>$</td><td>68</td><td>D</td><td>100</td><td>d</td></tr>
<tr><td>37</td><td>%</td><td>69</td><td>E</td><td>101</td><td>e</td></tr>
<tr><td>38</td><td>&</td><td>70</td><td>F</td><td>102</td><td>f</td></tr>
<tr><td>39</td><td>'</td><td>71</td><td>G</td><td>103</td><td>g</td></tr>
<tr><td>40</td><td>(</td><td>72</td><td>H</td><td>104</td><td>h</td></tr>
<tr><td>41</td><td>)</td><td>73</td><td>I</td><td>105</td><td>i</td></tr>
<tr><td>42</td><td>*</td><td>74</td><td>J</td><td>106</td><td>j</td></tr>
<tr><td>43</td><td>+</td><td>75</td><td>K</td><td>107</td><td>k</td></tr>
<tr><td>44</td><td>,</td><td>76</td><td>L</td><td>108</td><td>l</td></tr>
<tr><td>45</td><td>-</td><td>77</td><td>M</td><td>109</td><td>m</td></tr>
<tr><td>46</td><td>.</td><td>78</td><td>N</td><td>110</td><td>n</td></tr>
<tr><td>47</td><td>/</td><td>79</td><td>O</td><td>111</td><td>o</td></tr>
<tr><td>48</td><td>0</td><td>80</td><td>P</td><td>112</td><td>p</td></tr>
<tr><td>49</td><td>1</td><td>81</td><td>Q</td><td>113</td><td>q</td></tr>
<tr><td>50</td><td>2</td><td>82</td><td>R</td><td>114</td><td>r</td></tr>
<tr><td>51</td><td>3</td><td>83</td><td>S</td><td>115</td><td>s</td></tr>
<tr><td>52</td><td>4</td><td>84</td><td>T</td><td>116</td><td>t</td></tr>
<tr><td>53</td><td>5</td><td>85</td><td>U</td><td>117</td><td>u</td></tr>
<tr><td>54</td><td>6</td><td>86</td><td>V</td><td>118</td><td>v</td></tr>
<tr><td>55</td><td>7</td><td>87</td><td>W</td><td>119</td><td>w</td></tr>
<tr><td>56</td><td>8</td><td>88</td><td>X</td><td>120</td><td>x</td></tr>
<tr><td>57</td><td>9</td><td>89</td><td>Y</td><td>121</td><td>y</td></tr>
<tr><td>58</td><td>:</td><td>90</td><td>Z</td><td>122</td><td>z</td></tr>
<tr><td>59</td><td>;</td><td>91</td><td>[</td><td>123</td><td>{</td></tr>
<tr><td>60</td><td><</td><td>92</td><td>\</td><td>124</td><td>|</td></tr>
<tr><td>61</td><td>=</td><td>93</td><td>]</td><td>125</td><td>}</td></tr>
<tr><td>62</td><td>></td><td>94</td><td>↑</td><td>126</td><td>~</td></tr>
<tr><td>63</td><td>?</td><td>95</td><td>←</td><td>127</td><td>delete</td></tr>
</table>

Table 5.1b Extended Binary-Coded Decimal Interchange Code

64	space	132	d	200	H
74	[	133	e	201	I
75	.	134	f	208	}
76	<	135	g	209	J
77	(	136	h	210	K
78	+	137	i	211	L
79	|	145	j	212	M
80	&	146	k	213	N
90	]	147	l	214	O
91	$	148	m	215	P
92	*	149	n	216	Q
93	)	150	o	217	R
94	;	151	p	224	\
95	¬	152	q	226	S
96	-	153	r	227	T
97	/	161	~	228	U
106	!	162	s	229	V
107	,	163	t	230	W
108	%	164	u	231	X
109	_	165	v	232	Y
110	>	166	w	233	Z
111	?	167	x	240	0
121	`	168	y	241	1
122	:	169	z	242	2
123	#	192	{	243	3
124	@	193	A	244	4
125	'	194	B	245	5
126	=	195	C	246	6
127	"	196	D	247	7
129	a	197	E	248	8
130	b	198	F	249	9
131	c	199	G		

If we want to arrange characters according to a particular ranking or ordering relationship, however, the coding system must be considered. The customary procedure[1] is to consider the seven- or eight-bit pattern representing a character as a binary integer. The characters are then ranked according to the numerical order of the corresponding integers. For example, in ASCII, the letter A is represented by the same bit pattern as the integer 65, and B corresponds to 66; hence A will be considered *less than* B.

In arranging a dictionary or telephone book, there is little disagreement about ranking as long as we are dealing entirely with capital letters and blanks. (However, many catalogs and directories use complicated special rules. For instance, all names beginning with Mc may be "respelled" for purposes of ranking.) It is obvious that

[1] This procedure is "customary" only because the need to compare *characters* was not recognized until long after computers had been available with convenient means for comparing *numerical* values.

a blank must rank "less than" any letter of the alphabet, so that BROWN will precede BROWNING, for example. Both the ASCII and EBCDIC codes preserve this ranking. However, in ASCII the numerical digits rank *less* than the letters (but higher than blank), while in EBCDIC the digits have *greater* rank than any other characters. Most characters used in ordinary punctuation rank just greater than the blank in both codes, but the ranks of the individual symbols within this group do not agree. And there is considerable inconsistency in the ranking of special mathematical and other symbols.

A more serious difference between the codes is that upper- and lower-case letters are represented by different bit patterns and hence have different effective ranks (a is considered higher in rank than Z in ASCII, while A is above z in EBCDIC). For most practical comparisons, however, the upper- and lower-case versions of a letter must be the *same* rank. To see this, consider a group of names such as Defoe, De Forest, Degas, de Gaulle.

Therefore, we can rank characters in the customary way—by interpreting the code patterns as binary integers— *only* in rather simple problems, such as those in which only capital letters are used and in which there are only a few mathematical symbols. In other cases, we must work out a different way of assigning ranks to each character, and then we must compare the ranks rather than the coded characters themselves.

Strings Characters rarely occur in isolation in computer applications; they most often occur in groups that form words and lines of text. We call a group of consecutive characters a *string*. To refer to the sequence of characters, considering the string as a unit, we use a *string name*. A string name looks like a variable; however, it does not refer to a single cell but to whatever amount of space is needed to store the string. The computer must be informed that the string name is not an ordinary variable, by means of a type declaration of the form

```
string string name[limit] length
```

Here, the limit is an integer that tells the maximum number of characters the string may contain at any one time. And "*length*" is a variable to be associated with the string—at any time, the value of this variable is the number of characters currently contained in the string. For mnemonic purposes, we may adapt some special convention for relating the string name and the length variable, for instance,

```
string Title[72] LTitle
```

where we use the string name preceded by L for the length variable. However, the length variable could have been given any other name.

String Constants A string constant, like a string name, does not refer to a single cell but to whatever amount of space is needed to store the string. A string constant is written by enclosing characters between quotation marks:

'ABC'   '123'   '$7.49'   '(A + X)'   '#&%'

The length of a string constant is simply the number of characters between the quotation marks. The constants above have lengths 3, 3, 5, 7, and 3 respectively. Note that blanks are counted in determining the length. This is illustrated further in the following examples:

String	*Length*
'(A + X)'	7
'(A+X)'	5
' '	1

The first two strings have different lengths and, in fact, are different strings. The third example is a string of length one, consisting of the single character *blank*.

A string constant must be distinguished from a cell name, such as a variable or a numeric constant.

String	*Variable*
'TotalHrs'	TotalHrs
'ABC123'	ABC123
'X'	X

String	*Numeral*
'123.45'	123.45
'9'	9
'0.1e5'	0.1e5

The bit pattern used for representing an integer will not, in general, be the same as the pattern for the string consisting of the same digits. For instance, a seven-bit binary representation of the integer 9 is 0001001, while in ASCII the one-character string '9' is represented as 0111001 and in EBCDIC it is 11111001. (The agreement of the last 4 bits in all three codes is not quite coincidental.)

Assignment Instructions for Strings An assignment instruction of the form

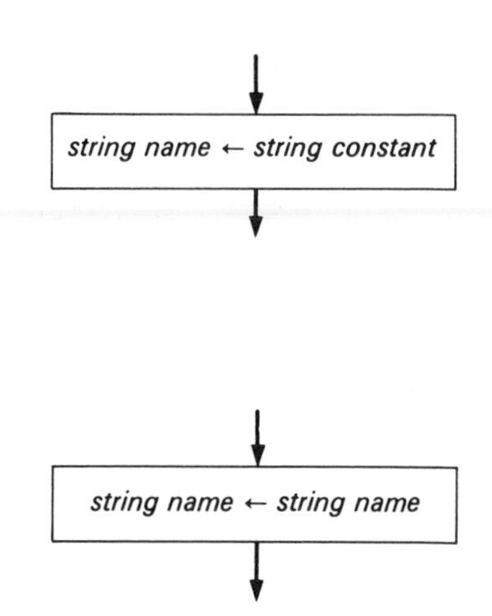

or

will store the string referred to on the left in the space allocated to the string name on the right, provided that the current length of the string to be stored does not exceed the limit specified in the declaration of the string name on the left. The length variable will be set to the proper value, that is, to the length of the string named on the right.

Consider the following program segment as an example:

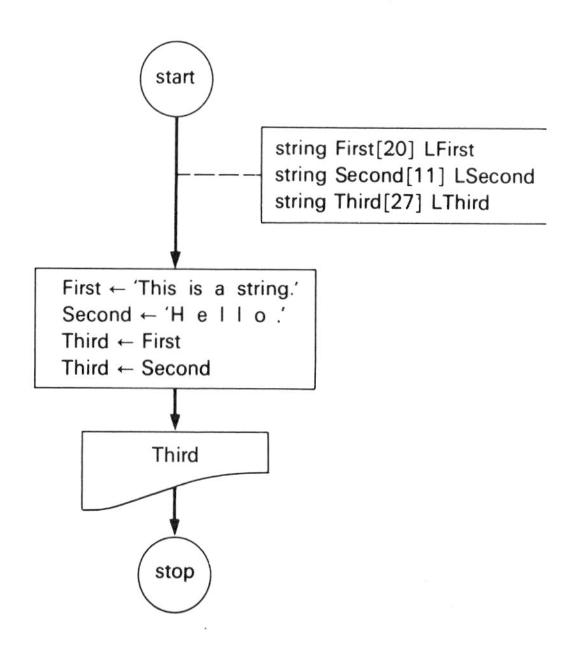

The length of the constant string assigned to First is 17 characters, within the length limit specified in the declaration. Similarly, the length of the constant string assigned to Second is 11 characters. The assignment of the string named First to the string variable Third automatically sets the length variable LThird to the value 17. The final assignment fetches the characters currently in the string named Second and stores them as the new characters of the string named Third, adjusting the value of the length variable LThird at the same time.

Null Strings The length of a string may be *zero*. Surprisingly enough, such a string is useful for some purposes. A string whose length is zero is called a *null string*. Such a string has a name, and its length has the definite value zero, although the string does not contain any characters. In effect, this provides a way to keep a string name "alive" when it is not being actively used to store characters. The representation of a null string, as a string constant, consists of two adjacent quotation marks:

''

Joining Two Strings A *join* operation, designated by the operator ||, is provided for combining the characters of two strings to form a longer string.

string name || string name

is a *string expression* referring to a string composed of a combination of the characters of the two specified strings.[2] The operands in this operation may also be string constants or, in fact, any string expressions. For example, the "value" of the string expression

'Now is the time' || 'for all good men to come.'

is the same as that of the string constant

'Now is the time for all good men to come.'

The length of the combined string is the *sum* of the lengths of the separate strings. If either operand of the *join* operator is a null string, the result will be the same as the other operand. Note that the rule for adding lengths works in this case also, because the length of the null string is zero.

String Input and Output Read and print instructions can include string names, along with numerical variables, in the input or output list. Print instructions can include string expressions as well. Input data corresponding to a string name must be enclosed in quotation marks, in the same form as a string constant. (The number of characters of input data must not exceed the limit specified in the string declaration.) A print instruction referring to a string will simply cause the characters currently contained in the string to appear on the printer. The following program will read two strings, join them, and print the resulting string:

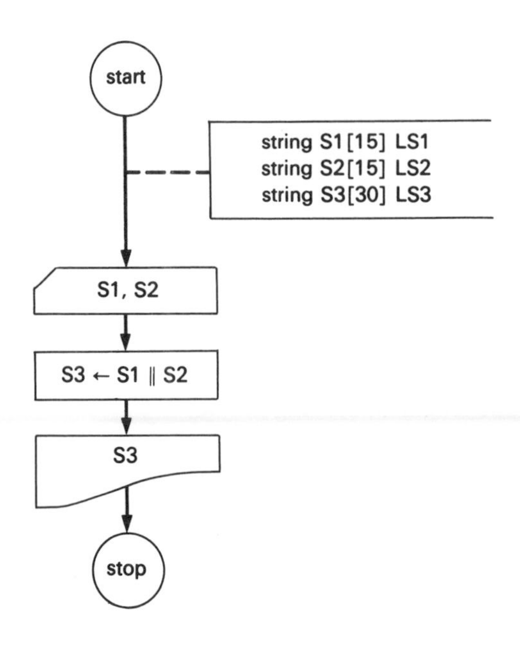

[2] The term *concatenate* or *catenate* is often used for this operation.

The same result would be printed if we omitted the assignment instruction and wrote

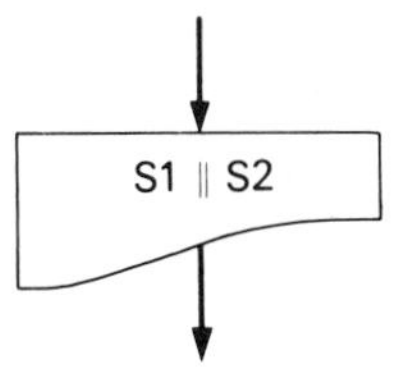

Consider another example:

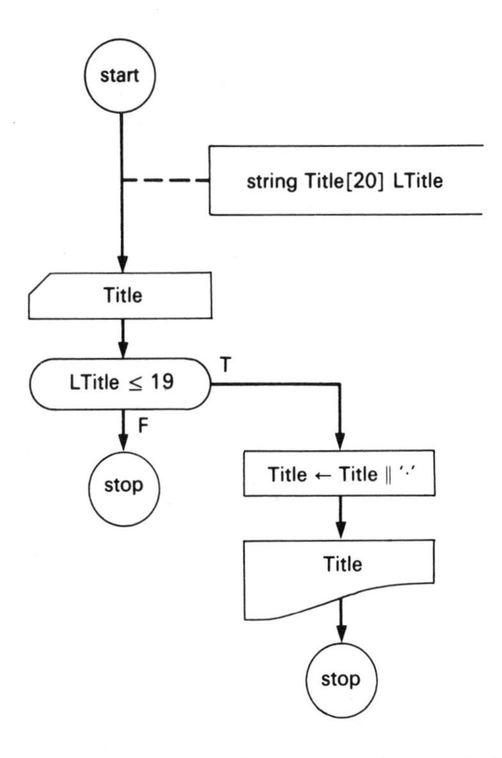

If the input string is not more than 19 characters long, a period will be adjoined to the end of it.

Note also the difference between the following two print instructions:

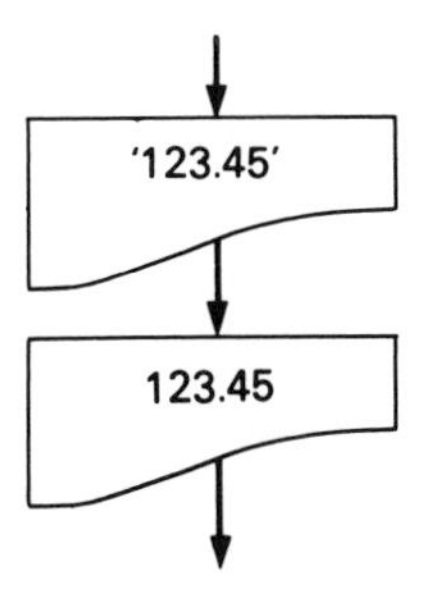

Of course, the bit patterns used in storage for the string constant in the first case and for the numerical constant in the second case would be entirely different; nevertheless, the appearance of the printed results might be identical. However, it is important to realize that the printed appearance of the numerical constant is controlled by the format *conventions* built into the output section of the computer system, but the printing of the string constant is controlled entirely by the *programmer.* For numerical constants representing very large or very small values, floating decimal external form may be used automatically; that is, it may be impossible to print *numerically* a constant such as 0.000000000321 in this fixed decimal external form. However, the string constant '0.000000000321' would, of course, be printed exactly as specified.

Single quotation marks must be used to enclose input strings, so that the string length can be determined. Output strings, on the other hand, will be printed without the enclosing quotation marks. This causes a slight difficulty in the case of strings that begin or end with a sequence of blanks. It will generally be impossible to distinguish, from the appearance of the output, between areas on the paper where "blank" characters from the string have been "printed" and areas where nothing has been printed at all.

An Example: "Personalized" Letters Some commercial mailing firms "personalize" letters that are to be sent to all persons on a mailing list. The letter contains some fixed text, along with a smaller amount of information that is different for each addressee. This information is sometimes just the name and address, but it may also use other information in the file from which the mailing list was selected.

For example, the Alumni Association may be soliciting contributions from those who donated to its fund drive during the previous year. The fixed text might be:

> Dear (*name*),
>
> Now that you are living in (*city*) you may not realize how much your Alma Mater has changed since you graduated in (*year*). Your contribution, last year, of (*amount*) has helped to make possible a number of important capital improvements around the campus, which are described in the enclosed folder. You realize, (*name*), how much your own Department of (*field*) will profit from these same improvements. You will want to contribute even more this year, we are sure, so that you can feel a glow of pride when you come from (*city*) to visit the campus at the time of the next reunion of your class of (*year*).

For each addressee, the name, city, amount of last year's contribution, year and field of graduation, are read into the computer and are inserted in the proper places as the letter is printed.

Comparing Strings As we have noted, it is possible to tell whether two bit patterns are *equal* without interpreting the pattern. Thus, a conditional instruction may be based on the following assertion:

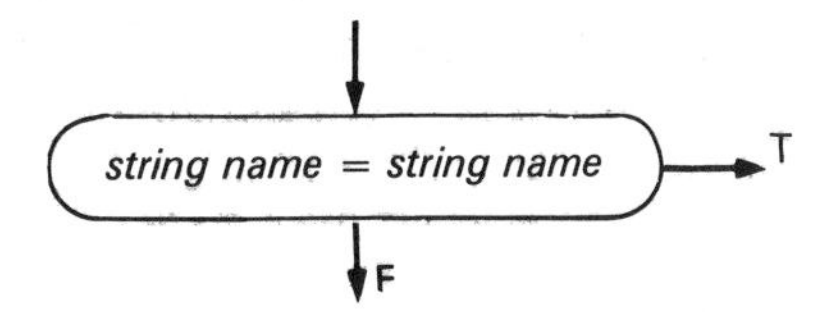

The assertion is *true* only if the two strings mentioned have the same current *length* and consist of the same sequence of characters. Comparisons involving the relations >, <, ≤, or ≥, however, will be based on the interpretation of the bit patterns as binary numbers; this may not give the desired ranking if the string includes both upper- and lower-case letters, or digits, or certain mathematical symbols.

When two strings are unequal in *length*, but the characters in the shorter string exactly match the beginning of the longer string, the shorter string is considered "less than" the longer string. (In effect, we may think of the shorter string as being extended with null characters, which rank even lower than blanks.) It follows from this rule that a null string is "less than" any other string.

An Example: Inserting an Item in a List We have a list of names of the employees of Blank Manufacturing Company, in alphabetical order. A new employee is hired, and we want to print a corrected list of employees, inserting the new name at the correct point. We put the new name ahead of the old list in the reader, so that it will be the first string read. Then we read the old list, and compare each name with the new name before printing it. Assume that the artificial name ZZZZZ appears at the end of the old list as a sentinel, to indicate that all the names have been read.

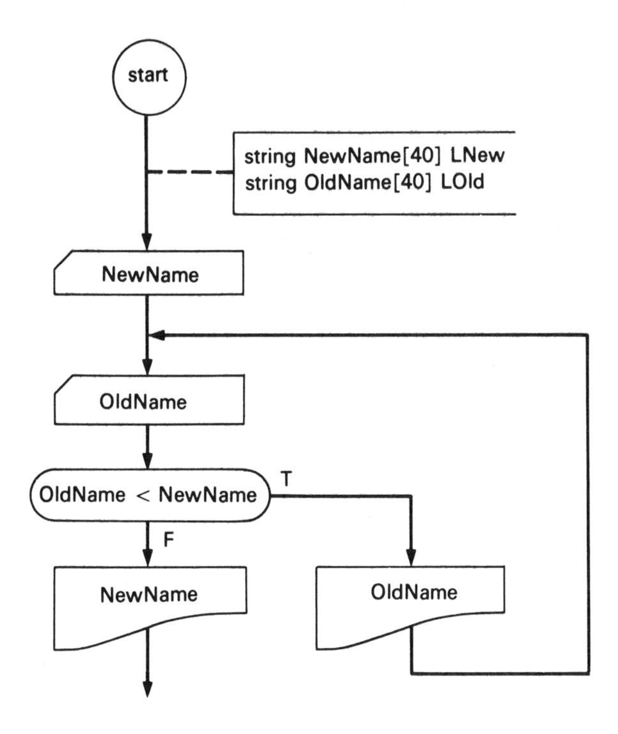

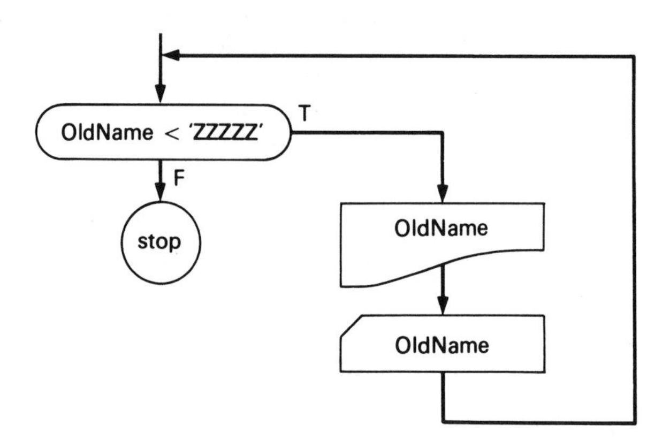

Note that the program continues to execute the first loop, comparing each OldName with the NewName, so long as each OldName is less in rank than the name sought. Then it begins comparing the names it reads with the constant string 'ZZZZZ' while printing the remainder of the old list. Note that this program will work even if the new name is ahead of all the old names, or even if it is larger in rank than any old name except the sentinel.

String Lengths Although a string length variable is included in the declaration, none of the foregoing operations make any explicit reference to its value. However, these operations automatically refer to it whenever necessary. The print and comparison operations use it to determine how many characters are involved, and the read and assignment operations give this variable the correct new value. The join operation automatically uses the lengths of the operand strings to determine the length of the combined string, and this length will be associated with the string in any further operations (such as assignment of the joined string).

Flowchart Strings form a separate non-numeric data type and must be declared. The declaration includes a string name, an integer giving the maximum length of the string, and an associated variable to be used for storing the current length of the string:

```
string string name[limit] length
```

For example,

```
string Title[20] LTitle
string S1[15] LS1
string S2[24] LS2
```

A string constant is written by enclosing the characters between single quotes:

```
'ABC'     '123'  '$7.49'  '(A + X)'  '#&%'
'(A+X)'   ' '    'T'      '9'        '0.1e5'
```

The length of a string constant is simply the number of characters between the quotation marks.

An assignment instruction may be written with a string name on the left and a *string expression* on the right. A string expression is a string name, a string constant, or a function such as *join* that produces a string as its "value." In any assignment operation, an error will occur if the length of the string expression on the right exceeds the limit in the declaration for the string name on the left.

A *null string* is any string whose current length is zero. The string constant consisting of two adjacent quotation marks,

```
''
```

represents a null string.

The *join* operation, designated by the operator ||, has two strings as operands. A longer string is formed by adjoining the second operand string to the end of the first.

Read and print instructions can include string names, along with numerical variables, in the input or output list. Print instructions can include string expressions as well. Input data corresponding to a string name has the same form as a string constant; again, during input care must be taken not to exceed the string length limit.

Strings may be compared for equality or inequality by an ordinary conditional instruction, using an assertion of the form

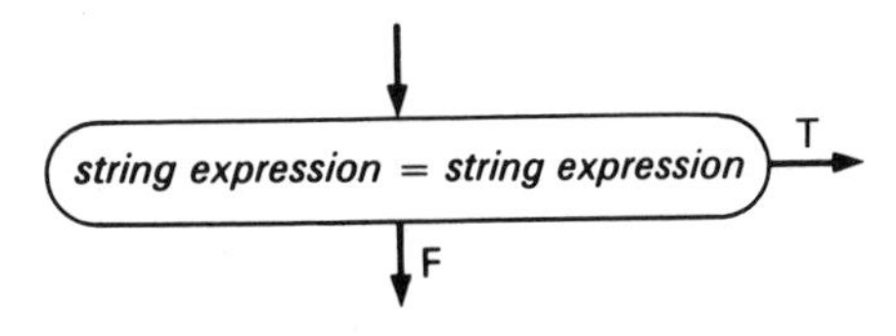

String comparisons involving the relations $>$, $\geq$, $<$, or $\leq$ are based on the interpretation of the bit patterns as binary numbers. This may not give the desired ranking if both upper- and lower-case letters, or numerical digits, or certain mathematical symbols are contained in the strings. When the characters in a shorter string exactly match the beginning of a longer string, the shorter string is considered "less than" the longer one. A null string is "less than" any non-null string.

Basic A string name in Basic looks like an ordinary variable followed by a dollar sign. That is, it is either a single capital letter followed by a dollar sign or a letter and a digit followed by a dollar sign:

```
X$     R3$    A$     T$
A1$    Q$     V1$    F1$
```

A string constant is written by enclosing the characters between *double* quotes:

```
"ABC"      "123"   "$7.49"   "(A + X)"
"(A+X)"    " "     "0.1E5"   "#&%"
```

An assignment instruction may be written with a string name on the left and a string name or a string constant on the right.

A *null string* is any string whose current length is zero. The string constant consisting of two adjacent double quotation marks,

""

represents a null string.

There is no *join* operation in Basic.

READ, INPUT, and PRINT instructions can include string names, along with numerical variables, in the input or output list. PRINT instructions can include string constants as well. Input data corresponding to a string name has the same form as a string constant.

String comparisons involving the relations $=$, $\neq$, $>$, $\geq$, $<$, or $\leq$ are based on the interpretations of the bit patterns as binary numbers. This may not give the desired ranking in cases that involve numerical digits or mathematical symbols.

Fortran Fortran does not provide convenient features for manipulating strings, primarily because the language was developed long before the importance of character manipulation was recognized. We present here facilities that are available in Standard Fortran; besides this, we exhibit some additional features that are provided in one important nonstandard Fortran "dialect"—Watfiv (Cress et al., 1970).

Standard Fortran provides for reading and printing strings. A constant string to be *printed*, for some purpose such as a heading, is included directly in a format statement. A Fortran string constant consists of a numeral giving the length, followed by the letter H, followed by a string of exactly the length indicated by the numeral; for example:

```
3HABC        3H123     5H$7.49
5H(A+X)      1H        1HT
7H(A + X)    3H#&%
1H9          5H0.1E5
```

In all these cases, the H and the digits to the left of it are *not* part of the string itself. To print the string 'TITLE' along with some numeric values, we could use the following format statement:

```
89 FORMAT (I10, 5HTITLE, 3F15.3)
```

A character may be read into the cell named by an ordinary variable by means of the A1 format. The following program will read one character into each of six cells:

```
   READ (5, 88) I, J, K, L, M, N
88 FORMAT (6 A 1)
```

The same characters, preceded by a constant string, could be printed as follows:

```
   WRITE (6, 87) I, J, K, L, M, N
87 FORMAT (6H HELLO, 6 A 1)
```

Most versions of Fortran permit more than one character to be read into or printed from a cell—the format specification A*n* is permitted, where *n* may have any value from 1 up to some limit such as 4, 6, or 10, depending on the version.

Although character data may be read or printed in Standard Fortran, and character constants may be used in limited ways, there is no provision in Standard Fortran for string variables. However, character data may be read into a cell whose name is a *numerical* variable, and this variable may then be used in expressions and instructions. Thus, short strings of characters, not exceeding the number of characters that will fit in a cell, may be moved about using *assignment* instructions. It is possible (but awkward) to manipulate longer "strings" of characters that have been stored in consecutive cells. *Comparison* of short strings (stored in single cells) can also be accomplished using numerical variables. Such comparisons will, of course, be based on the interpretation of the bit patterns as "numerical values" of the variables. For practically all computers, this comparison will give the desired result in most cases, if the name of the cell used to store the characters is an *integer* variable. However, numerical comparison of *real* variables involves a more complicated way of interpreting the bit patterns stored in the cells and is incompatible on many computers with the desired ranking of the bit patterns as alphabetical strings. Therefore, it is generally recommended that characters to be compared be read into cells having *integer variable* names.

Nonstandard dialects of Fortran provide additional features for the manipulation of strings of characters. Most provide an alternative form for string constants, consisting of the characters of the constant string written between single quotes:

```
'ABC'     '123'     '$7.49'     '(A + X)'
```

String constants in this form may be mixed with other string constants in the standard (nH) form, in the same program.

One widely used Fortran dialect, called Watfiv, provides a CHARACTER data type declaration. Such a declaration lists a string variable, followed by an asterisk and then by the string length. (The length of a string in Watfiv is *fixed* and does not change; however, all the characters from some point on may be blanks.) String variables may be read, printed, assigned, or compared. The alternative form for string constants is also permitted in Watfiv.

```
   CHARACTER TITLE*20
   CHARACTER S1*15
   CHARACTER S2*24, S3*24
   READ (5, 97) S3
97 FORMAT (6 A 4)
   S2 = S3
   S3 = 'ABCDEFGHIJKLMNOPQRSTUVWX'
   IF (S2 .NE. S3) STOP
   WRITE (6, 98) S2
98 FORMAT (1X, 6 A 4)
   STOP
   END
```

String comparisons are based on the interpretation of bit patterns as binary numbers. This may not give the desired ranking in cases that involve numerical digits or mathematical symbols.

The length of a CHARACTER variable is fixed by the declaration and remains unchanged throughout the computation. For this and other reasons, functions producing "values" of CHARACTER type (such as the "join" function) are not provided in Watfiv. A "string expression" must be either a variable of CHARACTER type or a string constant.

PL/1 The string features in our flowcharts have been patterned after those available in PL/1. The terminology is slightly different in PL/1, however. A string declaration has the following form:

```
declare string name character(limit)
  varying;
```

For example,

```
declare TITLE character(20) varying;
declare S1 character(15) varying;
declare (S2, S3) character(24) varying;
```

A string constant is written by enclosing the characters between single quotes:

```
'ABC'     '123'  '$7.49'  '(A + X)'  '#&%'
'(A+X)'   ' '    'T'      '9'        '0.1e5'
```

A string expression may be a string name, a string constant, or any two string expressions joined by the *concatenation operator*, ‖. This symbol indicates that a longer string is to be formed by joining the second string expression to the first. For example, we could add a period to the end of a variable string by writing

```
TITLE ‖
```

Get list and put skip list instructions can include string names, along with numerical variables, in the input or output list. Put skip list instructions can include string expressions as well. Input data corresponding to a string name has the same form as a string constant; again, during input care must be taken not to exceed the string length limit.

PL/1 does not provide a separate variable for the current length of a string; instead, a length function is provided. This function has a string expression as its argument, and its value is an integer giving the current length of the argument string:

```
length (string expression)
```

For example

```
length ('$7.49')
```

has the value 5. The length of a string produced by the concatenation operator is the sum of the lengths of the component strings.

A null string is any string whose current length is zero. The string constant consisting of two adjacent quotation marks:

```
''
```

represents a null string.

String comparisons involving any of the six relations are based on the interpretation of the bit patterns as binary numbers. This may not give the desired ranking if both upper- and lower-case letters, or numerical digits, or certain mathematical symbols are contained in the strings.

Words to Remember

character	length of string	join
ASCII	constant string	string comparison
EBCDIC	null string	string expression
string		

*Exercises**

1. Modify the payroll program of Section 3.7 to read and print the employee's *name* as well as the other information.

Name	*Emp. No.*	*Total Hours*	*Base Rate*
Clinton, D.	15625	40	$1.60
Pinckney, C.	24389	25	$2.70
Gerry, E.	42879	45	$4.31
Calhoun, J.	50653	40	$9.75
Clay, H.	59319	40	$3.25
Webster, D.	68921	42	$4.31
Bryan, W.	97336	20	$1.60
Zwettler, X.	97972	40	$4.31

2. (no F) As part of an instructional dialog, the student's response is to be compared with certain correct and incorrect responses that were anticipated by the author of the dialog. Write a program to read a string, and compare it with each of five strings: R1, R2, R3 (right answers expected) and W1, W2 (wrong answers expected). Print the constant string 'You are right.' or 'You are wrong.' as indicated, if the input string matches any of the five expected answers. If no match is found, print the constant string 'Please reformulate your answer.'

   For example, the student may be responding to the question "Who was president of the United States in 1864?" The anticipated correct answers might be 'Lincoln', 'Honest Abe', and 'Abraham Lincoln'. Anticipated wrong answers might be 'Andrew Johnson' and 'U. S. Grant'.

	Expected Right Answers	*Expected Wrong Answers*
a.	Lincoln	Andrew Johnson
	Honest Abe	U. S. Grant
	Abraham Lincoln	
b.	Baltimore	Los Angeles
	Orioles	Dodgers
	American League	
c.	Armstrong	John Glenn
	Neil Armstrong	Aldrin
	N. Armstrong	

* Some of these exercises use features that are not available in some of the languages. A "no B" (Basic), "no F" (Fortran), or "no W" (Watfiv) indicates that the language does not include the required features. PL/1 has the necessary features for working all the exercises.

3. (no B, no F, no W) Write instructions to join together the three constant strings 'Now is the time', 'for all good men to come', and 'to the aid of the party'. Then read another string and join it on to the *left* of the string just formed.

   Variable string
   a. Today is my birthday.
   b. We are out of potato chips.
   c. All bad men have gone away.

4. Write one or more formal rules, similar to those of Section 2.10 (see also Section 2.10, exercise 8) defining *string expression*. Include the "join" operation.
5. (no B, no F, no W) Write a program to print the "personalized" letters described in the example. Use a separate print instruction for each line to be printed. Thus you must join together the constant strings and the strings read as data into a single string to be printed for each line of the letter.

Name	*City*	*Amount*	*Year*	*Field*
Miss Hopper	Hollywood	$100	1925	journalism
Mr. Keaton	Hollywood	$100	1930	drama
Mr. Grosvenor	Baddeck	$400	1910	journalism
Mr. Beebe	San Mateo	$400	1937	journalism
Mr. Sloan	New York	$1,700	1910	economics
Mr. Hofmann	New York	$1,700	1915	art
Mr. Nimitz	San Francisco	$2,000	1920	engineering
Mr. Bassett	St. Louis	$2,800	1966	engineering

6. Comment on the idea of "personalized" letters, from the humanistic standpoint. Are such letters more or less personal than other approaches that might be used? Also consider privacy and other potential problems that might arise.
7. (no F) Change the "name insertion" program to a "name deletion" program, to be used when an employee resigns. Read the name to be deleted, and print all the names on the "old" list that do not match this name. (Should any special action be taken in case no match occurs?) Begin with the list of employee names given as data for exercise 1 and perform the following operations: (a) delete "Clay, H."; (b) delete "Clinton, D."; (c) delete "Zwettler, X."; (d) "Brian, W." (incorrect spelling).

5.2 *How Can We Manipulate the Individual Characters of a String?*

Although for many purposes an entire string can be manipulated as a unit, it is often desirable to subdivide the string or to perform operations on the individual characters. To do this, we *name* the separate characters by writing the string name, followed by an expression enclosed in brackets indicating the consecutive number of the desired character within the sequence of characters that form the string. For example, if the string named Title appears in a string declaration with a maximum length of 20 characters, the tenth character would be specified as

Title[10]

If the string named Come consists of the 65 characters

'Now is the time for all good men to come to the aid of the party.'

the positions where the letter o appears would be indicated as follows:

```
Come[2]  in the word "Now"
Come[18] in the word "for"
Come[26] in the word "good"
Come[27] in the word "good"
Come[35] in the word "to"
Come[38] in the word "come"
Come[43] in the word "to"
Come[53] in the word "of"
```

The expression in brackets, which is called the *subscript*, acts as a "pointer" to a particular character in the string (see Figure 5.2a). Note that the name of the string, together with the value of the subscript, is used to identify a particular character in the string. Thus, any integer *expression* may be used for the subscript, including as the most common case an integer *constant*. (When noninteger expressions are used, we assume that the value is *rounded* if necessary to the nearest integer.) There is no preferred way of writing a subscript. Any subscript expression having the correct *value* can be used, along with a string name, to refer to a character. (Care should of course be taken to insure that the subscript value does not exceed the length limit for the string.)

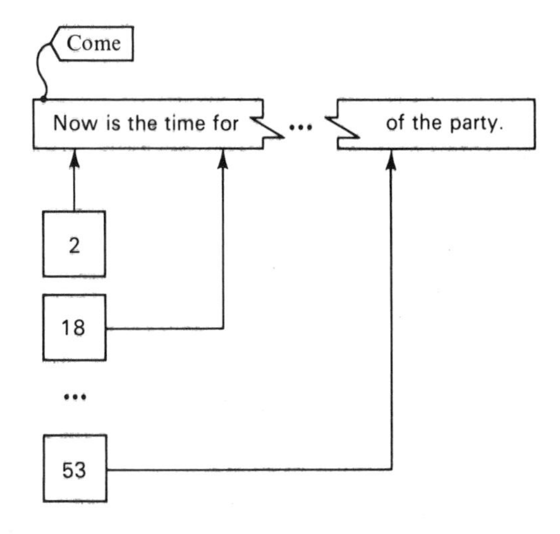

5.2a. *The subscript value indicates (points to) a particular character of the string.*

Substrings A substring is any sequence of *consecutive* characters that form a portion of a string. We can use a count-controlled loop to form a new string by copying a substring, if we know the position in the string where the substring begins and the length of the substring.

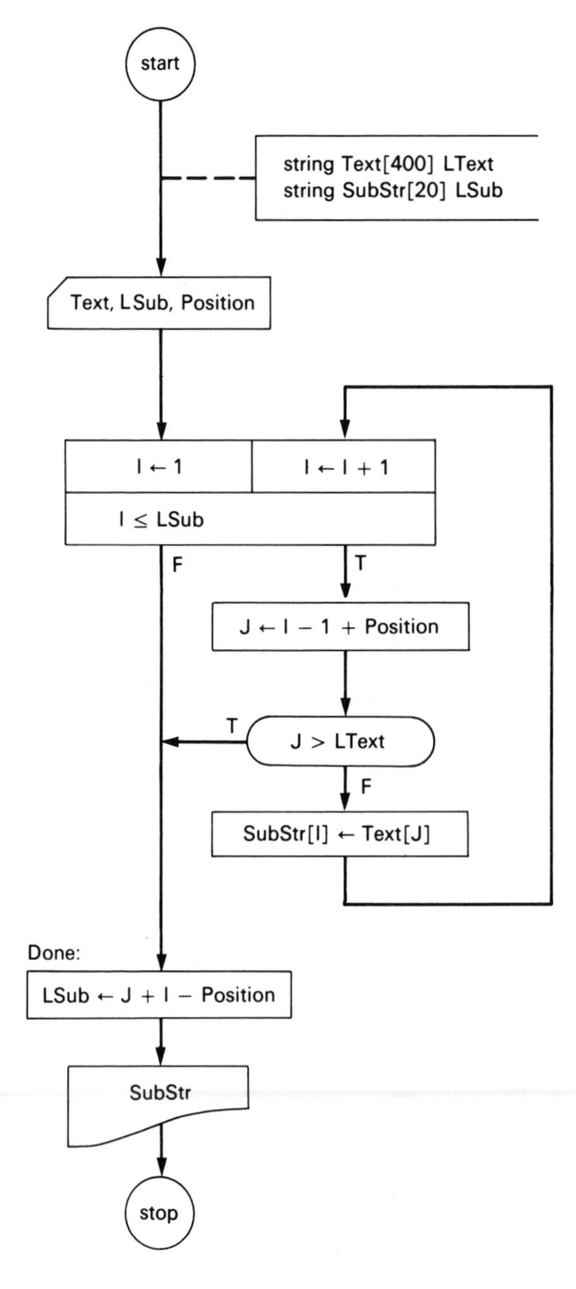

The operations in the previous section, which considered the entire string as a unit, took care of the string length automatically. When we manipulate the individual characters of a string, however, we must also keep track of any changes in the length of the string and modify the value of the associated length variable accordingly. This is illustrated in the last three boxes of the foregoing example. The value of LSub, the length variable corresponding to the string name SubStr, is set to the correct value, so that the proper number of characters will be printed.

Pattern Matching A common problem, in work with strings of text, is the following: Given a string and a "pattern" (usually a shorter string), find out whether the pattern occurs in the given string and, if so, where. We say that a match occurs if some substring of the given string exactly matches the pattern. Here is a pattern-matching program.

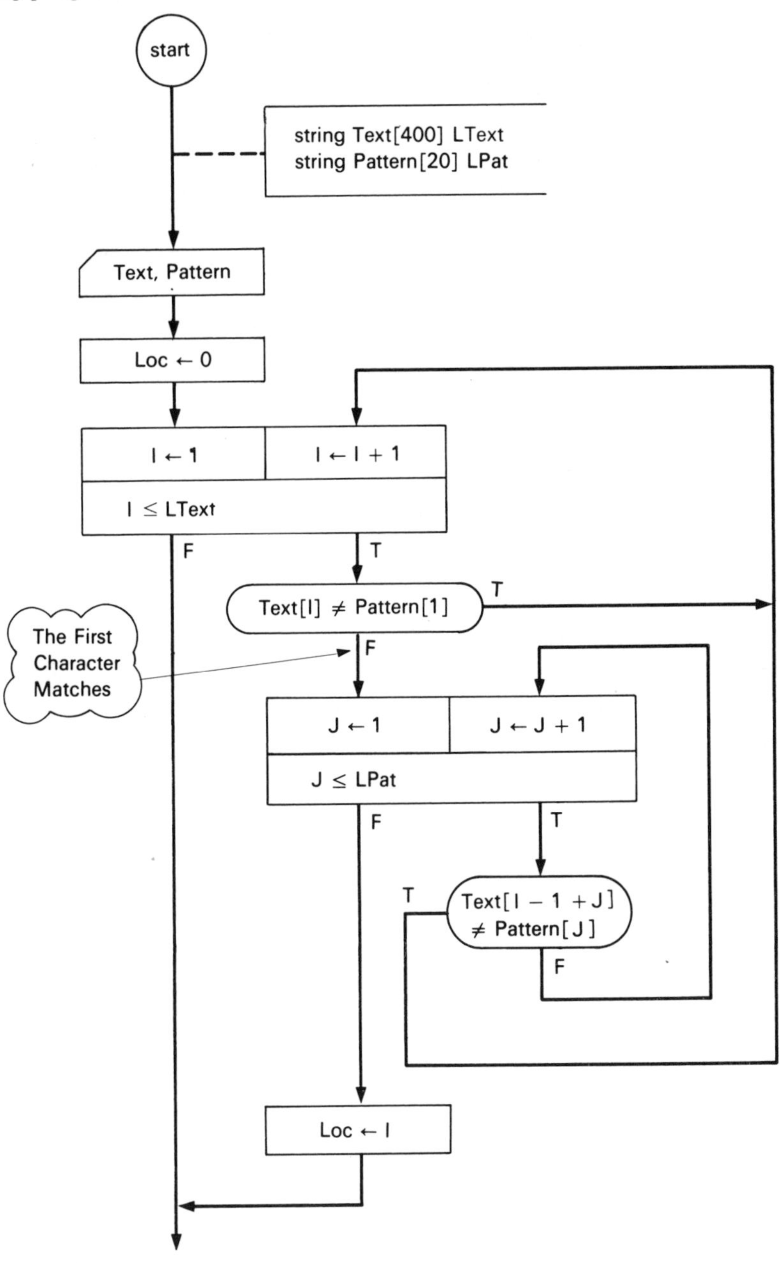

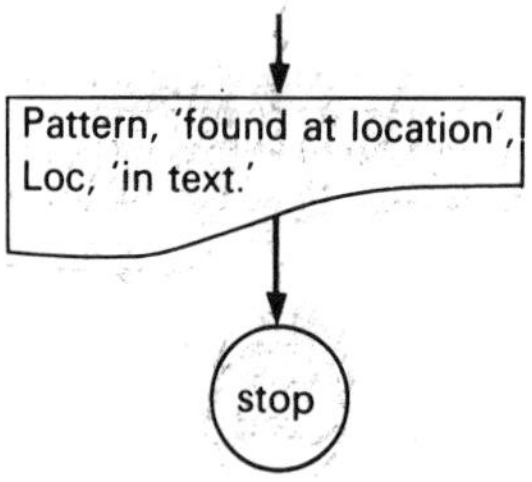

Pattern, 'found at location',
Loc, 'in text.'

stop

The pattern-matching strategy consists of looking for the *first* character of the pattern among the characters of the given string, and then looking for consecutive characters that match after the first character is found. If the pattern match never succeeds, the final value of Loc will be zero.

Packed and Unpacked Strings A cell, as we have seen, is the amount of space inside a computer that is used to store one number. Similarly, we now define a *byte* as the amount of space used to store one character. Thus, one cell is large enough to contain several bytes. (For example, a 32-bit cell on an IBM 360 or 370 computer consists of four eight-bit bytes.) Although the machine-oriented instructions on many modern computers can address individual bytes, earlier machines could not refer directly to a unit of storage smaller than a cell.

On computers whose smallest addressable unit of storage is larger than one byte, a dilemma arises as to how the characters of a string ought to be stored. To be able to address each character directly, we must "unpack" the string, storing one character in each cell. But to minimize the amount of space needed to store a string, we must "pack" the string, storing as many characters in each cell as it will hold. Of course, this dilemma evaporates where byte addressing is possible, because the individual characters can be referenced directly even when they are stored in their most compact form. However, even on machines with byte addressing, the older languages such as Fortran still exhibit vestiges of the distinction between packed and unpacked character representations. An awareness of this distinction will help in understanding the following discussion of character manipulation features in the various languages.

Flowchart In our flowcharts, we assume either that byte addressing is available or that strings are stored in unpacked form. Thus a string name, along with a subscript value, designates an individual character of the string. Instructions that reference the separate characters in this way can be freely combined with instructions that prescribe an operation (such as input, output, assignment, or comparison) involving the entire string.

(On a machine without byte addressing, it might be worthwhile to provide separate pack and unpack functions. If a long string was retained in the computer without being actively manipulated during a certain phase of the computation, it could be compressed into a smaller amount of space and expanded when it again became active.)

The form

string name[*subscript expression*]

designates an individual character of a string. If the value of the subscript expression is not an exact integer, it is automatically *rounded* to the nearest integer, so that the consecutive number of the designated character within the string is given by

$\lfloor$*subscript expression* + 0.5$\rfloor$

Note that the *name* of the string, together with the *value* of the subscript expression, designates the desired character.

When the individual characters of a string are manipulated, the string may become longer or shorter. If this happens, it will be necessary to adjust the value of the *length* variable associated with the string name.

Copying a substring:

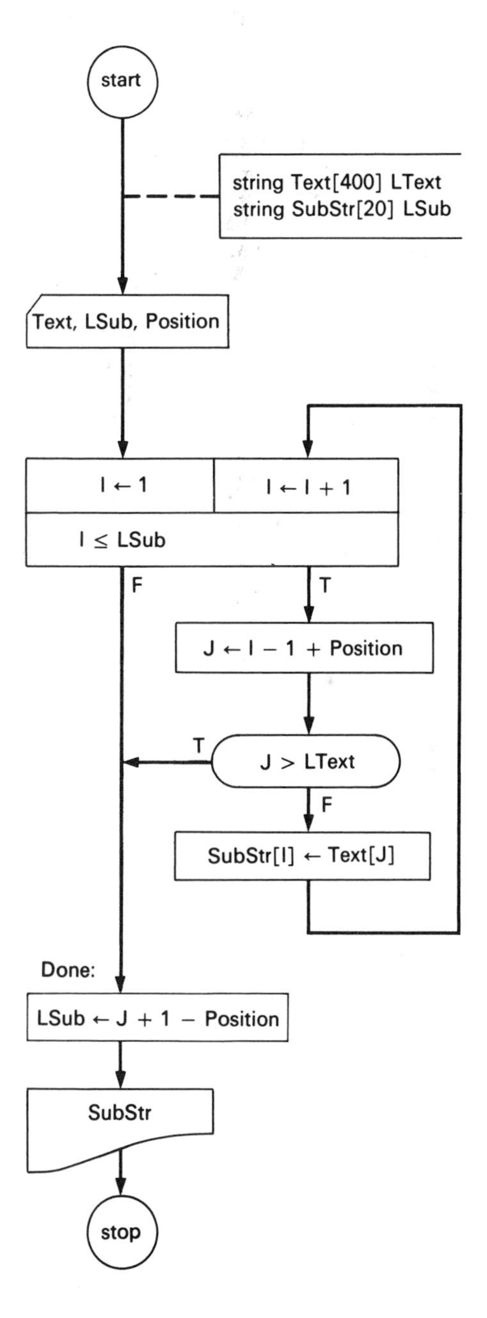

Pattern matching:

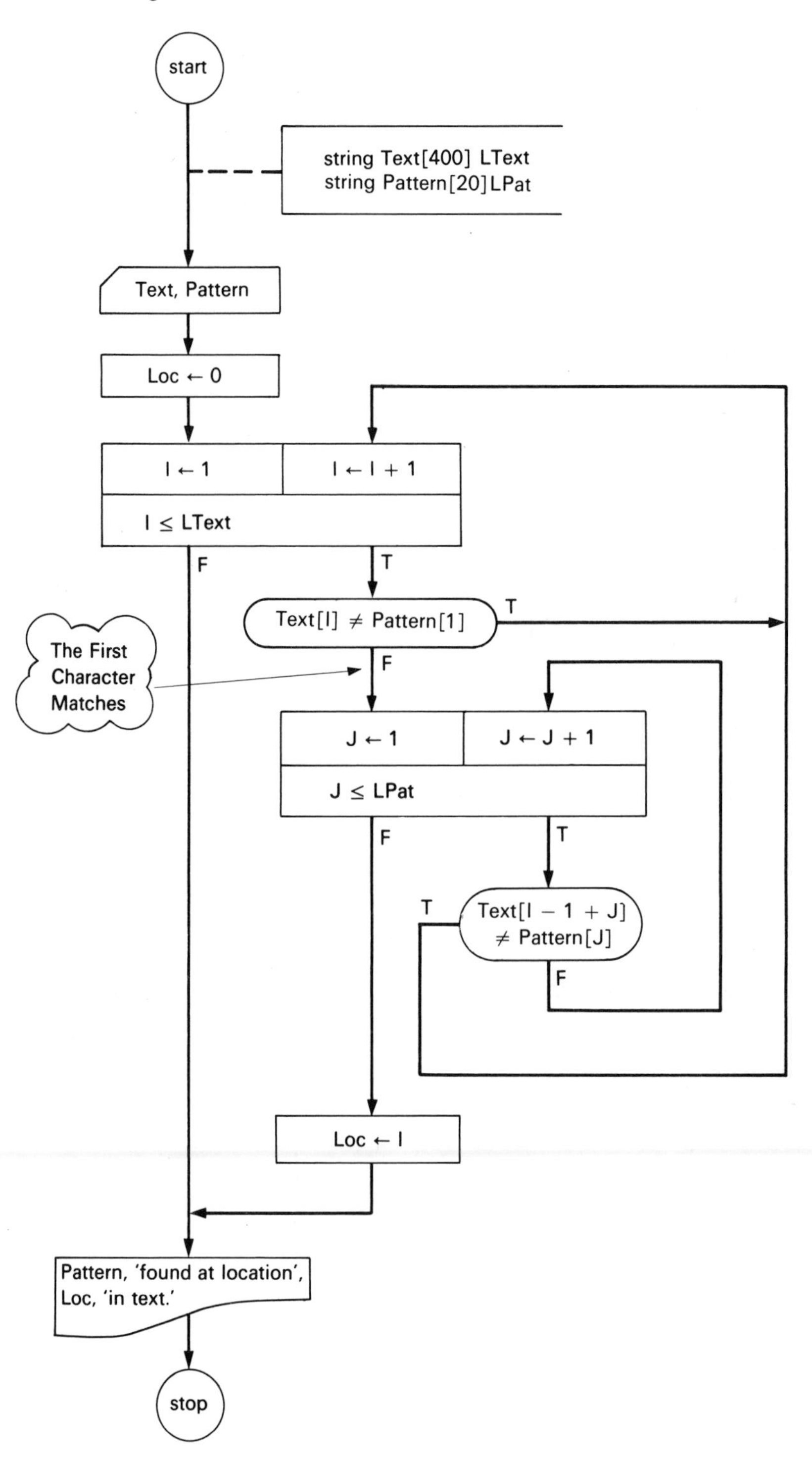

Basic Although almost all versions of Basic provide some means for referring to the individual characters of a string, the details vary widely. The version we describe is specified by Kemeny and Kurtz (1971). (Other versions are listed in Appendix B.)

String operations such as input, output, assignment, and comparison assume that the string is packed. However, a special feature is provided for unpacking a string and converting it to numeric form at the same time. A list of *numeric* variables may be associated with a string. This list must be given a name, which must be declared in a DIM (dimension) statement:

```
10 DIM T[400]
20 DIM S[20]
```

The CHANGE instruction unpacks the string and places the numeric codes for the characters of the string into consecutive positions in the numeric list:

```
30 INPUT T1$, S1$
40 CHANGE T1$ TO T
50 CHANGE S1$ TO S
```

The numeric list name may now be used with a subscript to designate the codes corresponding to the characters of the string. The CHANGE operation automatically places the length of the string in "position 0" of the list. This value may be referred to by the list name with a subscript expression having the value 0.

Copying a substring:

```
10  DIM T[400]
20  DIM S[20]
30  INPUT T1$, P, L
40  CHANGE T1$ TO T
50  FOR I = 1 TO L
60     LET J = I - 1 + P
70     IF J > T[0] THEN 100
80     LET S[I] = T[J]
90     NEXT I
100 LET S[0] = J + 1 - P
110 CHANGE S TO S1$
120 PRINT S1$
900 END
```

Note that the CHANGE operation can be used in either of two ways. At line 40 it is used to unpack the string T1$ into the numeric list T, and at line 110 it is used to pack the numeric list S into the string S1$. In the second form, where the string name *follows* the word TO, the length must be previously stored at "position 0" of the list.

Pattern matching:

```
10  DIM T[400]
20  DIM P[20]
30  INPUT T1$, P1$
40  CHANGE T1$ TO T
50  CHANGE P1$ TO P
60  LET L = 0
70  FOR I = 1 TO T[0]
80     IF T[I] <> P[1] THEN 130
85     REM: FIRST CHAR. MATCHES.
90     FOR J = 1 TO P[0]
100       LET K = I - 1 + J
105       IF T[K] <> P[J] THEN 130
110       NEXT J
120    LET L = I
125    GO TO 140
130    NEXT I
140 PRINT P$, "FOUND AT LOCATION",
145 PRINT L, "IN TEXT."
150 END
```

Fortran Character data read or printed with format code A 1 will be stored in unpacked form. Numerical variables are used in Standard Fortran as the names of cells where characters may be stored. (However, most computers require that such names be of *integer* type.) Input, output, assignment, and comparison operations may be specified by these names. A group of consecutive cells can be used to store the individual characters of a string, if these are read into the cells by an A 1 format code. A DIMENSION declaration is used to specify the length of a string stored in this way.

Copying a substring:

```
      DIMENSION ITEXT(80)
      DIMENSION ISOME(20)
      READ (5, 97) ITEXT
97    FORMAT (80 A 1)
      READ (5, 81) IP
81    FORMAT (8 I 10)
      IF (IP .GT. 60) STOP
      DO 90 I = 1, 20
        J = I - 1 + IP
        ISOME(I) = ITEXT(J)
90      CONTINUE
100   WRITE (6, 98) ISOME
98    FORMAT (1X, 20 A 1)
      STOP
```

Pattern matching:

```
      DIMENSION ITEXT(80)
      DIMENSION IPAT(7)
      READ (5, 97) ITEXT
97    FORMAT (80 A 1)
      READ (5, 96) IPAT
96    FORMAT (7 A 1)
      L = 0
      DO 130 I = 1, 74
        IF (ITEXT(I) .NE. IPAT(1)) GO TO 130
C       THE FIRST CHARACTER MATCHES.
        DO 110 J = 2, 7
          K = I - 1 + J
          IF (ITEXT(K) .NE. IPAT(J))
     1    GO TO 130
110       CONTINUE
        L = I
        GO TO 140
130     CONTINUE
140   WRITE (6, 98) IPAT, L
98    FORMAT (1X, 7 A 1,
     1   19H FOUND AT LOCATION ,
     2   I2, 9H IN TEXT.
      STOP
```

Use of the Watfiv dialect, which simplifies string assignment and string comparison, is not compatible (except on machines with addressable bytes) with operations on the individual characters of the string. This is because the operations in Watfiv work with packed strings, while the individual characters can be manipulated only if the string is *unpacked*.

PL/1 Substring copying and pattern matching functions, called substr and index, respectively, are available in PL/1. When he uses these functions, the user does not know how the individual characters are manipulated. While most users consider this an advantage, the serious student will profit from the insight he gains by studying the explicit procedures for performing these operations that are needed with the other languages.

Copying a substring:

```
COPY: procedure options (main);
   declare TEXT character(400) varying;
   declare SOME character(20) varying;
   get list (TEXT, POSITION, LSUB);
   SOME = substr (TEXT, POSITION, LSUB);
   put skip list (SOME);
   end COPY;
```

Pattern matching:

```
MATCH: procedure options (main);
   declare TEXT character(400) varying;
   declare PATTERN character(20) varying;
   declare LOC fixed;
   get list (TEXT, PATTERN);
   LOC = index (TEXT, PATTERN);
   put skip list (PATTERN,
      'FOUND AT LOCATION',
      LOC, 'IN TEXT.');
   end MATCH;
```

Words to Remember

subscript	byte	packed string
substring	byte addressing	unpacked string

Exercises

1. Write a program to read the paragraph in Section 3.1 beginning "In the great temple at Benares" and count all the occurrences of the word "the". Do not count the pattern "the" when it occurs as part of another word, such as "*the*se" or "o*the*r".
2. Write a program to determine how many words there are in a given text. Apply the program to the paragraph in exercise 1, Section 4.1.
3. Write a program to copy a string, except that any blanks at the beginning are to be omitted. Strings:
   a. " Now is the time to come."
   b. "   The Science of Computing"
   c. "No blanks at the beginning."
4. Write a program to search a string containing an arithmetic expression to see if there are any unbalanced parentheses. Keep track of the level of nesting, starting with zero at the beginning of the string. Increase the level by one whenever a left parenthesis is encountered and decrease it by one whenever a right parenthesis is found. If the level is ever less than zero, or if it is not zero when the end of the string is reached, then the string contains unbalanced parentheses.
   a. $3 + (KO + Mi)$
   b. $((A \div B) + (C \times D)) \div (E + (F \div G)$
   c. $(A \div B) + (C \times D)) \div (E + (F \div G)$

5.3 What Is a List?

Bit patterns representing numerical values or other information, as well as characters, can be grouped into an ordered set or *list*.[3] The only essential distinction between a string and a list is that the *length* of a list remains fixed, while a string may become shorter or longer during a computation. Besides this, most programming languages provide different facilities for list operations than for string operations. We

[3] The term *list* is used in computer science to designate a broader class of data structures than we are considering in this section; the structures studied here should accordingly be called *sequential linear lists* (Knuth, 1968). However, the simpler term is widely used in the humanities and in texts on Basic. In applications from the fields of science, engineering, and mathematics, on the other hand, the term *vector* (or *linear array*) is normally used to designate the structures that we are calling *lists*.

do not usually operate on a list as a *whole*; most often, we manipulate the individual elements of the list in a count-controlled loop, and we often use the length of the list as the upper limit of the loop.

The items of a list are stored in separate cells and are referred to by *subscripted variables.* A subscripted variable consists of a list name followed by a subscript, which is an expression enclosed in brackets. The value of the subscript expression (rounded to the nearest integer, if necessary) indicates the consecutive number of the desired item within the list. The list name must appear in a declaration, along with an integer constant giving the length of the list.

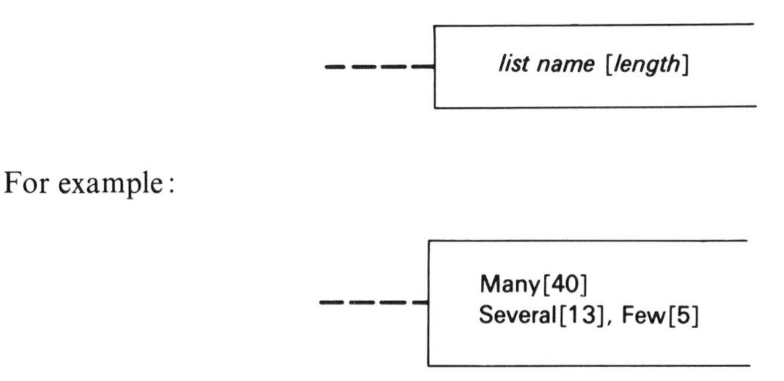

The value of the subscript expression must not be negative and must not be larger than the length appearing in the declaration.

Why should a group of numbers be combined into a list? The motivation for this can readily be seen by analogy to strings. It is advantageous to use a list when (a) the same sequence of operations is to be performed on all of the numbers; *and* (b) all of the numbers must be kept available.

If the first number is no longer needed after the operations on the second number have begun, then an ordinary unsubscripted variable should be used; this variable can take on each value in turn.

Sorting Among the most important data manipulation tasks performed on lists is that of sorting. Many methods have been invented for sorting data with computers; half a book (Knuth, 1973) has been devoted to the subject (see also Martin, 1971). A simple measure of the efficiency of a sorting method is the number of *comparisons* it requires, given in terms of the number of items, "M". Some simple general-purpose sorting methods require approximately M^2 or $M^2/2$ comparisons. Probably the most efficient of these methods is "sorting by insertion," which requires approximately $M^2/4$ comparisons. The motivation for this algorithm is suggested by the way a person might arrange a hand of cards at bridge. We at first assume that the items to be arranged in order are available in the reader, to be read one at a time as needed.

We read the first item, storing it at V[1]. Then, we read the next item—but we read it into a temporary cell because we want to store the first two items in the correct ascending order. We compare the new item with the number at V[1]; if the new item is larger, we store it at V[2]. Otherwise we move the contents of V[1] to V[2] and store the new item at V[1] (see Figure 5.3a). In either case, the two numbers stored so far are in the correct ascending order, and it is time to read the third item.

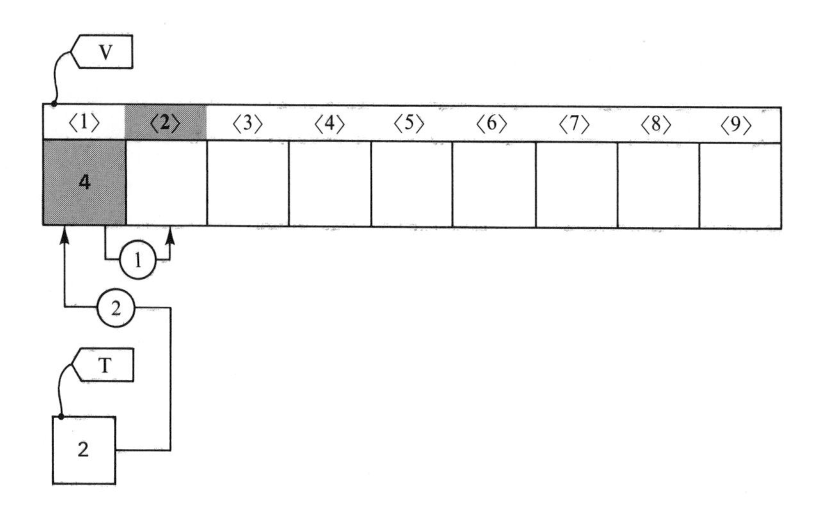

5.3a. *The new item is compared with* V[1]. V[1] *is larger in value, so it is moved to* V[2] *and the new item is stored at* V[1].

Again we store the new item in a temporary cell, rather than directly in V[3]. We want the largest of the three items read so far to be stored in V[3]. We can tell whether the new item is the largest, by comparing it with V[2], because we have guaranteed that the items previously stored are already in ascending order. If the new item is larger than the contents of V[2], we store the new item at V[3] and proceed to read the next item. However, if V[2] is larger, we move V[2] to V[3], thus achieving our partial goal of storing the largest of the three items in V[3], and making cell V[2] available. But now we must compare the new item with V[1]. If the new item is larger, we store it at V[2]; otherwise we move the contents of V[1] to V[2] and store the new item at V[1] (see Figure 5.3b). In either case, the three numbers stored so far are in ascending order, and we are ready to read the fourth item.

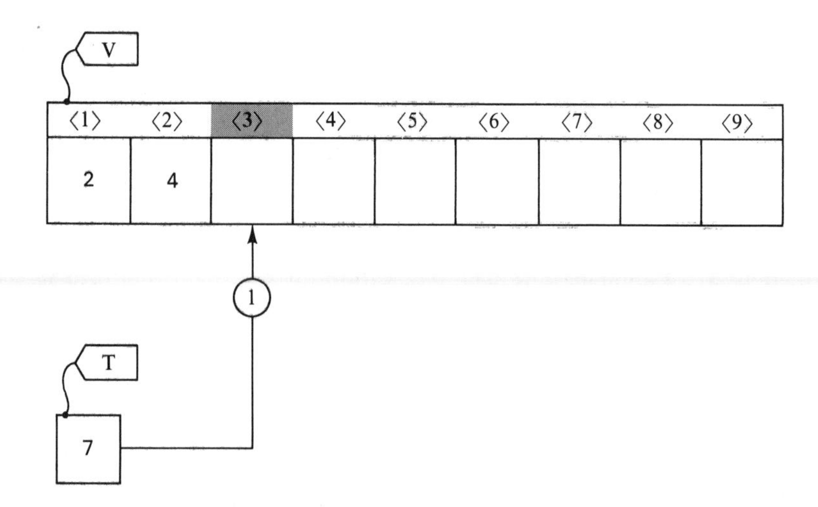

5.3b. *The new item is compared with* V[2]. *The new item is larger in value, so it is stored at* V[3]. V[1] *and* V[2] *are not moved.*

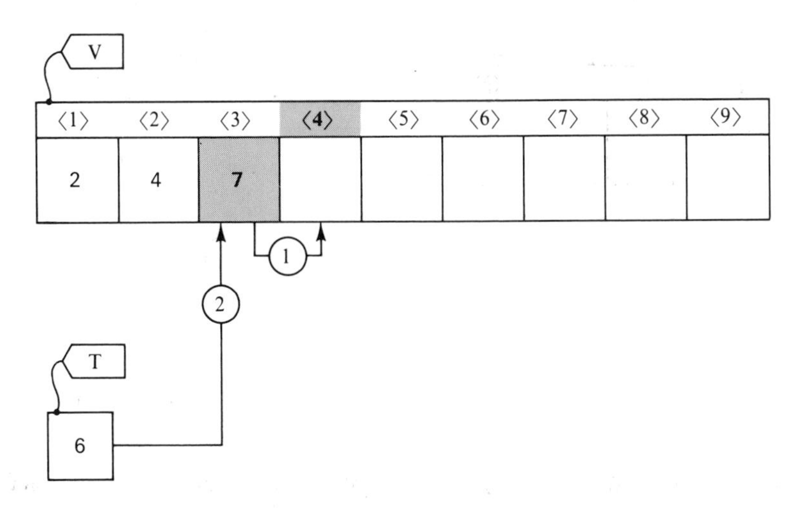

5.3c. *The new item is compared with* V[3], *and all cells to the left, until a value smaller than the new item is found. Values to the right of that one are moved up and the new item is inserted in proper sequence.*

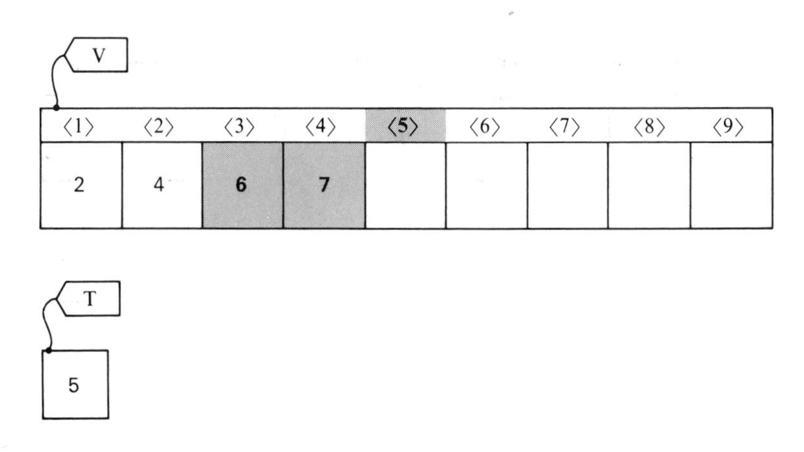

5.3d. *Can you draw the arrows to complete this picture?*

A pattern is beginning to emerge (see Figures 5.3c and 5.3d). When we read the fourth item, we shall want to store it in a temporary cell and then compare it with the items previously stored, in *reverse* subscript sequence, until we find an item that is *smaller* than the new item. Each time we come to a previously stored item that is larger than the new item, we move it to the cell with the next higher subscript, thus making its former location available. As soon as we come to a previously stored item that is smaller than the new item, we store the new item in the available cell; then we read another new item. The following flowchart shows how we read and dispose of the fourth item. Note that the value of I decreases from 3 to 1 as the value of K increases from 1 to 3.

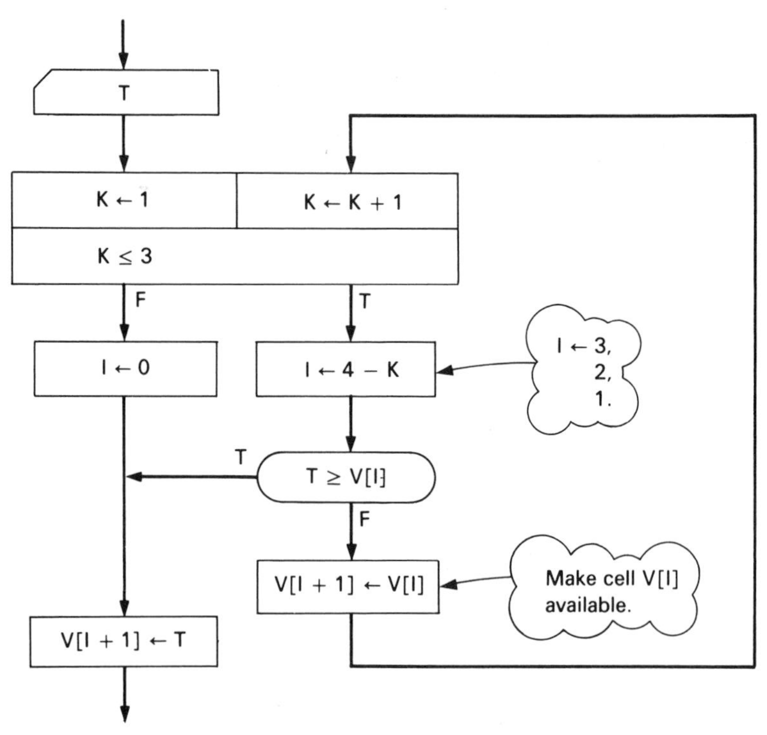

The same flowchart can be generalized, if we replace the upper limit of the loop by a variable (say, J), and change the formula for I from 4 − K to J + 1 − K. We can then use an outer loop to control J. Note that the maximum value of J is one *less* than the number of items to be sorted—the fourth item in the foregoing example was placed in the correct position when the value of J was 3. This results from the fact that the first item to be read was placed directly in V[1] and did not need to be compared with any of the other items.

The following flowchart reads and sorts a list of 40 items.

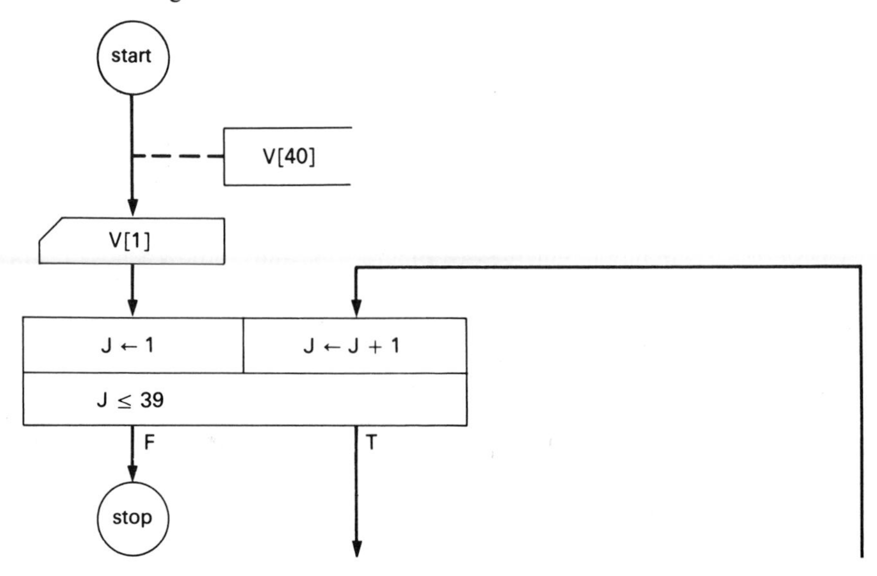

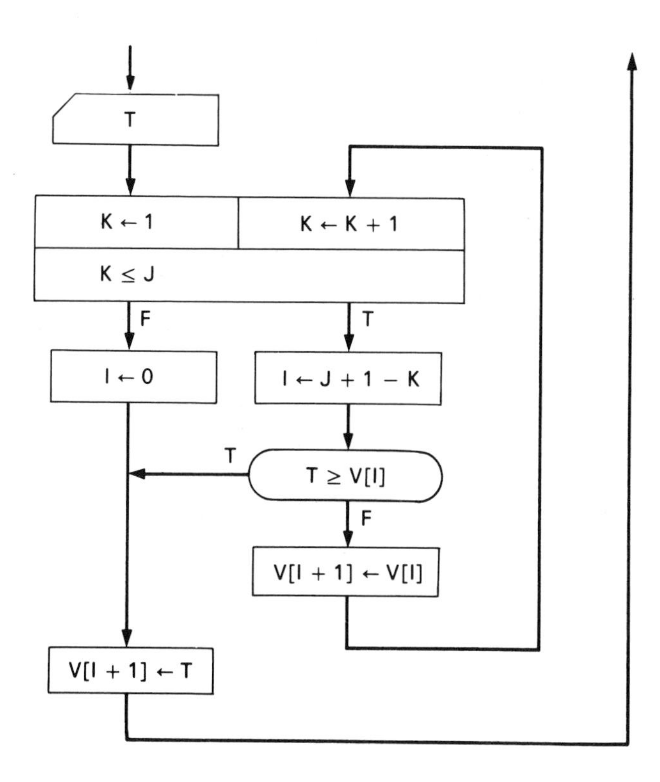

The process of "sorting by insertion" is somewhat easier to understand when we assume that the items are to be read as they are sorted. Much more common in actual application, however, is the case in which the variables to be sorted are *already* in storage in the list V. A minor change adapts the algorithm to this case. Instead of *reading* the next item, we move it to T from V[J + 1].

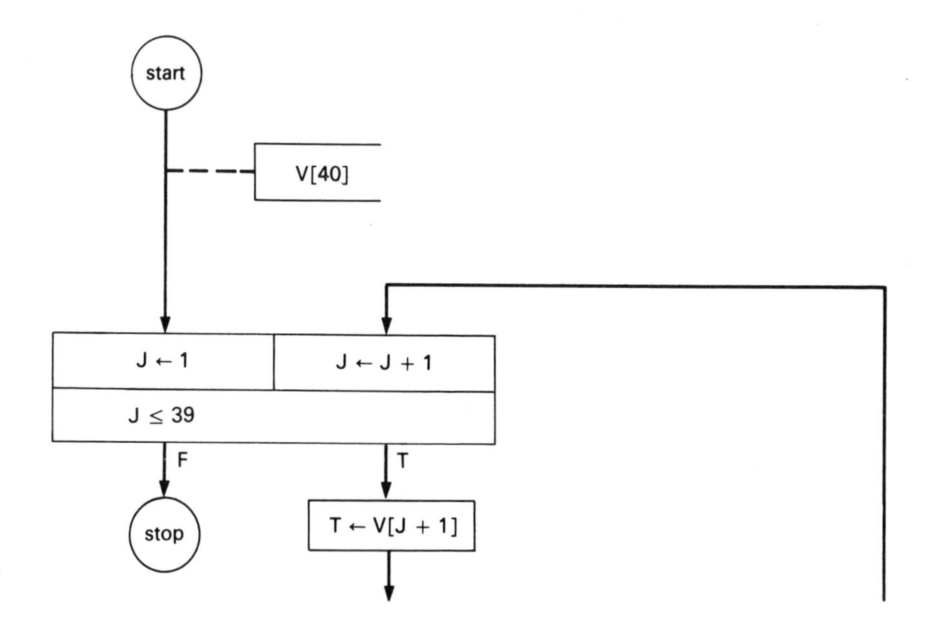

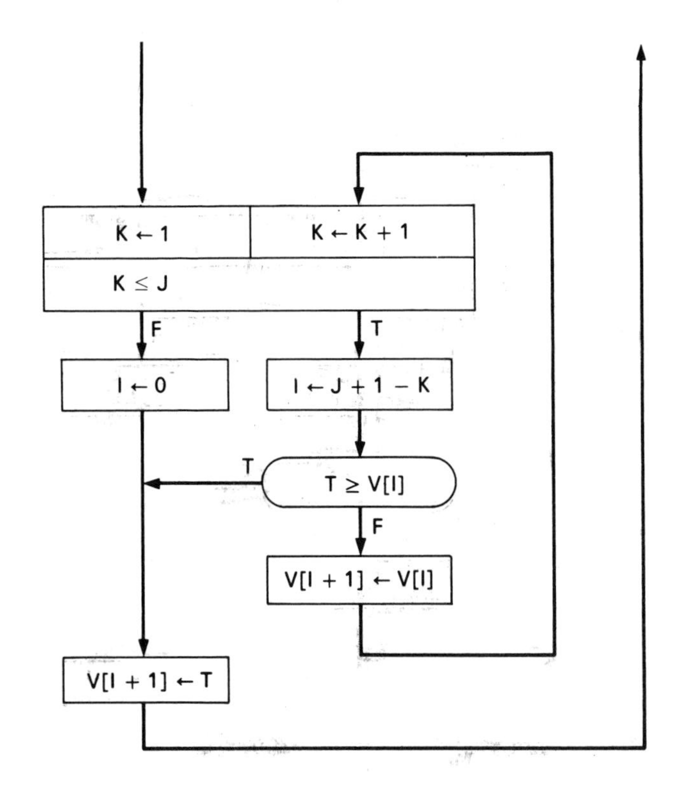

Print Line Control Up to this point, we have almost ignored questions of the arrangement of the printed page. We have merely assumed that all the variables in the output list of a given print instruction will be printed on the same line, and that a new line will start with each new print instruction. For printing a list of subscripted variables, however, the programmer must be able to specify the number of items to appear on each line. To print all the subscripted variables of a list, a count-controlled loop may be used, which will print one subscripted variable for each iteration of the loop. We need a way to override the usual convention that would start a new line for each new print instruction. Accordingly, we establish the new convention of adding an *extra* comma to the end of a print list, to *suppress* skipping to a new line after the list is printed. For example,

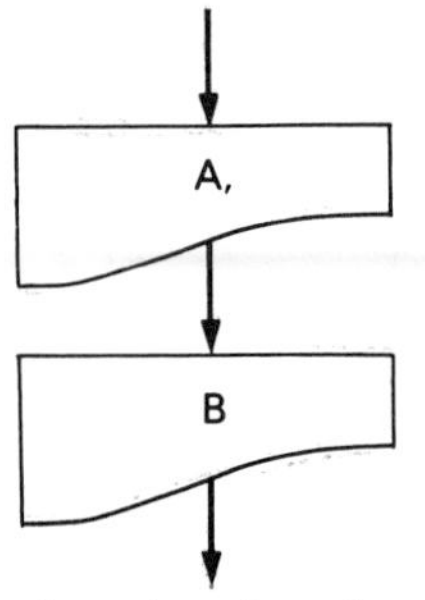

will cause the values of A and B to be printed on the same line, just as if we had written

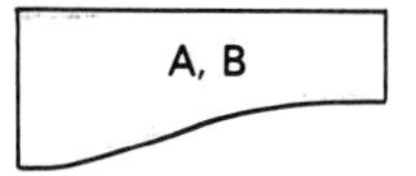

On the other hand,

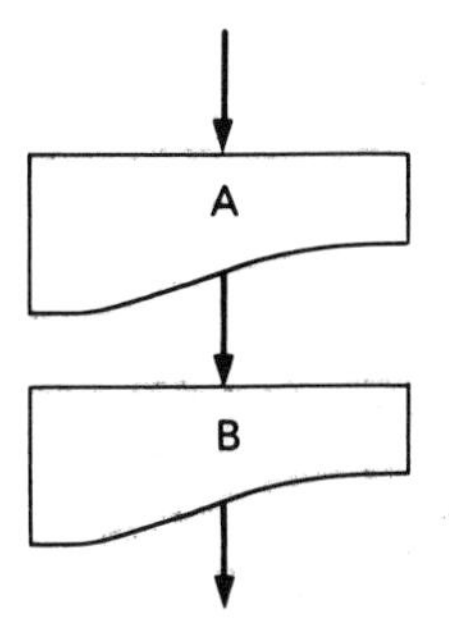

will cause A and B to be printed on separate lines, because there is not an extra comma at the end of the first print list.

Using this convention, we can write a loop to print several subscripted variables on the same line:

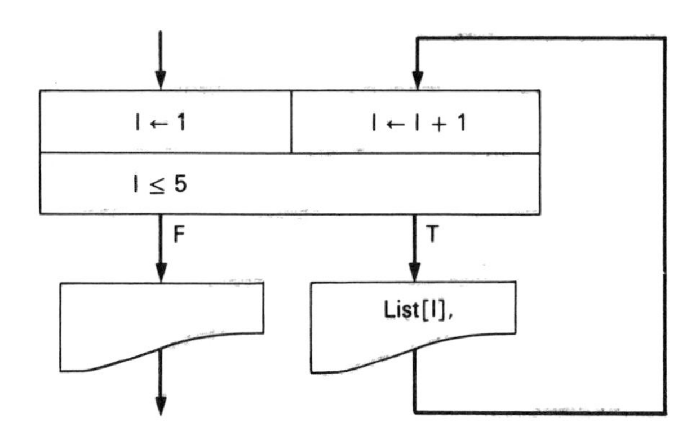

The effect of this loop is the same as if we had written:

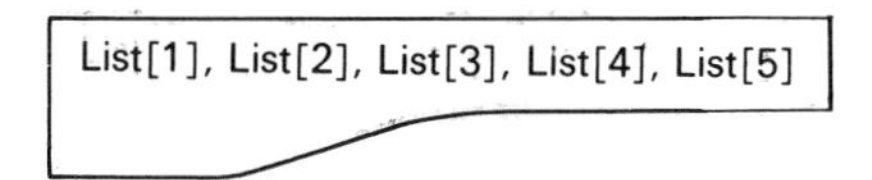

Note that in this latter version there is no comma after List[5] so a skip to the next print line will occur after the fifth subscripted variable has been printed. For this reason, in the loop version we need an extra print step with an "empty" output list after the end of the loop. This step prints nothing, but skips to the next print line because there is no comma.

Flowchart A list name must appear in a declaration giving the *length* of the list:

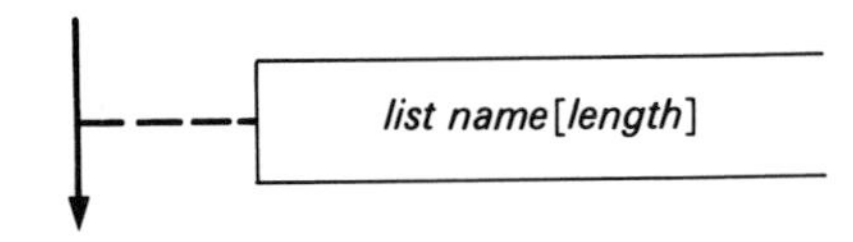

For example:

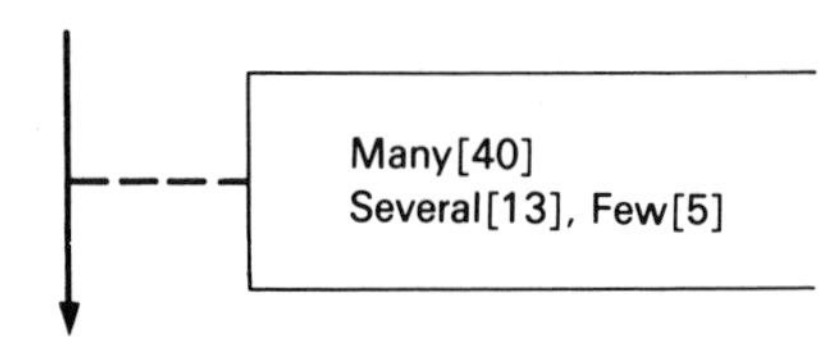

An individual list item is referenced by a *subscripted variable*, which consists of the list name followed by a subscript expression enclosed in brackets. The list name, along with the value of the subscript expression, specifies an item of the list.

Sorting:

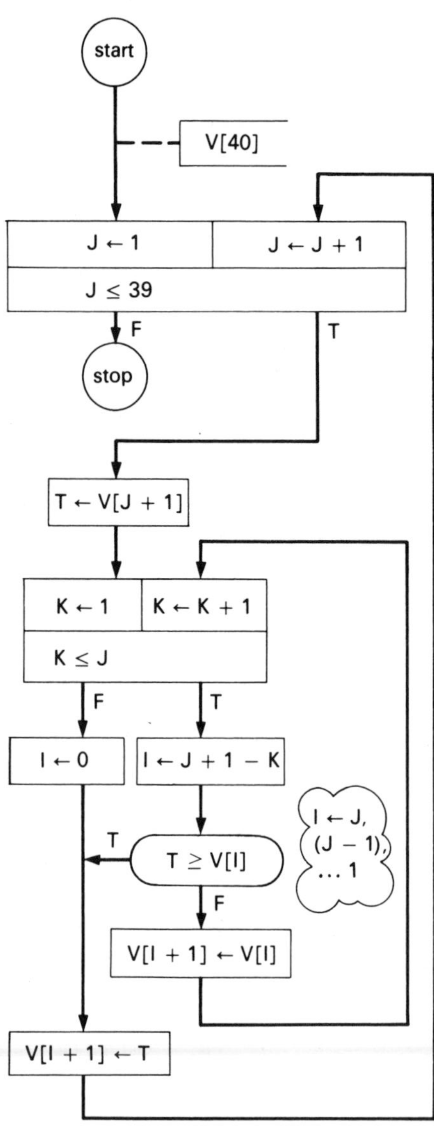

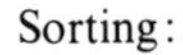

Basic A list name must appear in a DIM statement, along with the length of the list:

```
10 DIM M[40]
20 DIM S[13], F[5]
```

An individual list item is referenced by a *subscripted variable*, which consists of the list name followed by a subscript expression enclosed in brackets. The list name, along with the value of the subscript expression, specifies an item of the list.

Sorting:

```
120 DIM V[40]
125 FOR J = 1 TO 39
130   LET T = V[J + 1]
135   FOR K = 1 TO J
140     LET I = J + 1 - K
142     REM: I = J, (J - 1), . . . , 1
145     IF T >= V[I] THEN 165
150     LET V[I + 1] = V[I]
155     NEXT K
160   LET I = 0
165   LET V[I + 1] = T
170   NEXT J
175 END
```

In Basic, an extra comma at the end of a print list *suppresses* the skip to a new line that would normally occur at the end of execution of a PRINT instruction.

```
50 FOR I = 1 TO 5
60   PRINT LIST[I],
70   NEXT I
80 PRINT
```

Fortran Although the string handling features of Fortran are somewhat awkward, its features for manipulating lists are excellent. A list name must appear in a DIMENSION declaration, which specifies the length of the list:

DIMENSION *list name* (*length*)

For example:

```
DIMENSION M(40)
DIMENSION S(13), F(5)
```

If a type declaration for the same list name occurs in the program, the length specification may be incorporated into this type declaration rather than appearing in a separate DIMENSION declaration:

```
REAL M(40)
INTEGER S(13), F(5)
```

An individual list item is referenced by a *subscripted variable*, which consists of the list name followed by a subscript expression enclosed in parentheses. The subscript expression must be of *integer* type, and its value must be greater than zero but not greater than the length of the list.

Sorting:

```
      DIMENSION V(40)
      DO 250 J = 1, 39
        T = V(J + 1)
        DO 235 K = 1, J
          I = J + 1 - K
C         I = J, (J - 1), . . . , 1
          IF (T .GE. V(I)) GO TO 245
          V(I + 1) = V(I)
235       CONTINUE
        I = 0
245     V(I + 1) = T
250     CONTINUE
      STOP
```

The instruction I = J + 1 − K is necessary because the form of a subscript in Standard Fortran is somewhat restricted. A subscript must be of integer type, and it must be a constant, a variable, or an expression of one of the simple forms:

variable ± *constant*
constant * *variable*
constant * *variable* ± *constant*

A *list segment* notation is available in Fortran for printing all or part of a list:

```
   READ (5, 81) (LIST(I), I = 1, 5)
81 FORMAT (1X, 8 I 10)
```

If there are more items in the list than can be printed on one line, the format specification will be repeated as many times as necessary until the entire list has been printed. Each time the end of the format is reached, the printer will skip to a new line. Hence, in the following example, there are 40 numbers to be printed; eight of them will appear on each line.

```
   DIMENSION SOME(40)
   WRITE (6, 83) (SOME(I), I = 1, 40)
83 FORMAT (1X, 8 F 15.3)
```

The list segment notation can also be used for input. Because Fortran input is record-oriented, each new READ instruction and each repetition of the format specification causes the next input data item to be taken from the beginning of a new record in the reader. The following READ instruction will take eight items from each of five data records:

```
   READ (5, 82) (SOME(I), I = 1, 40)
82 FORMAT (8 F 10.0)
```

PL/1 A declaration for a list is just like an ordinary variable declaration, except that it includes lower and upper *subscript bounds* in parentheses:

declare *list name* (*lower*:*upper*) float;

In our examples, the lower bound will invariably be 1. Two or more declarations of lists and ordinary variables may be combined:

```
declare MANY(1:40) float;
declare (SEVERAL(1:13), FEW(1:5),
  X, SUM) float;
```

An individual list item is referenced by a *subscripted variable*, which consists of the list name followed by a subscript expression enclosed in parentheses. The list name, along with the value of the subscript expression, specifies an item of the list. The value of this expression must lie between the lower and upper subscript bounds.

Sorting:

```
INSERT: procedure options (main);
  declare (V(1 : 40), T) float;
  declare (J, K, I) fixed;
  do J = 1 to 39;
    T = V(J + 1);
    do K = 1 to J;
      I = J + 1 - K;
      /* I = J, (J - 1), . . . , 1 */
      if T >= V(I) then go to CONT;
      V(I + 1) = V(I);
      end;
    I = 0;
  CONT: V(I + 1) = T;
    end;
  end INSERT;
```

A list segment notation is available in PL/1 to simplify the reading and printing of lists:

```
declare J fixed;
get list ((SOME(J) do J = 1 to 40));
put skip list ((SOME(J) do J = 1 to 40));
```

Also, it is possible to suppress skipping to a new line before printing, merely by writing put list instead of put skip list.

Words to Remember

list	subscript	comparison
dimension declaration	subscripted variable	sort

Exercises

1. Write a program to find the largest number in a list named V, of length 10. One method is to keep a variable whose value is the "largest found so far." Start by giving this variable the value of V[10]; then use a count-controlled loop to compare this value with each of V[1] to V[9], giving it a new value whenever a larger number is found.

a.	19,	13,	5,	27,	1,	26,	31,	16,	2,	9
b.	2,	1,	4,	3,	6,	5,	8,	7,	10,	9
c.	19,	13,	8,	27,	3,	14,	7,	16,	17,	6

2. Write a program to find the total of the values of all the items in a list. (Use the data in exercise 1.)
3. Write a program to total separately the positive and the negative items in a list.
   a. 9, 3, −5, 17, −9, 16, 21, 6, −8, −1, −8, −9, −6, 7, −4, 5, 8, 7, 10, 9, 4, −3, 6, 7, −4
   b. 102, −18, 13, 52, 22, 14, 95, −4, 15, 22, 84, −10, 4, 33, 2, 18
   c. 16.7, −12.2, −6.4, −3.2, 50.9, 17.5, −9.3, 23.3, −0.4, −6.5
4. Write a program to read 40 items, sort them, and print
   a. the largest and the smallest;
   b. the tenth, twentieth, and thirtieth largest (the quartiles);
   c. the fourth, eighth, twelfth, . . . , thirty-sixth largest (the deciles).

   Use the following items: −0.9, 6.7, 11.6, 615.7, 77.8, 63.4, −2.2, 6.1, 0.1, −14.7, 163.9, 49.6, 420.8, 217.0, 63.1, 194.9, 27.7, 571.8, 41.2, 22.0, 46.1, 302.5, 149.5, 1341.8, 24.8, 102.5, 409.4, 1.6, 37.3, 406.7, 0.1, 178.1, 623.8, 0.1, 37.2, 25.8, 95.8, 98.4, 110.4, 94.7.

5.4 *What Are Some Other Ways of Using Lists?*

There is a natural connection between lists and indexed loops. Lists are used when the same operations are to be performed on all the variables in a group. In the most common case, each variable is to be processed in sequence. We write an indexed loop, and use the loop index (counter) as a subscript in the block. For example, to print a list:

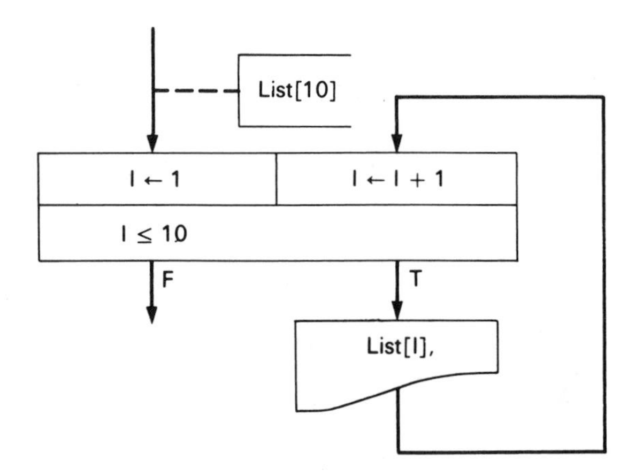

The loop index I is a natural variable for use as the subscript. Generally, the subscript will consist just of the loop index, if *every* subscripted variable in the list is to be processed in sequence. The required sequence of *subscript* values begins at 1 and increases in the same way as the index of the loop (Figure 5.4a).

In other applications, however, the variables in a list are to be processed in some other order. Suppose we want to print items 4 through 10 of a list, using an indexed loop. The subscript must start at 4 when the index is 1; thus it will reach 10 when the index is 7.

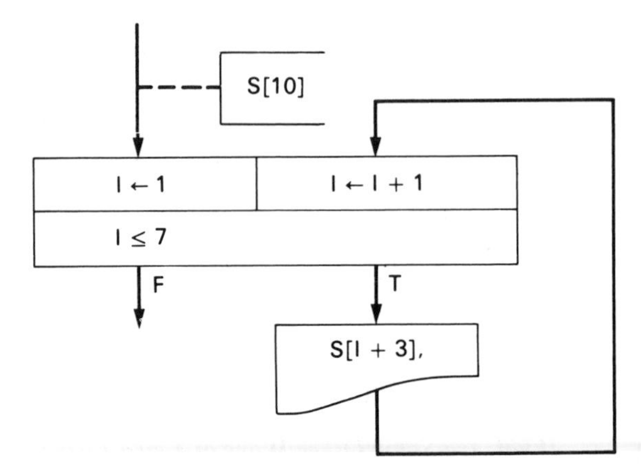

Here, a "shifted index" is used as a subscript. The subscript expression is formed by adding a "shift constant" to the index. This pattern is used when a portion of a list is to be processed in sequence, but the starting point is not at the subscript value 1. In general, the number of times the block is to be executed is used as the upper limit of the loop. The amount of the shift is determined by considering what value the subscript should have when the index is at 1 (Figure 5.4b).

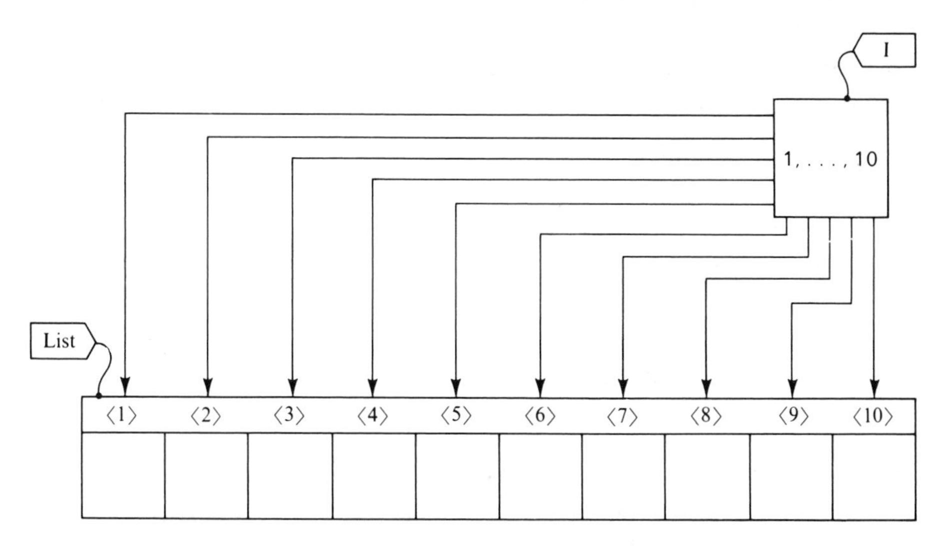

5.4a. *As the index variable* I *takes on successive values from one to ten, it acts as a pointer to each subscripted variable in turn.*

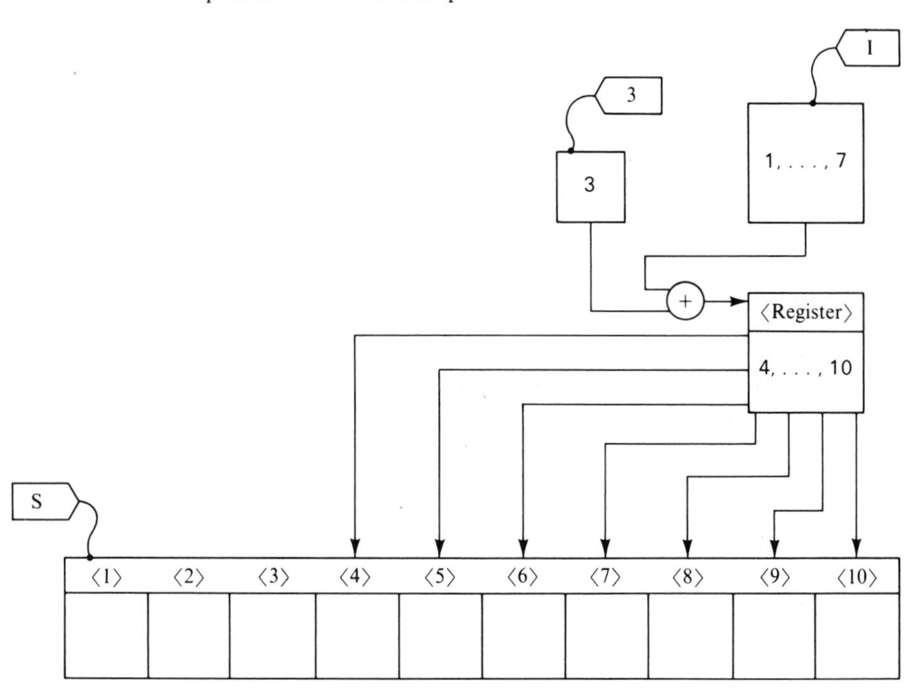

5.4b. *As the index variable* I *takes on successive values from one to seven, the numerical value in the register goes from four to ten. Thus, the register acts as a pointer to each of the subscripted variables,* S[4] *through* S[10], *in turn. With other expressions, the register can be made to take on decreasing values, or values spaced at equal intervals.*

Another pattern, which we saw in the sorting program in Section 5.3, involves a *decreasing* sequence of subscript values. In a case of this kind, the subscript expression will involve the negative of the index. In the sorting program we have the index

K and the subscript value J + 1 − K. This problem could have been written using J + 1 − K directly in the subscript expression:

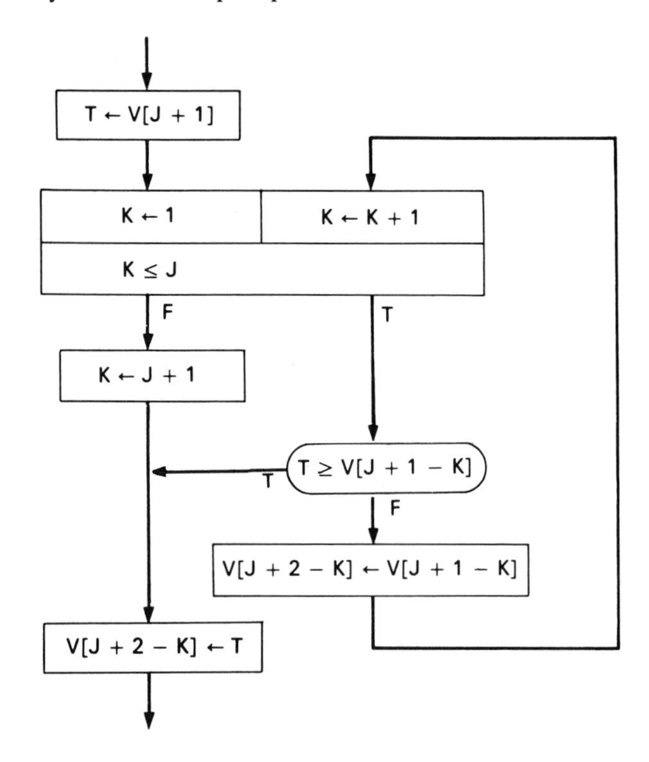

Subscripts spaced at equal intervals may sometimes be required; for instance, we may wish to repeat a process using a sequence of subscripts such as (3, 7, 11, 15, 19). Again, we first determine how many repetitions are required (in this case, five). The spacing interval enters the subscript expression as a *multiplier* of the index; then, as before, we shift if necessary so that the first subscript value will be correct:

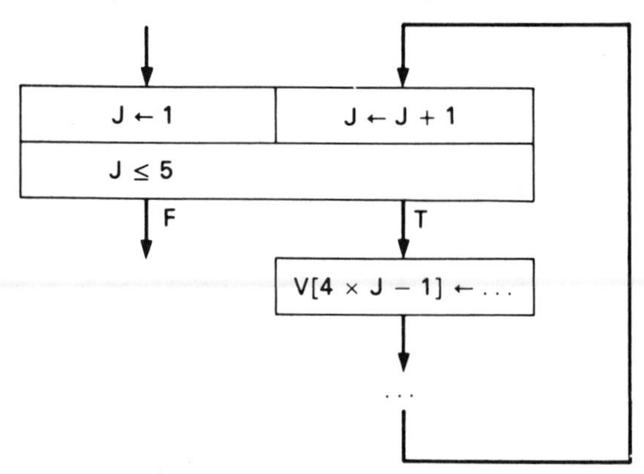

In still other cases, the relation between the subscript and the index is more complicated or irregular—or there may be no relation at all. Or several different lists may be involved, with subscripts that vary in entirely different ways relative to the index. The following program copies a list, deleting all the negative numbers:

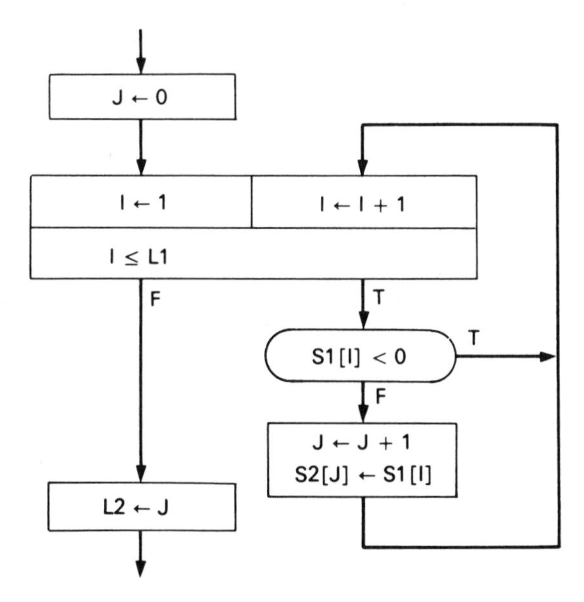

Here the subscript I is identical to the index, but the value of J increases irregularly and unpredictably, depending on the data (Figure 5.4c).

A sequence of subscript values for one list may be stored in another list, as in the following example. The Blank Manufacturing Company pays a bonus of $5.00 per week to all employees who live more than 20 miles from the factory. During the payroll calculation, the gross pay of each employee is stored in a list named GrossPay, using his employee number as a subscript. Then the bonus is added to the gross pay of all employees on the "bonus list." This list, named BE, consists of 17 employee numbers:

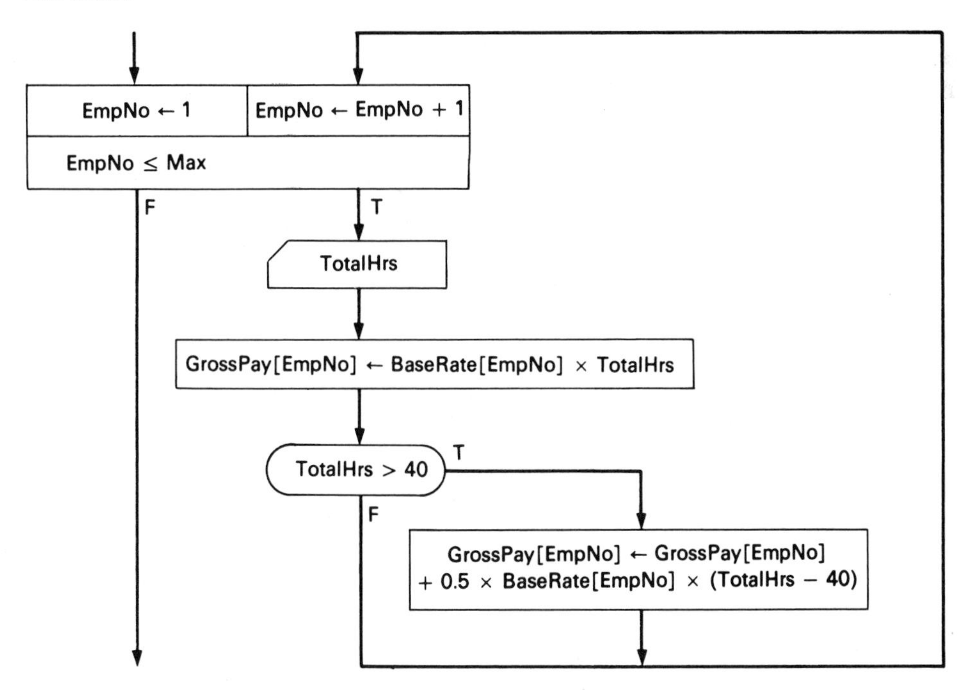

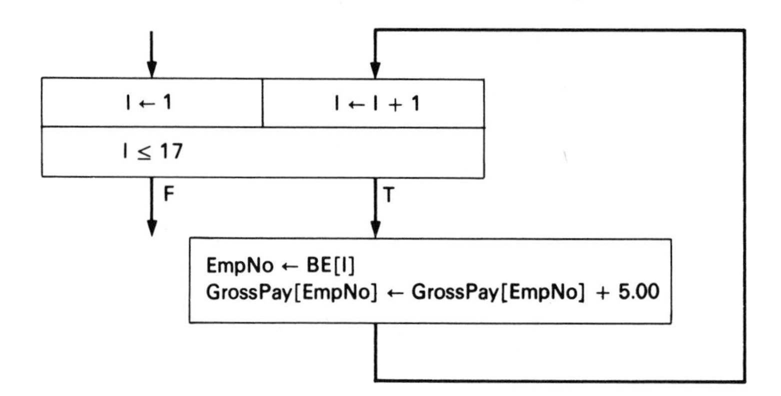

In the second loop, the variable EmpNo is used as a subscript, but the index is I. The relation between the index and the subscript is defined by the bonus list, BE:

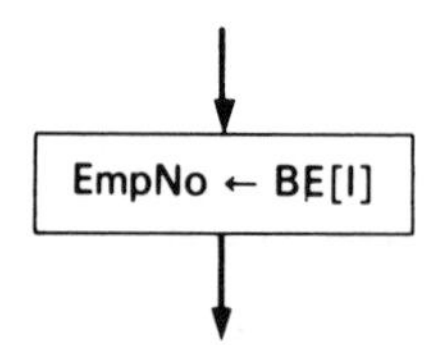

In effect, the bonus list is a list of pointers to various entries in the two lists named GrossPay and BaseRate (Figure 5.4d).

A pointer is a variable whose value is used indirectly in a calculation. Very commonly, as in this case, the value stored in the pointer cell is used as a subscript to identify one of the variables in some other list. The pointer value does not enter directly in the arithmetic expression determining the result of the calculation but is used, in effect, as part of the name of another cell. For this reason, a pointer is often called a *reference variable*.

Some loops have an index that is never used inside the main body of the loop. This index controls the number of times (or the maximum number of times) the loop is executed. Consider, for example, a loan amortization problem, where we begin by reading the original principal, PO, the monthly payment amount, M, and the annual interest rate, R. We first compute the monthly interest rate, R ÷ 12.

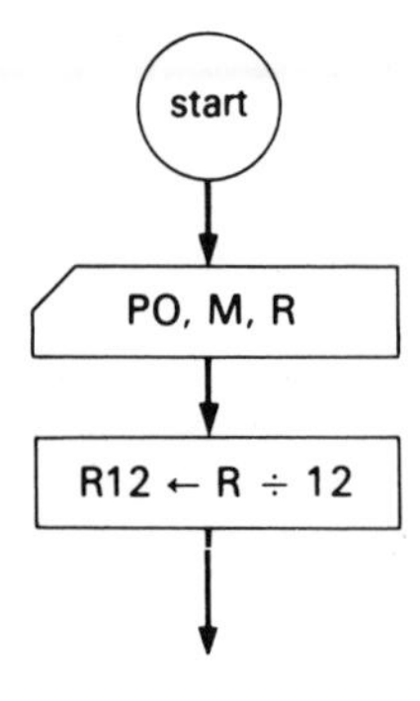

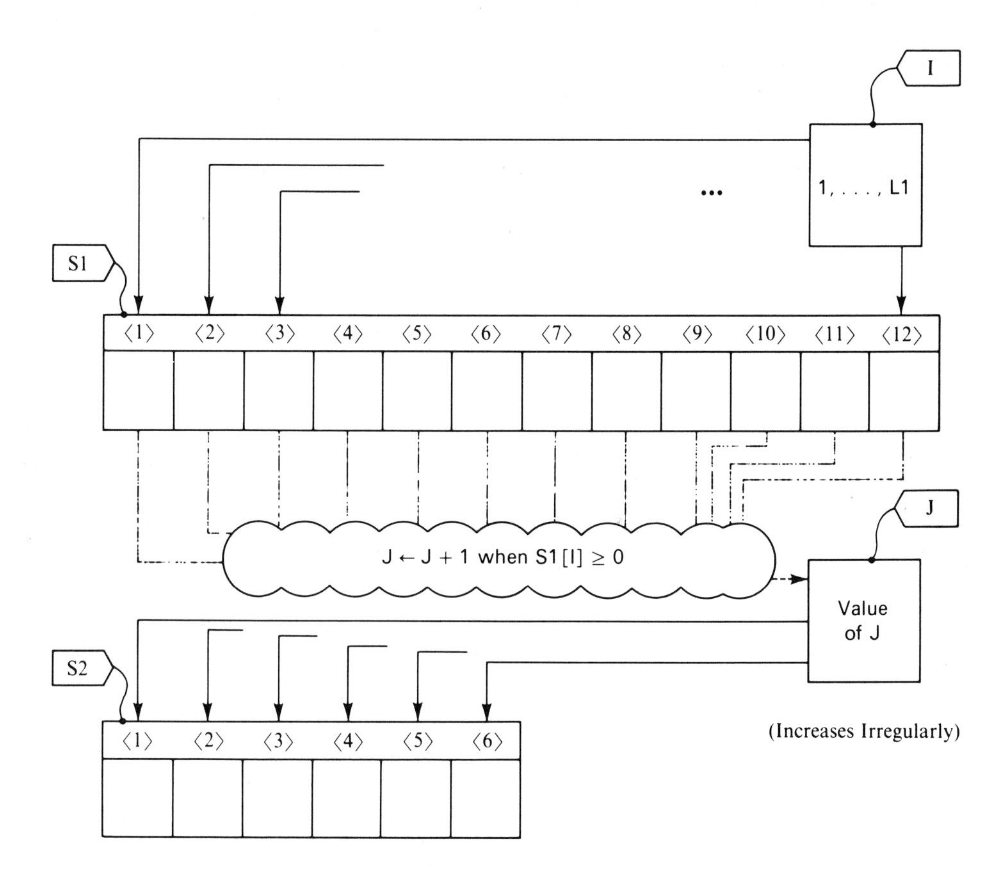

5.4c. The variable I, *used as a pointer to the list* S1, *increases regularly because* I *is the index of a count-controlled loop. On the other hand,* J, *which points to the various subscripted variables of* S2 *in turn, increases irregularly, depending on the values found in the list* S1.

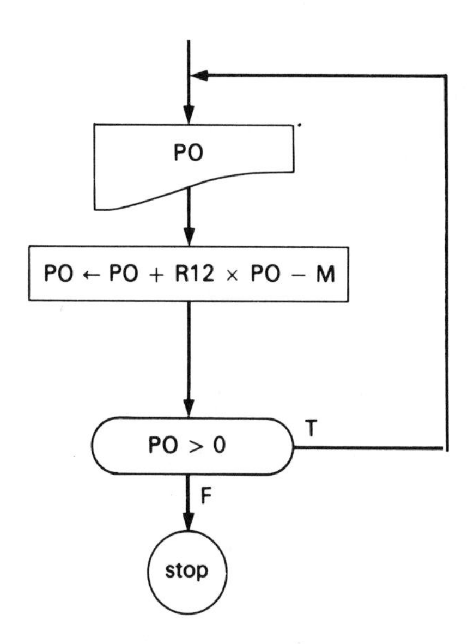

However, we may inadvertently give the program some strange data values for PO, M, or R, for which the principal may decrease very slowly because R12 × PO is almost as large as M. Let us insert in the program the assumption that no loan runs more than 50 years (600 months). An easy way to do this is to use an indexed loop:

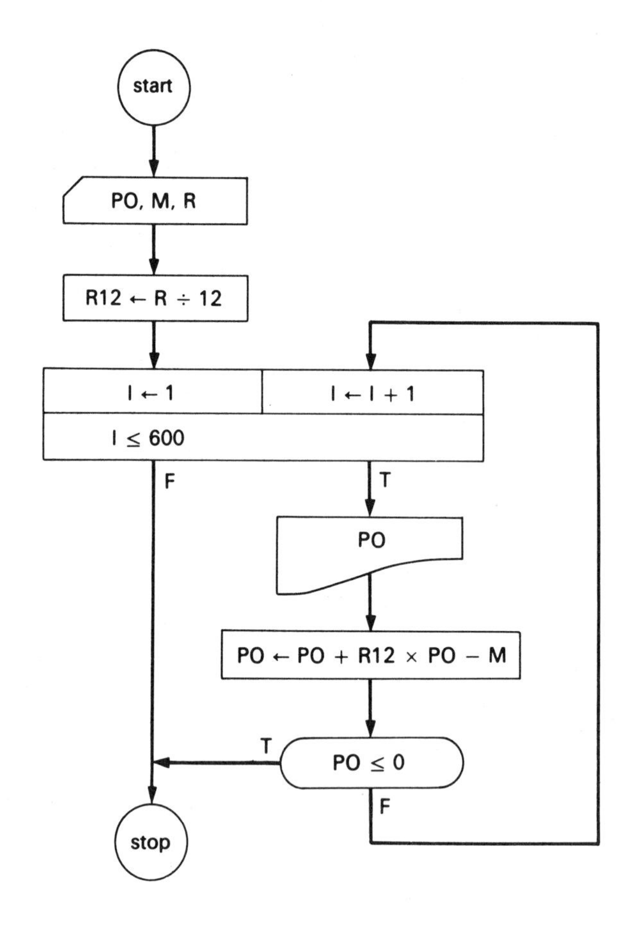

In this program the index is never used inside the block, and we hope the upper limit will never be reached. There is no relation between index and subscript, because there is no subscript. The index is used only as a safety factor to avoid wasting time when data errors occur.

Searching Problem: We have a list, X, of 1000 values stored in *ascending* order:

X[1] ≤ X[2] ≤ X[3] ≤ . . . ≤ X[1000]

We read a data item, Y, and we want to find the *nearest* value to Y among the X values. Specifically, we want to find the value of a subscript, I, such that X[I] is larger

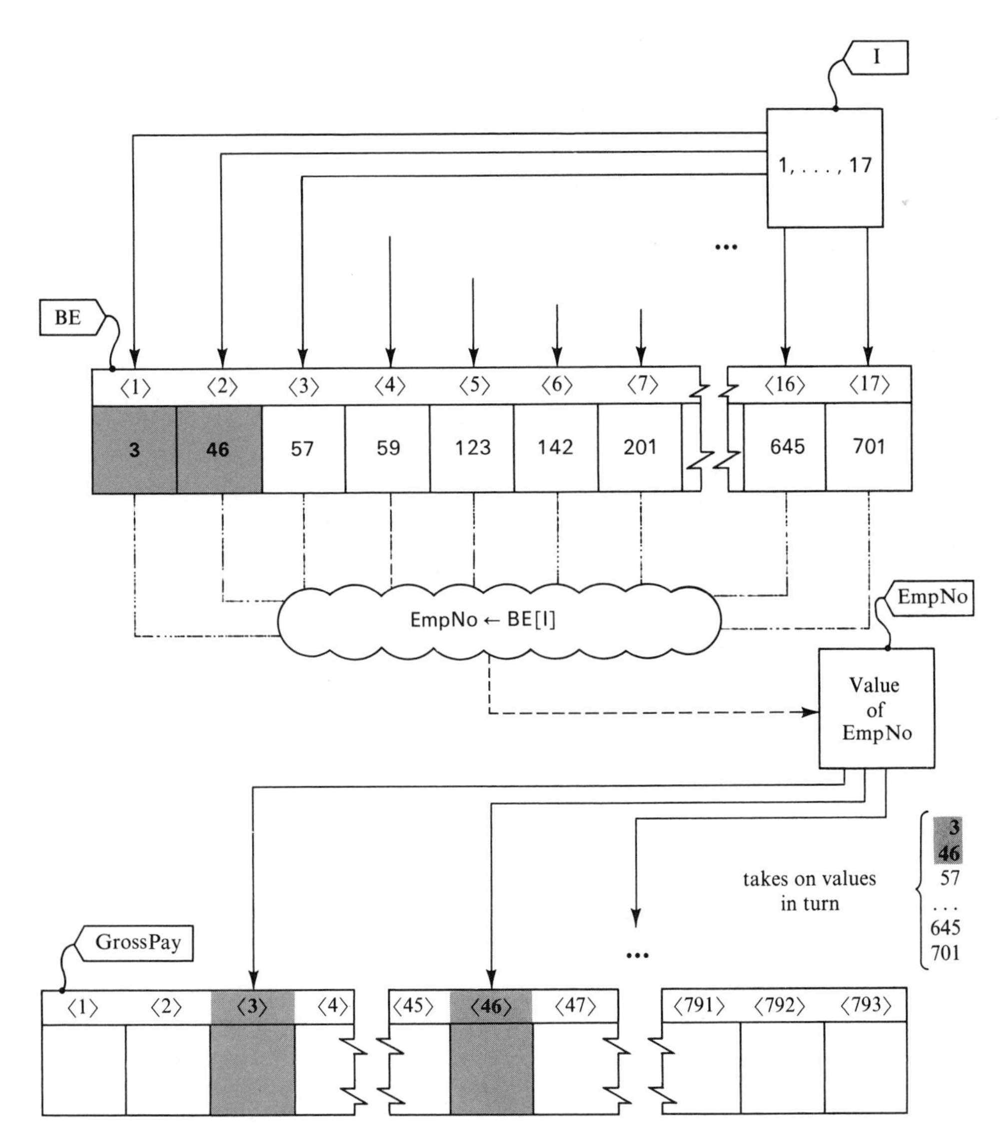

5.4d. *The numbers in the list* BE *(the Bonus Employee list) are stored successively in the cell* EmpNo, *and are used as pointers to cells in the list* GrossPay, *indicating which employees should have a bonus added to their gross pay.*

than Y (or equal to it), while X[I − 1] is less than Y in value:

$$X[I] \geq Y > X[I-1]$$

A straightforward way of solving this problem is to use a linear search, examining each X value in turn until we find one that is larger than Y or equal to it:

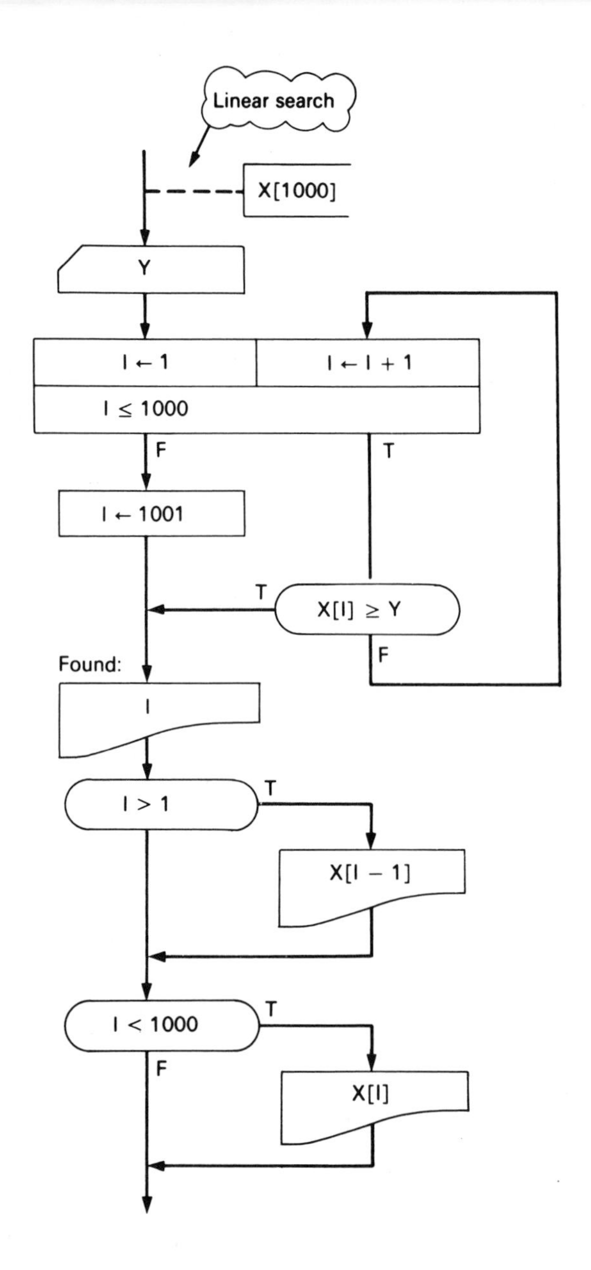

If an entry in the X list agrees exactly with the desired value Y, then the variable I will point to that entry when the loop is terminated and the instruction labeled Found is reached. Otherwise, I will point to the first X entry that exceeds Y in value.

The linear search procedure requires, on the average, 500 comparisons for 1000 items (or $M/2$ comparisons for M items). A more efficient method, known as binary search, requires only about 20 comparisons for 1000 items (or $2 \times \log_2 M$ comparisons for M items). In the binary search procedure (Price, 1971) we examine the middle item first, to see if the desired item is in the lower or upper half. The half that contains the item is then subdivided, and a test determines which quarter of the items contains the item sought. Continuing in this way, we narrow the search down to half as many items each time, until the search converges on a single item (Figures 5.4e, 5.4f, 5.4g, 5.4h). In this method, as in linear search, the items are assumed to be in order before the search begins.

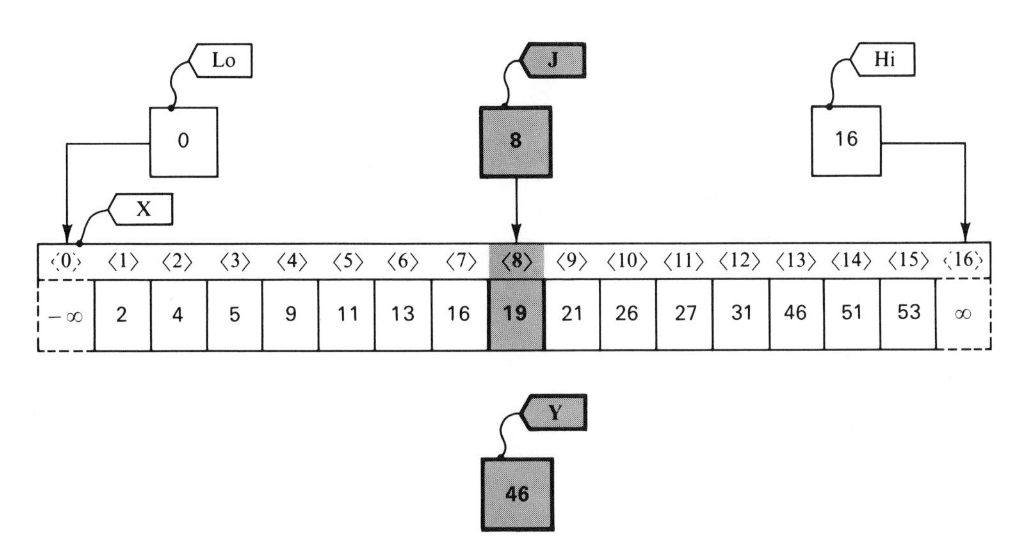

5.4e. Hi *and* Lo *pointers are initially set to point to "phantom" items, so that the algorithm will still work even when the desired item is beyond the range of the list entries.* J *is set to point to an item in the middle of the list, and the value of* X[J] *is compared with that of* Y. *Since 19 is less than 46, the* Lo *pointer will be reset. The desired item must be in the top half of the list.*

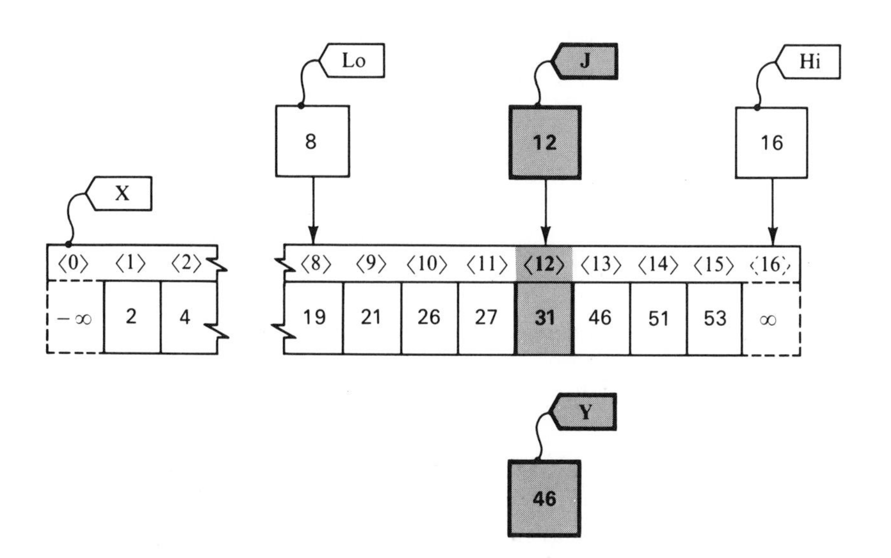

5.4f. J *is now set to point to the middle of the top half of the list.* X[J] *again has lower value than* Y, *so* Lo *will be reset. The search has been narrowed to the upper quarter of the original list.*

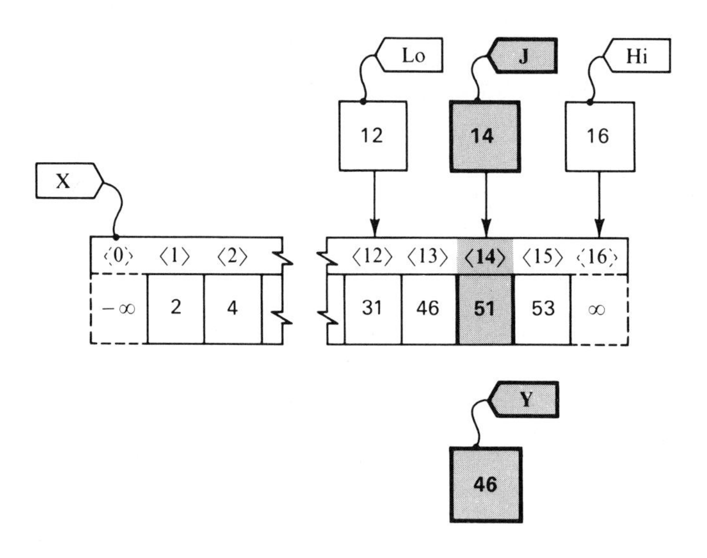

5.4g. *This time, the item in the middle of the remaining group has a larger value than* Y, *so* Hi *will be reset.*

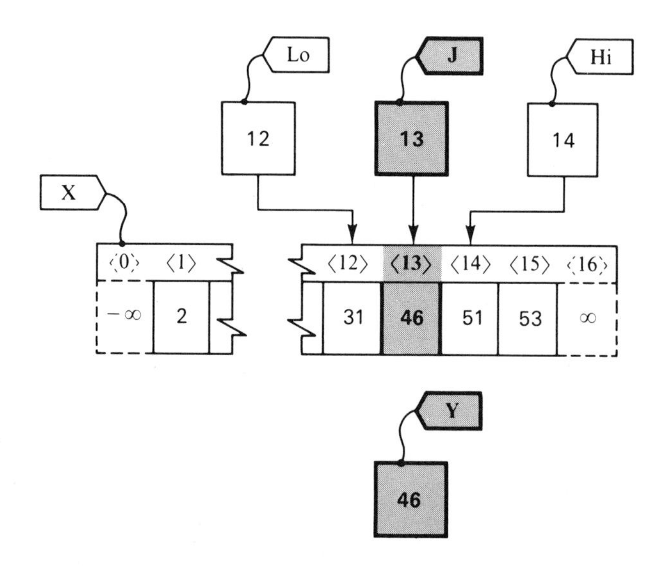

5.4h. *The middle one of the remaining items is again compared with* Y *and, in this case, an exact match is found. If the search had been for some* Y *value not on the list (say, 43), one more step would have been taken (resulting in 12 and 13 for the values of* Lo *and* Hi, *respectively).*

If one entry in the X list agrees exactly with the desired value, Y, then both variables Hi and Lo will point to that entry when the loop is terminated and the instruction labeled Found is reached. Otherwise, Hi and Lo will differ by exactly one in value. It will always be true, throughout the search, that

$$X[Lo] < Y < X[Hi]$$

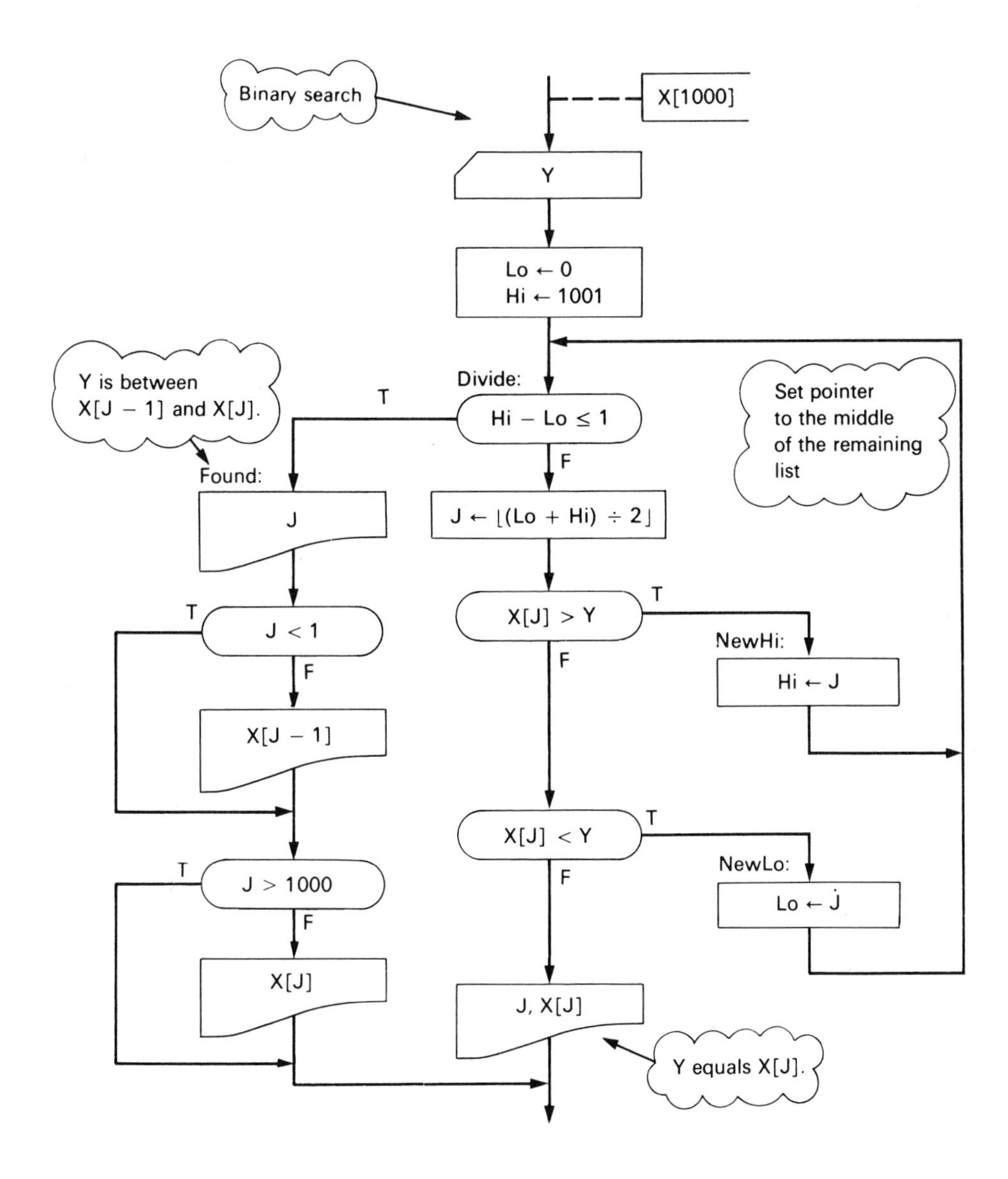

The only exception that can occur is at the final step, if one X entry agrees exactly with Y in value. Therefore, if Found is reached with Hi and Lo having values that differ by one, it must be that the value of Y *does not* occur in the list, but is bracketed between the values of X[Hi] and X[Lo].

If several list items match Y, the binary search may end up pointing to a different one of these identical items than the one found by the linear search. (Of course, the requirement that the list be in order guarantees that if there are two or more identical items they will occur at consecutive positions.) The linear search will find the one with the lowest subscript; but the one found by the binary search will be determined by the way the search subdivides the list.

For an excellent discussion of binary searching and related algorithm, see Knuth (1973, pp. 406–419).

Flowchart The count-controlled loops in our flowcharts always use an index whose value starts at 1 and increases in steps of 1. The philosophy is to complicate the subscript as much as necessary to keep the indexing simple. The *number of times* a loop is executed will thus be exactly equal to the value of the limit expression.

Printing the fourth through tenth items of a list:

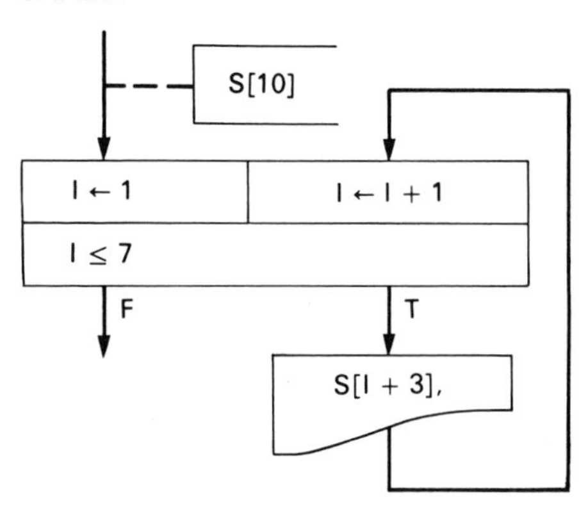

Decreasing subscript values:

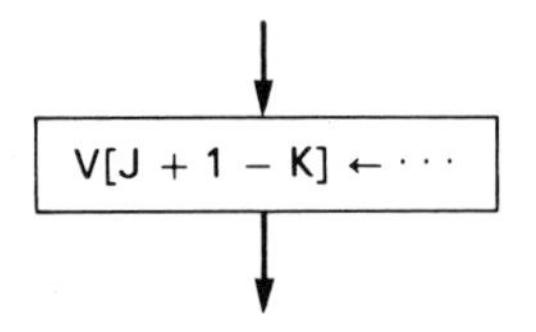

Subscript sequence (3, 7, 11, 15, 19):

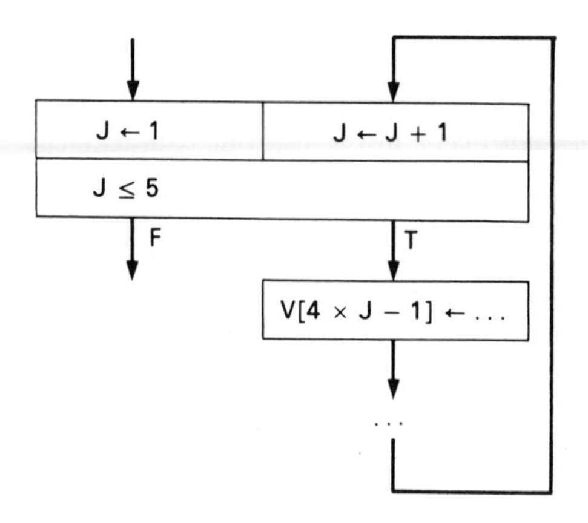

Irregular subscript sequence:

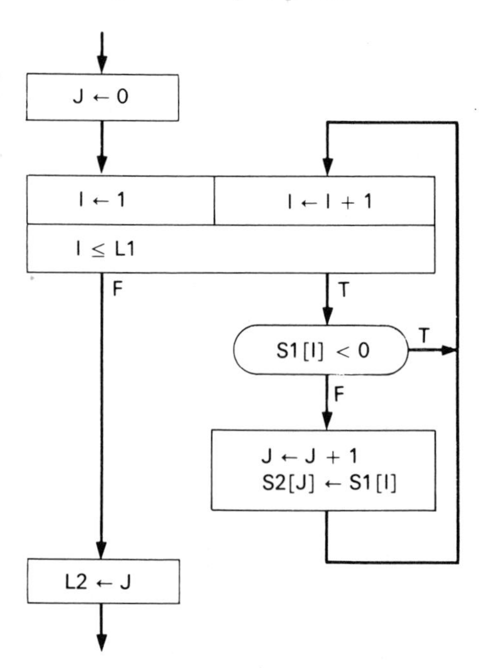

Subscript sequence stored in a list of pointers:

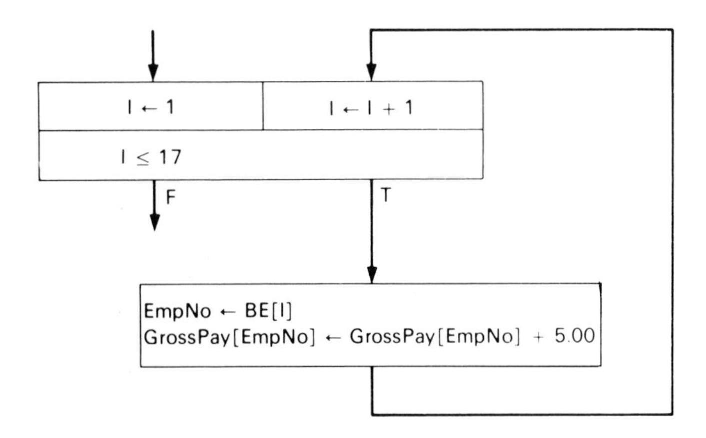

Index without subscript:

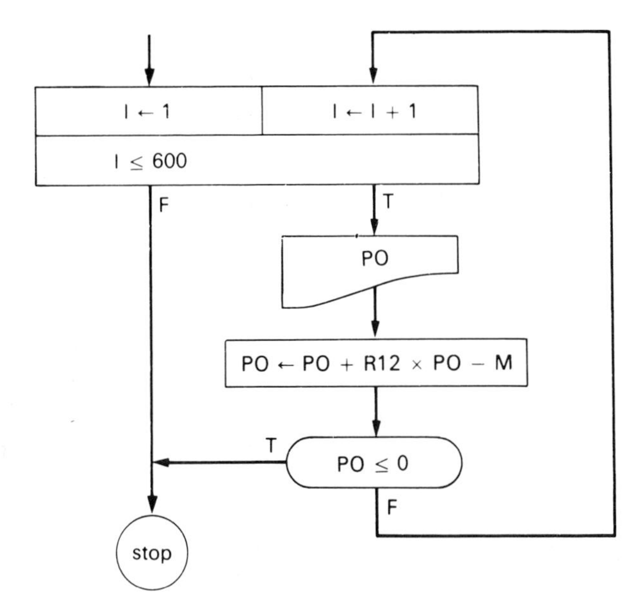

Binary search:

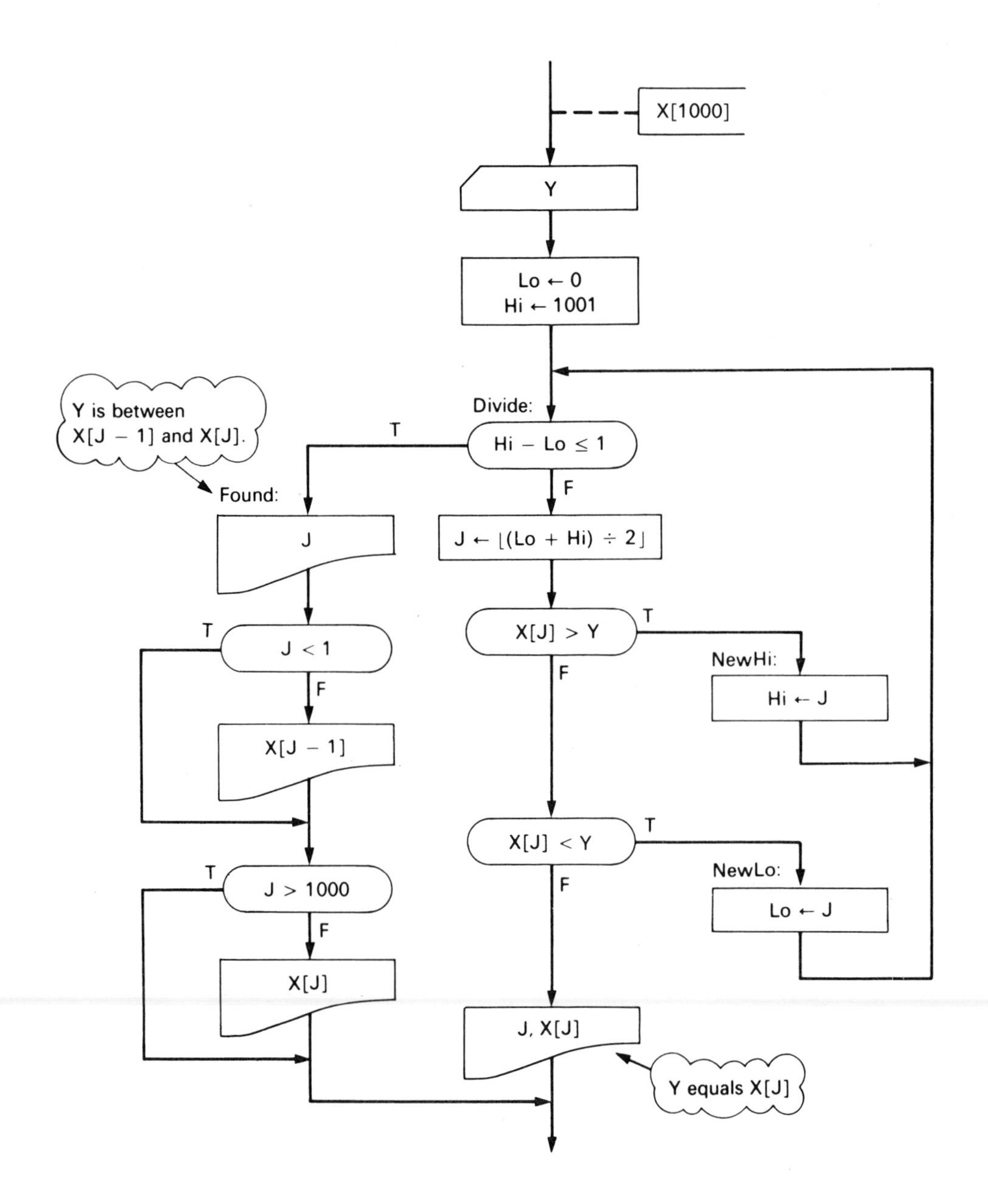

Basic In Basic, we can write indexed loops beginning at 1, as in the foregoing flowchart examples. If we do not specify any step size, the index will increase by 1 each time.

Printing the fourth through tenth items of a list:

```
200 DIM S[10]
210 FOR I = 1 TO 7
220     PRINT S[I + 3]
230     NEXT I
```

This maintains the philosophy of complicating the subscript expression as much as necessary to keep the indexing simple.

An alternative approach is to impose part of the burden on the index and to keep the subscript expression somewhat simpler. This is possible in Basic. We may write

```
110 FOR I = A TO B STEP C
```

If STEP C is omitted, the step size will be 1. Note that I must be a *variable*, while A, B, and C may be any *expressions*. The previous example may be rewritten as:

```
200 DIM S[10]
210 FOR I = 4 TO 10
220     PRINT S[I]
230     NEXT I
```

A decreasing subscript is achieved with a negative step value:

```
410 FOR I = J - 1 TO 1 STEP -1
420     IF T >= V[I] THEN 460
430     LET V[I + 1] = V[I]
440     NEXT I
450 LET I = 0
460 LET V[I + 1] = T
```

The subscript sequence (3, 7, 11, 15, 19) is obtained by using 4 as the step size and 3 as the initial value:

```
510 FOR J = 3 TO 19 STEP 4
520     LET V[J2] = ...
        ...
580     NEXT J
```

In general, when we write FOR I = A TO B STEP C, the variable I takes on values A, A + C, A + 2 * C, and so on. The loop is executed as long as the value of I is *between* A and B, inclusive.

Irregular subscript sequence:

```
620 LET J = 0
630 FOR I = 1 TO L1
640     IF S1[I] < 0 THEN 670
650     LET J = J + 1
660     LET S2[J] = S1[I]
670     NEXT I
680 LET L2 = J
```

Subscript sequence stored in a list of pointers:

```
745 FOR I = 1 TO 17
750     LET E = B[I]
755     LET G[E] = ...
760     NEXT I
```

Index without subscript:

```
830 FOR I = 1 TO 600
840     PRINT P
850     LET P = P + R1 * P - M
860     IF P < 0 THEN 880
870     NEXT I
880 REM
```

Binary search:

```
200 DIM X[1000]
205 LET L = 0
210 LET H = 1001
215 INPUT Y
220 IF H - L <= 1 THEN 270
222 REM: SET POINTER TO MIDDLE
223 REM: OF REMAINING LIST.
225 LET J = INT ((L + H) / 2)
230 IF X[J] > Y THEN 250
235 IF X[J] < Y THEN 260
237 REM: Y EQUALS X[J].
240 PRINT J, X[J]
245 GO TO 900
250 LET H = J
255 GO TO 220
260 LET L = J
265 GO TO 220
267 REM: Y IS BETWEEN
268 REM: X[J - 1] AND X[J]
270 PRINT L, H
275 IF L < 1 THEN 285
280 PRINT X[L]
285 IF H > 1000 THEN 295
290 PRINT X[H]
295 REM
900 END
```

Fortran In Fortran, we can write indexed loops beginning at 1. If we do not specify any step size, the index will increase by 1 each time.

Printing the fourth through tenth items of a list:

```
      DIMENSION S(10)
      DO 230 I = 1, 7
        WRITE (6, 83) S(I + 3)
83      FORMAT (8 F 15.3)
230     CONTINUE
      STOP
```

This keeps the indexing simple while complicating the subscript expression. In standard Fortran, if the subscript expression is more complicated than

constant × *variable* {±} *constant*

the expression must be evaluated separately and assigned to a variable, which is then used as the subscript:

```
      K = I - 1 + J
      ...S(K)...
```

An alternative approach is to impose part of the burden on the index, and keep the subscript expression somewhat simpler. This alternative is particularly attractive in standard Fortran, because of the restriction to simple subscript expressions. We may write

DO *label* I = L1, L2, L3

where L1 is the initial value, L2 is the limit, and L3 (optional) is the step size. Note that I must be a variable, while L1, L2, and L3 may be variables or constants. Also note that in Fortran, as contrasted to Basic, PL/1, and most other languages, the loop is executed at least *once*[4] even if the value of L1 is greater than that of L2.

[4] The Fortran Standard does not specify what happens when the value of L1 is greater than that of L2. However, virtually all versions of Fortran assign the value of L1 to I and then execute the loop once before comparing the values of I and L2. Fortran users universally regard this as the *de facto* standard, and consider the published standard defective at this point.

The previous example may be rewritten as:

```
      DIMENSION S(10)
      DO 230 I = 4, 10
        WRITE (6, 83) S(I)
83      FORMAT (8 F 15.3)
230     CONTINUE
      STOP
```

A negative step cannot be used in Fortran; an expression involving −K must be assigned to the subscript variable:

```
      DO 440 K = 2, J
        I = J - K + 1
        IF (T . GT. V(I)) GO TO 460
        V(I + 1) = V(I)
440     CONTINUE
      I = 0
460   V(I + 1) = T
      STOP
```

A positive initial value and step may be used, however:

```
      DO 580 J = 3, 19, 4
        V(J) = ...
        ...
580     CONTINUE
      STOP
```

Irregular subscript sequence:

```
      J = 0
      DO 670 I = 1, L
        IF (S1(I) .LT. 0.0) GO TO 670
        J = J + 1
        S2(J) = S1(I)
670     CONTINUE
      L2 = J
```

Subscript sequence stored in a list of pointers:

```
      INTEGER EMPNO, BE
      DO 760 I = 1, 17
        EMPNO = BE(I)
        GP(EMPNO) = ...
760     CONTINUE
```

Index without subscript:

```
      DO 870 I = 1, 600
         WRITE (6, 83) PO
83       FORMAT (F 15.2)
         PO = PO + R * PO - AM
         IF (PO .LT. 0) STOP
870      CONTINUE
      STOP
```

Binary search:

```
    DIMENSION X(1000)
81  FORMAT (8 I 10)
82  FORMAT (8 F 10.0)
83  FORMAT (8 F 15.3)
    INTEGER HI
    LO = 0
    HI = 1001
    READ (5, 82) Y
220 IF (HI - LO .LE. 1) GO TO 270
C   SET POINTER TO MIDDLE
C   OF REMAINING LIST.
    J = (LO + HI) / 2
    IF (X(J) .GT. Y) GO TO 250
    IF (X(J) .LT. Y) GO TO 260
C   Y EQUALS X(J)
    WRITE (6, 81) J
    WRITE (6, 83) X(J)
    STOP
C   NEW HI
250 HI = J
    GO TO 220
C   NEW LO
260 LO = J
    GO TO 220
C   Y IS BETWEEN X(J - 1) AND X(J).
270 WRITE (6, 81) LO, HI
    IF (LO .GE. 1) WRITE (6, 82) X(LO)
    IF (HI .LE. 1000) WRITE (6, 82) X(HI)
    STOP
```

PL/1 It is possible to write indexed loops beginning at 1. If no step size is specified, 1 is assumed. For instance, a program to print the fourth through tenth items of a list could be written as:

```
declare S(1 : 10) float;
declare I fixed;
do I = 1 to 7;
   put list (S(I + 3));
   end;
```

This keeps the indexing simple while complicating the subscript expression. An alternative approach is to impose some or all of the burden on the index, and correspondingly to simplify the subscript expression. We may write

```
do I = L1 to L2 by L3;
```

where L1 is the initial value, L2 is the limit, and L3 (optional) is the step size. Note that I must be a variable, while L1, L2, and L3 may be variables or constants. The previous example may be rewritten as:

```
declare S(1 : 10) float;
declare I fixed;
do I = 4 to 10;
   put list (S(I));
   end;
```

A negative step can be used in PL/1, to obtain a decreasing subscript:

```
do I = J - 1 to 1 by -1;
   if T >= V(J) then go to CONT;
   V(I + 1) = V(I);
   end;
I = 0;
V(I + 1) = T;
```

Subscript sequence (3, 7, 11, 15, 19):

```
do J = 3 to 19 by 4;
   V(J) . . . ;
   end;
```

Irregular subscript sequence:

```
J = 0;
do I = 1 to L1;
   if S1(I) <=0 then go to SKIP;
```

```
      J = J + 1;
      S2(J) = S1(I);
    SKIP: end;
    L2 = J;
```

Subscript sequence stored in a list of pointers:

```
    do I = 1 to 17;
    EMPNO = BE(I);
    GROSSPAY(EMPNO) = ....;
    end;
```

Index without subscript:

```
    do I = 1 to 600;
       put skip list (PO);
       PO = PO + R12 × PO - M;
       if PO=0 then go to LAST;
       end;
    LAST:...
```

Binary search:

```
FIND: procedure options (main);
     declare (Y, X(1:1000)) float;
     declare (LO, HI, J) fixed;
     get list (Y);
     LO = 0; HI = 1001;
DIVIDE: if HI - LO <= 1 then go to FOUND;
     /* Set pointer to middle
        of remaining list.*/
     J = floor ((LO + HI) / 2);
     if X(J) > Y then go to NEWHI;
     if X(J) < Y then go to NEWLO;
     /* Y equals X(J).*/
     put skip list (J, X(J)); go to L2;
NEWHI: HI = J;
     go to DIVIDE;
NEWLO: LO = J;
     go to DIVIDE;
     /* Y is between X(J - 1) and X(J).*/
FOUND: put skip list (J);
     if J < 1 then go to L1;
     put skip list (X(J - 1));
L1: if J > 1000 then go to L2;
     put skip list (X(J));
L2: end FIND;
```

Words to Remember

shift reference variable pointer

Exercises

1. Read a list of five items and a list of 60 items. For each item in the first list, determine how many times (if any) it occurs in the second list. List of 60 items: 10, 11, 21, 12, 13, 15, 28, 13, 11, 14, 25, 19, 14, 13, 27, 10, 17, 27, 14, 11, 15, 26, 11, 17, 18, 25, 13, 18, 11, 29, 10, 19, 29, 18, 17, 15, 22, 17, 19, 16, 25, 11, 16, 17, 23, 10, 13, 13, 26, 19, 15, 24, 19, 13, 12, 25, 17, 12, 19, 21.
   Lists of 5 items:
   a. 10, 11, 12, 13, 14
   b. 14, 25, 15, 19, 23
   c. 11, 15, 20, 24, 29
   d. 13, 15, 10, 11, 17
2. In the previous exercise, read the list of five items and store them, but read only one of the 60 items at a time and compare it with each item of the first list. Can you see any advantage or disadvantage to this approach?
3. Write a program to read sets of data, each consisting of an item number (between 1 and 20) and an amount. Accumulate and print the total of the amounts for each item. Use a sentinel to indicate when the end of data is reached. *Do not* store the individual input data items in a list.

Number	*Amount*	*Number*	*Amount*	*Number*	*Amount*
9	42.1	15	17.7	9	−0.5
11	32.2	4	132.7	13	14.2
2	218.0	18	−0.7	17	213.9
12	26.0	1	11.8	6	−3.8
3	13.9	18	73.9	13	268.6
14	5.7	8	7.7	12	96.7
6	6.2	15	18.5	8	28.8
19	58.2	2	18.5	9	−7.8
4	25.3	16	36.0	20	5.7
12	−1.9	7	40.6	19	36.4
5	39.8	12	51.9	5	6.5
16	2.3	8	29.4	1	300.2
20	10.5	10	7.0		

4. Write a program to count the number of input data items falling into each of several ranges: 0 to 199, 200 to 299, 300 to 399, 400 to 499, 500 to 599, 600 and over. *Do not* store the individual input data items in a list. Use a sentinel to indicate when the end of the input data is reached. Print the total number of items in each range.
   a. 52, 143, 204, 344, 343, 307, 287, 332, 420, 416, 480, 517, 559, 518, 570, 644, 608, 563, 532, 533
   b. 314, 10, 116, 196, 1096, 141, 143, 27, 354, 359, 44, 49, 598
   c. 313, 221, 170, 275, 331, 62, 195, 348, 502, 276, 706

5. Write a program to find all of the prime numbers between 1 and 1000 using the following method, called the Sieve of Eratosthenes:
   a. Store the value 0 in all the cells of a 1000-cell list.
   b. Using a count-controlled loop, store 1 in each cell whose subscript is a multiple of 2, beginning at 4.
   c. Next, store a 1 in each cell whose subscript is a multiple of 3, beginning at 9.
   d. Continue, for each n, storing 1 in each cell whose subscript is a multiple of n, beginning at n^2. Stop when n^2 exceeds 1000.
   e. Short cut: Check to see if there is already a 1 at n^2. (For example, 16 and 36 will already have a 1 stored before they are reached by step *d.*) If there is a 1 at n^2, all multiples of n will also have a 1 already.
   f. After all the values of n have been used, until n^2 exceeds 1000, go back and print the subscript values of all cells still containing zeros. These are the prime numbers. (Can you prove it?)

5.5 *When Should Lists and Strings Be Used?*

Here are two different programs for computing Fibonacci numbers:

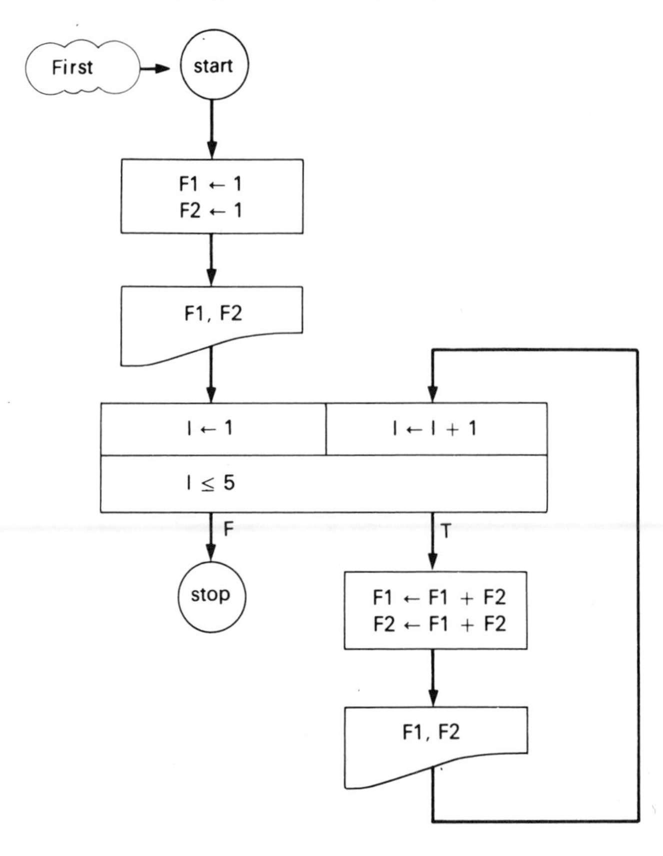

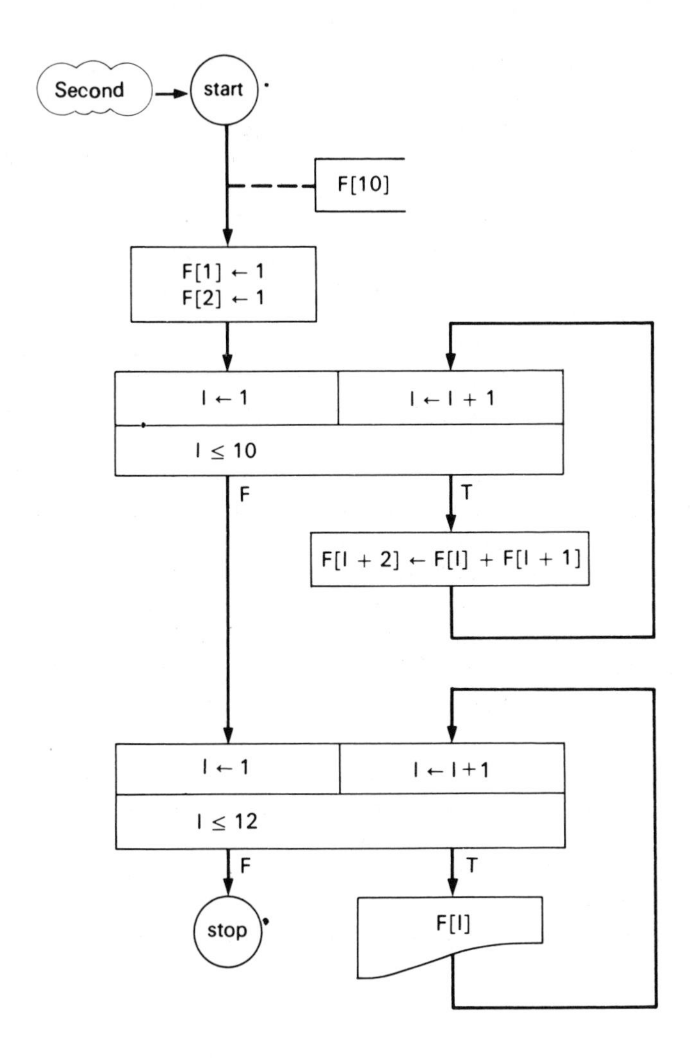

Both of these programs achieve the same result, printing the values of the first twelve numbers in the Fibonacci sequence:

1, 1, 2, 3, 5, 8, 13, 21, 34, 55, 89, 144

The essential difference between them is that the second version uses a list of 12 variables, while the first version uses only two variables. If we can achieve the same result without using a list, then the second version has the disadvantage of requiring additional space. The advantage of this version, however, is that the results all remain available after the calculation is finished. One way to use the results after they are all calculated is illustrated in the loop at the end of the second version, where the values are printed after they are all available. This is not a very convincing example of the need for lists, for we could have printed each number as it was calculated, as in the first version. However, we might want to print the numbers in *decreasing* order, in which case we could store them in a list.

In many common statistical problems we can choose whether to use a list. In most cases the use of a list can be avoided, although at first glance the problem seems to require them. The calculation of standard deviation, as usually described, is:

(a) Find the mean (average) of the data values.

(b) Subtract the mean from each data value to obtain the deviation; square each deviation and add the squares together to form the sum of the squares of the deviations.

(c) Take the square root of this sum. This is the standard deviation.

It would appear that the data must be stored, so that step (b) can be performed after all the data values have been used in step (a). But the formulas can be arranged (see Section 3.7, exercise 5) so that we accumulate the sum of the squares of the data values as they are read in, and use this sum to obtain the standard deviation by a slightly different formula. Thus we do not need to store the data. This means that we can write a program which is not limited by the number of data items to be used. Hundreds or thousands of values can be processed, and the statistical formulas can be applied to them, without requiring that space inside the computer be set aside to store them. The calculation can be performed using an amount of space which depends on the complexity of the statistical analysis but not on the number of items of data to be processed. There is a large class of statistical problems to which this idea applies.

String Manipulations with Subscripted Variables The "automatic" operations of reading, printing, joining, and comparing strings, available in many programming languages, can be studied in more detail if we write out the individual steps—using loops and subscripted variables—in much the same way as we write the steps for manipulating the items of a list.

Reading or printing a string merely requires a simple loop:

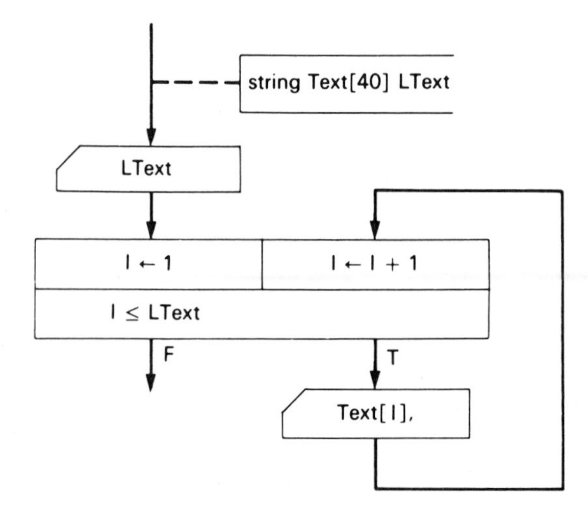

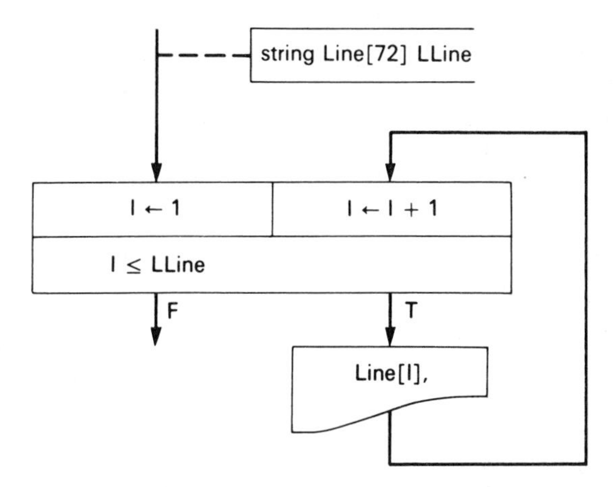

The form in which we prepare data for the first program may have to be somewhat different from the form used by the "automatic" string read feature. For example, the text may have to be preceded by an integer data value telling how many characters are to be read into the string, and the characters may have to be separated, with each character individually enclosed in quotation marks. Also the appearance of the characters printed by the second program may differ somewhat from that of characters printed "automatically."

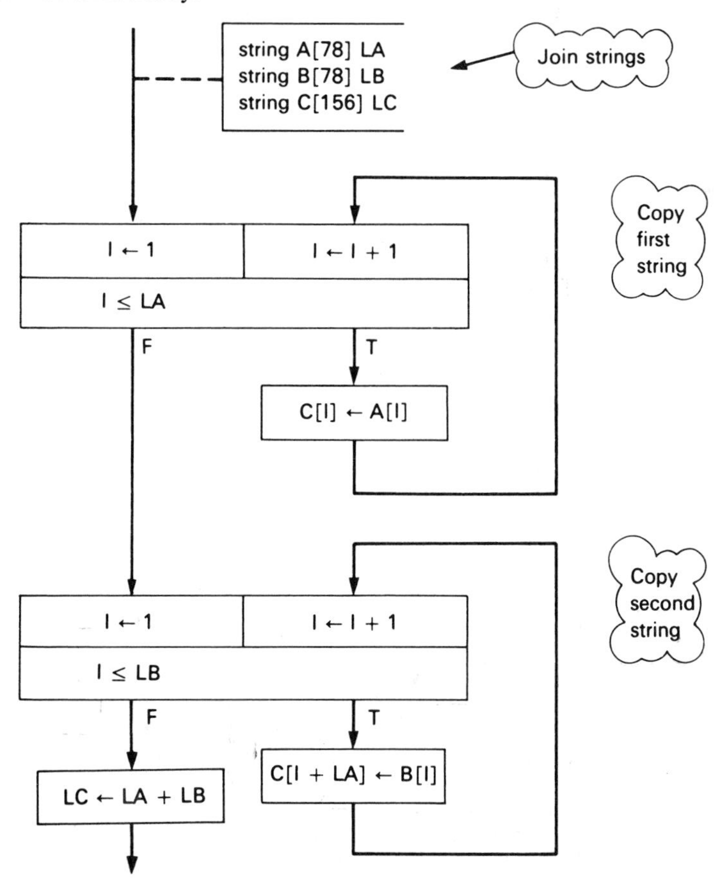

Joining strings is not especially complicated or difficult. Assuming that the lengths of the strings have been established, their characters are simply moved one at a time to become characters of the combined string. Notice the use of the subscript [I + LA] when the characters of B are being moved.

Comparing two strings, on the other hand, is a bit more intricate.

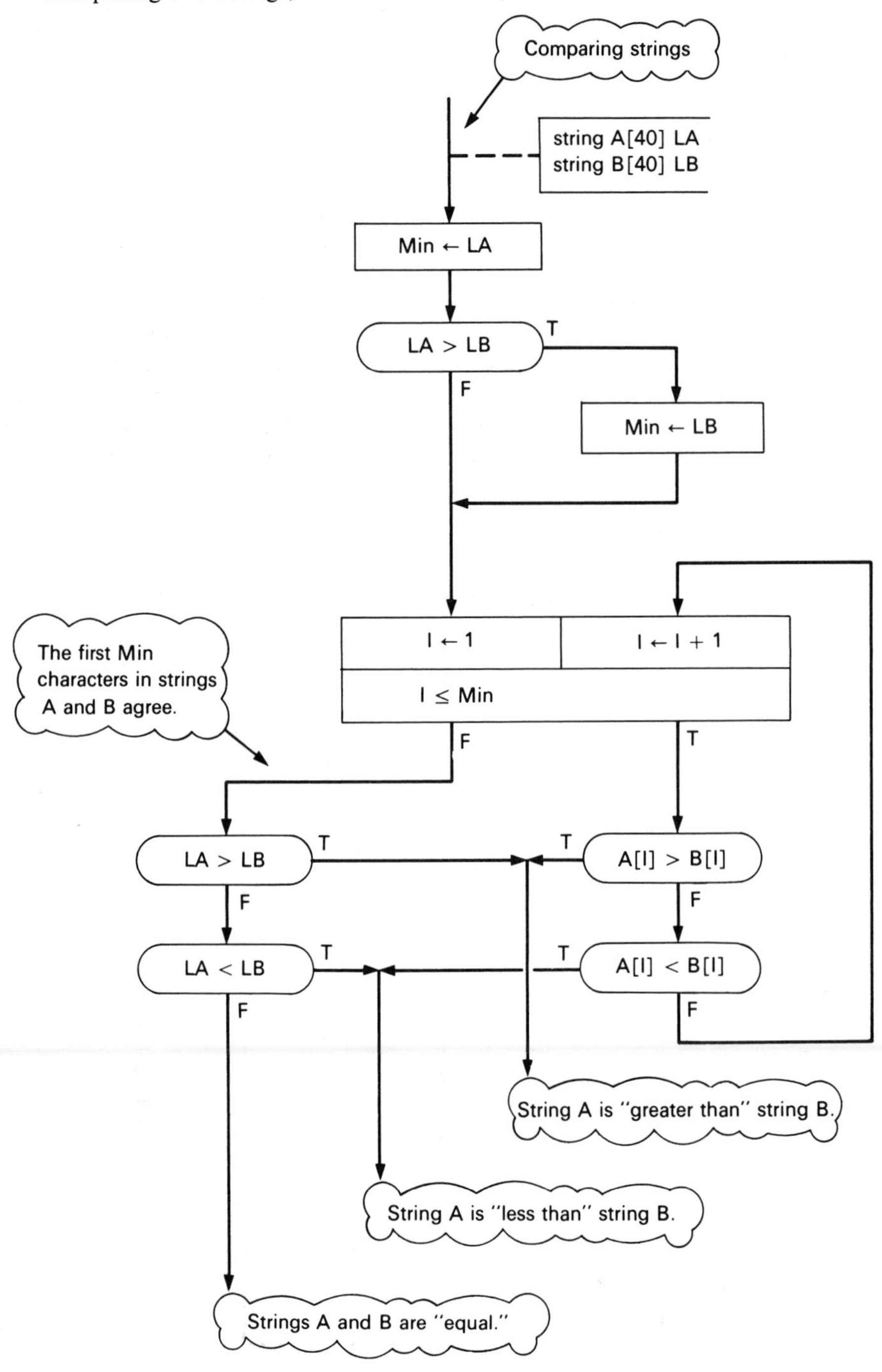

The value of the variable `Min` is set equal to the length of the shorter string; this is the number of characters to be compared. If all the characters in each string up to that point are equal, we compare the lengths and say that the shorter string is "less than" the longer. However, if we find a difference among the first `Min` characters, we immediately know how to rank the strings.

Correctly Ranking Character Strings As we saw in Section 5.1, an instruction involving a comparison of two strings of characters, in most programming languages, uses the bit patterns that represent characters inside the computer (intepreted as integers) to determine which alphabetic character or string is "less than" the other. For ASCII and EBCDIC codes, this gives the ranking we usually desire, provided that the characters are limited to capital letters, blanks, and simple punctuation.

In ASCII, the digits are "less than" the letters of the alphabet, while in EBCDIC the digits are "greater than" the letters. For some purposes, either of these arrangements may be satisfactory. However, a mixture of upper- and lower-case letters must be ranked as though both cases of the same letter had the same rank. For example, how can we sort a list of names such as the following (selected from Webster, 1958)?

Debussy, Claude
Debye, Peter
De Casseres, B.
Decatur, Stephen
Decazes, D.
De Chair, Sir Dudley
Defoe, Daniel
De Forest, Lee
Degas, Hilaire
de Gaulle, Charles
Dekker, Thomas
De Koven, Reginald
de Kruif, Paul
Delacroix, Ferdinand
De la Mare, Walter
Deland, Margaret
De La Rey, Jacobus
Delaroche, H. P.
de la Roche, Mazo
Delavigne, Casimir
De La Warr, Baron
Delcasse, Theophile
d'Erlanger, Baron F.
de Seversky, Alexander
Desmond, Shaw
de Soto, Hernando
Douglas, Stephen
Douglass, Frederick
Dubois, Paul
Du Bois, William

It is obviously necessary to consider upper- and lower-case letters as having the same rank; otherwise, all names beginning with *d* would be separated from names beginning with *D*, and a similar misarrangement would occur with upper-case letters inside a name. Also, before ranking the names, it is evidently necessary to "squeeze out" blanks (and apostrophes, as in *d'Erlanger*) when they occur within a name. On the other hand, *Douglas, Stephen* must precede *Douglass, Frederick*; therefore we must maintain an indication of the end of the surname.

A possible way to proceed, once we have decided on a set of alphabetization rules, is to generate a *different* string to be used for sorting purposes. This string will have to be generated character by character, according to the rules. For instance, we could change all lower-case letters to upper-case, and shorten the string if necessary by removing internal blanks and apostrophes. Any other strange characters would be replaced in some way by a character that has the desired ranking. After this has been done for each string in the list, we compare the *modified* strings to obtain the desired ranking. (A list, consisting of an *integer* "rank" value for each character in the string, could be generated on the same principle. Also, it might be possible to wait and convert the characters as they are being compared. This might save work if not all the characters need to be examined, and if each string is compared with only a few others.)

As part of the program to determine the desired ranking for a particular character of the string, it might be desirable to first convert the stored (ASCII or EBCDIC) pattern directly into an integer, and then use this integer as a pointer or subscript to find the desired item in a list of ranks:

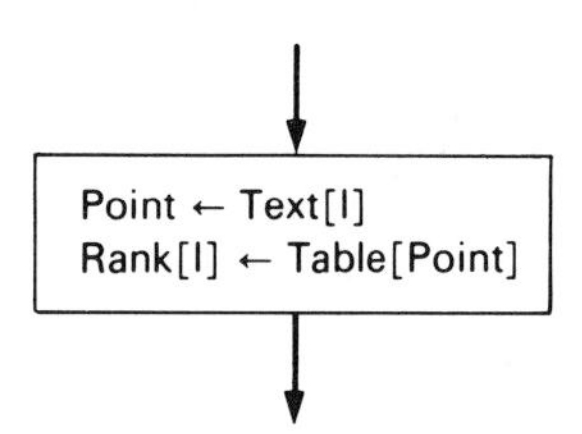

Clearly, the details of a program for correctly ranking strings of characters can be chosen by the programmer. The actual program will depend to a considerable extent on the character handling features available on the computer system to be used, and on the rules of ranking that are chosen.

Flowchart

Fibonacci sequence without list:

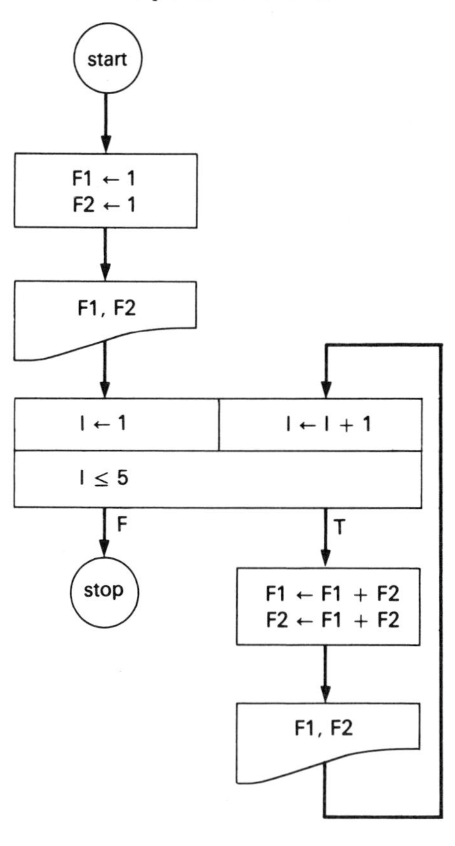

Fibonacci sequence with list:

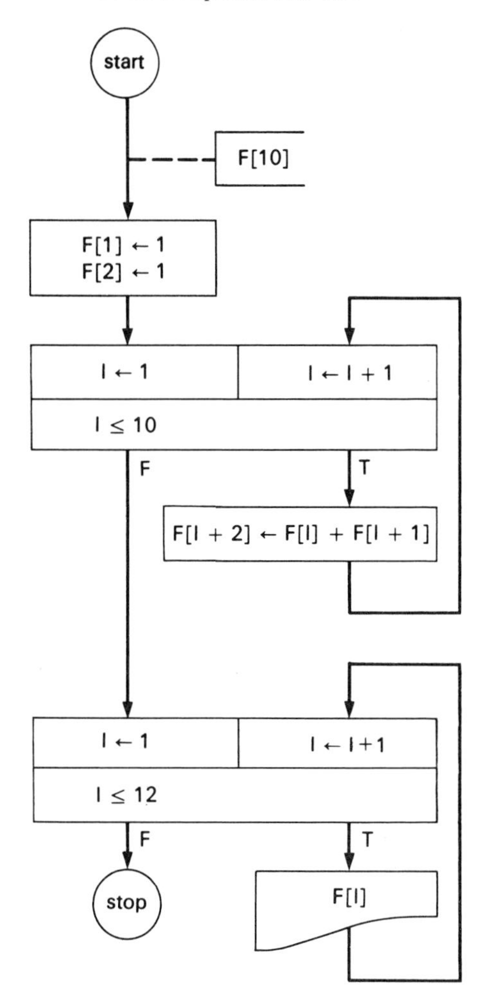

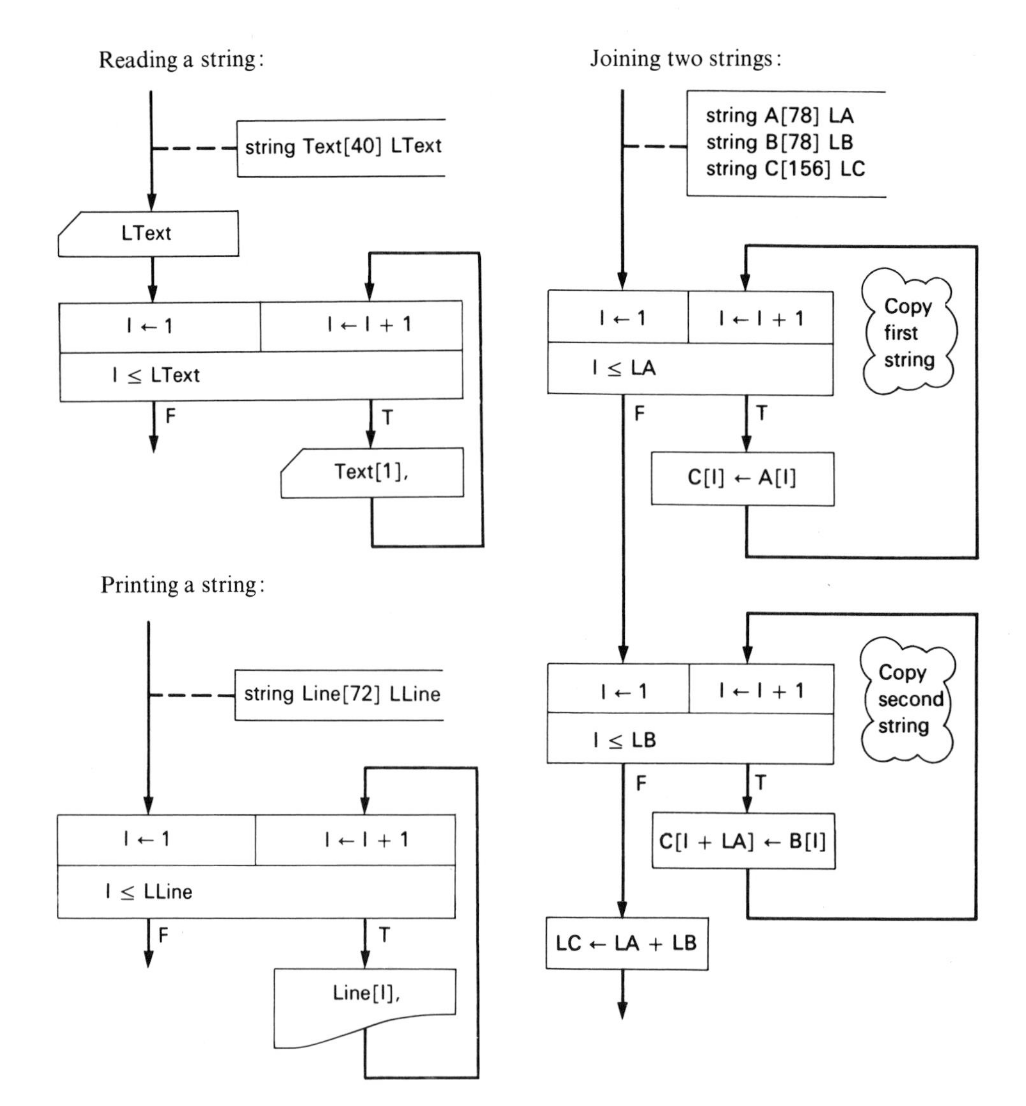

Reading a string:
string Text[40] LText
LText
I ← 1
I ← I + 1
I ≤ LText
F
T
Text[1],
Printing a string:
string Line[72] LLine
I ← 1
I ← I + 1
I ≤ LLine
F
T
Line[I],
Joining two strings:
string A[78] LA
string B[78] LB
string C[156] LC
I ← 1
I ← I + 1
I ≤ LA
Copy first string
F
T
C[I] ← A[I]
I ← 1
I ← I + 1
I ≤ LB
Copy second string
F
T
C[I + LA] ← B[I]
LC ← LA + LB

Comparing two strings:

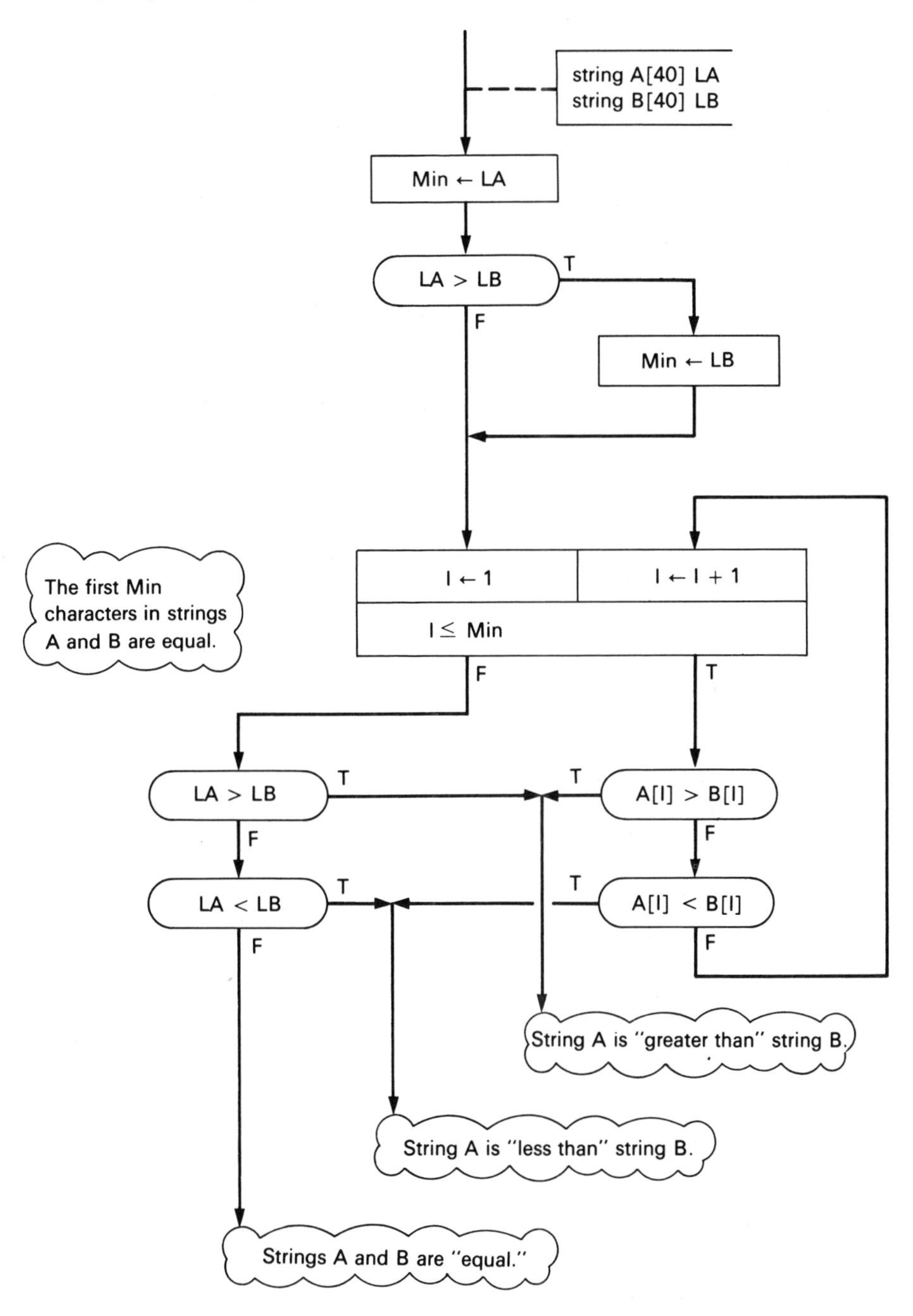

Basic

Fibonacci sequence with and without list:

```
110 REM ... FIRST
115 LET F1 = 1
120 LET F2 = 1
125 PRINT F1, F2
130 FOR I = 1 TO 5
140    LET F1 = F1 + F2
150    LET F2 = F1 + F2
160    PRINT F1, F2
170    NEXT I
180 END

210 REM ... SECOND
215 DIM F[12]
220 LET F[1] = 1
225 LET F[2] = 1
230 FOR I = 3 TO 12
240    LET F[I] = F[I - 2] + F[I - 1]
270    NEXT I
280 FOR I = 1 TO 12
285    PRINT F[I]
290    NEXT I
295 END
```

Joining two strings:

```
200 DIM A[78], B[78], C[156]
    ...
220 CHANGE A1$ to A
225 CHANGE B1$ TO B
227 REM: COPY FIRST STRING.
230 FOR I= 1 TO A[0]
235    LET C[I] = A[I]
240    NEXT I
242 REM: COPY SECOND STRING.
245 FOR I = 1 TO B[0]
250    LET C[I + A[0]] = B[I]
255    NEXT I
260 LET C[0] = A[0] + B[0]
265 CHANGE C TO C1$
    ...
```

Comparing two strings:

```
300 DIM A [40], B[40]
    ...
320 CHANGE A1$ TO A
325 CHANGE B1$ TO B
330 LET M = A[0]
335 IF A[0] < B[0] THEN 345
340 LET M = B[0]
345 FOR I = 1 TO M
350    IF A[I] < B[I] THEN 385
355    IF A[I] > B[I] THEN 395
360    NEXT I
362 REM: FIRST M CHARACTERS MATCH.
365 IF A[0] < B[0] THEN 385
370 IF A[0] > B[0] THEN 395
375 REM A IS EQUAL TO B.
    ...
385 REM: A IS LESS THAN B.
    ...
395 REM: A IS GRTR THAN B.
    ...
```

Note the use of the CHANGE operation, and the fact that the length of a string is referenced with the subscript value zero.

Fortran

Fibonacci sequence with and without list:

```
C     FIRST
      INTEGER F1, F2
      F1 = 1
      F2 = 1
      WRITE (6, 81) F1, F2
81    FORMAT (8 I 10)
      DO 170 I = 1, 5
        F1 = F1 + F2
        F2 = F1 + F2
        WRITE (6, 81) F1, F2
170     CONTINUE
      STOP

C     SECOND
      INTEGER F(12)
      F(1) = 1
      F(2) = 1
      DO 270 I = 3, 12
        F(I) = F(I - 2) + F(I - 1)
270     CONTINUE
      DO 290 I = 1, 12
        WRITE (6, 81) F(I)
290     CONTINUE
      STOP
```

Reading and printing a string:

```
      DIMENSION ITEXT(40)
      READ (5, 81) L
81    FORMAT (8 I 10)
      READ (5, 97) (ITEXT(I), I = 1, L)
97    FORMAT (80 A 1)
      WRITE (6, 98) (ITEXT(I), I = 1, L)
98    FORMAT (1X, 80 A 1)
      STOP
```

Joining two strings:

```
      DIMENSION IA(40), IB(40), IC(80)
      . . .
C     COPY FIRST STRING.
      DO 240 I = 1, LA
        IC(I) = IA(I)
240   CONTINUE
C     COPY SECOND STRING.
      DO 255 I = 1, LB
        J = I + LA
        IC(J) = IB(I)
255     CONTINUE
      LC = LA + LB
        . . .
```

Comparing two strings:

```
      DIMENSION IA(40), IB(40)
      . . .
      MIN = LA
      IF (LA . LT. LB) GO TO 345
      MIN = LB
345   DO 360 I = 1, MIN
        IF (IA(I) .LT. IB(I)) GO TO 385
        IF (IA(I) .GT. IB(I)) GO TO 395
360     CONTINUE
C     THE FIRST 'MIN' CHARACTERS MATCH
      IF (LA .LT. LB) GO TO 385
      IF (LA .GT. LB) GO TO 395
375   CONTINUE
C     A IS EQUAL TO B.
      . . .
385   CONTINUE
C     A IS LESS THAN B.
      . . .
395   CONTINUE
C     A IS GREATER THAN B.
      . . .
```

PL/1

Fibonacci sequence with and without list:

```
FIRST: procedure options (main);
  declare (F1, F2) float;
  declare I fixed;
  F1 = 1; F2 = 1;
  put skip list (F1, F2);
  do I = 1 to 5;
    F1 = F1 + F2;
    F2 = F1 + F2;
    put skip list (F1, F2);
    end;
  end FIRST;

SECOND: procedure options (main);
  declare F(1:12) float;
  declare I fixed;
  F(1) = 1; F(2) = 1;
  do I = 3 to 12;
    F(I) = F(I - 2) + F(I - 1);
    end;
  do I = 1 to 12;
    put skip list (F(I));
    end;
  end SECOND;
```

In PL/1, the only way to handle individual characters is to use a list of strings, with each string consisting of just one character. This makes it impossible to use the automatic features for handling the same group of characters as a single string. However, some instructions may be written to convert between a single string and a list of one-character strings. Such a process is rarely needed in practice, as the most frequently needed operations can be performed with the functions and operators provided in the language. However, it is instructive to see how this sort of manipulation could be done if desired.

Converting a string to a list of one-character strings:

```
UNPACK: procedure options (main);
  declare STRING character(40) varying;
  declare LIST(1:40) character(1);
  declare (LSTR, I) fixed;
  get list (STRING);
  LSTR = length (STRING);
  do I = 1 to LSTR;
    LIST(I) = substr (STRING, I, 1);
    end;
  . . .
  end UNPACK;
```

Converting a list of one-character strings to a single string:

```
PACK: procedure options (main);
  declare STRING character (40) varying;
  declare LIST (1:40) character(1);
  declare (LSTR, I) fixed;
  . . .
  do I = 1 to LSTR;
    substr (STRING, I, 1) = LIST(I);
    end;
  put skip list (STRING);
  end PACK;
```

In the second example, we see a construction peculiar to PL/1, called a *pseudo-variable.* This object is not a variable but is permitted on the left side of an assignment instruction with an expression on the right. The instruction

```
substr (STRING, I, 1) = LIST (I);
```

assigns a short string to replace some of the characters of a longer string, without changing any of the other characters of the longer string.

Exercise

Write a program to read a string of characters, and then copy this string to another string, deleting leading and trailing blanks. (Hint: find the first nonblank character and the last nonblank character in the original string.) Print the resulting string.

a. "   The quick brown fox jumps over the lazy dog.   "

b. "   The Science of Computing   "

Project

Write a program to keep track of the moves in a game of Kalah.[5] This "ancient game of mathematical skill" is played as follows (*Time*, 1963; see also Haggerty, 1964):

> Two players sit behind the two ranks of six pits on the board between them. Each pit contains three (for beginners) or six "pebbles" (which may be anything from matches to diamonds). Purpose of the game is to accumulate as many pebbles as possible in the larger bin (Kalah) to his right. Each player in turn picks up all the pebbles in any one of his own six pits and sows them, one by one, in each pit around the board to the right, including, if there are enough, his own Kalah, and on into his opponent's pits (but not his Kalah). If the player's last counter lands in his own Kalah, he gets another turn, and if it lands in an empty pit on his own side [only] he captures all his opponent's counters in the opposite pit and puts them in his Kalah together with the capturing pebble. It is not always an advantage for a player to go "out," since all the pebbles in the pits on the opposite side go into the opponent's Kalah. The score is determined by who has the most pebbles.
>
> In the game illustrated below Player *A* begins by moving the three pebbles in his pit A4, ending in his Kalah and thus earning another move, which he uses to play from pit A1, ending on empty pit A4 and thereby capturing *B*'s men. By similar moves and captures, *A*, by the end of move 6, becomes pebble-proud with eleven in his Kalah to a pathetic one in *B*'s. The sequence of moves in the sample game is (A4, A1)–B6–(A3, A2, A1)–(B4, B6, B1)–A4–B3–A6–(B2, B1). By the end of move 6, *A* is dangerously concentrated in the two pits, A5 and A6. Player *B*, seeding six pebbles on his own side, forces *A* to start distributing his hoard around the board. By the end of move 12, Player *A* still has twelve in his Kalah to five in *B*'s, but *B* moves the five pebbles in B2 and then has only to move the single pebble in his pit B1 to capture *A*'s seven remaining pebbles, ending the game and winning it by a score of 24–12.

Further observations concerning the rules of the game:

1. If a player's last counter lands in an empty pit on his own side, but the opponent's pit (opposite) is empty, so that there is nothing to capture, then the last counter does not go to his Kalah, but is placed in the last pit in the normal way.

[5] The name is a registered trade mark of the Kalah Game Co., Holbrook, Mass. This company markets a commercial version of the game. A satisfactory (utilitarian but perhaps not aesthetic) substitute can be made from an egg carton and a handful of beans.

2. If a player reaches his opponent's Kalah while sowing pebbles, he merely jumps over it and goes on sowing. The next pebble, which would have gone into the opponent's Kalah, goes into his own first pit.

In the computer program, call the players "number 1" and "number 2," and number the pits from 1 to 14, with 1 at the left hand of player 1. The first player's Kalah is pit 7, and the second player's Kalah is pit 14. The board is represented in the computer by a list of length 14.

A move is indicated by a pair of data values, giving the player number (1 or 2) and the pit number. If desired, a check may be included to see whether this player is entitled to move next, and whether he is moving from a nonempty pit on his side.

The heart of the program consists of instructions for taking the pebbles out of a pit and distributing them to the other pits in order of increasing subscript values. This sequence must be modified when the end of the board is reached, or when the opponent's Kalah is reached. Hence, there will be an "irregular" relation between the subscript and the number of pebbles.

At the end of this sequence, tests are made to see whether a capture occurs, and to determine which player moves next.

If you have access to an on-line computer, two players may enter their moves alternately at the console. Otherwise, prepare a game in advance and let the computer verify it and print a record of all the moves.

START:

K	6	5	4	3	2	1	B
0	3	3	3	3	3	3	
	3	3	3	3	3	3	0
A	1	2	3	4	5	6	K

Move 1: A4

K	6	5	4	3	2	1	B
0	3	3	3	3	3	3	
	3	3	3	0	4	4	1
A	1	2	3	4	5	6	K

Move 2: A1

PLAYER A CAPTURES 3 FROM B3

K	6	5	4	3	2	1	B
0	3	3	3	0	3	3	
	0	4	4	0	4	4	5
A	1	2	3	4	5	6	K

Move 3: B6

K	6	5	4	3	2	1	B
1	0	3	3	0	3	3	
	1	5	4	0	4	4	5
A	1	2	3	4	5	6	K

Move 4: A3

K	6	5	4	3	2	1	B
1	0	3	3	0	3	3	
	1	5	0	1	5	5	6
A	1	2	3	4	5	6	K

Move 5: A2

K	6	5	4	3	2	1	B
1	0	3	3	0	3	3	
	1	0	1	2	6	6	7
A	1	2	3	4	5	6	K

Move 6: A1

PLAYER A CAPTURES 3 FROM B5

K	6	5	4	3	2	1	B
1	0	0	3	0	3	3	
	0	0	1	2	6	6	11
A	1	2	3	4	5	6	K

Move 7: B4

K	6	5	4	3	2	1	B
2	1	1	0	0	3	3	
	0	0	1	2	6	6	11
A	1	2	3	4	5	6	K

Move 8: B6

K	6	5	4	3	2	1	B
3	0	1	0	0	3	3	
	0	0	1	2	6	6	11
A	1	2	3	4	5	6	K

Move 9: B1

PLAYER B CAPTURES 1 FROM A3

K	6	5	4	3	2	1	B
5	0	1	0	1	4	0	
	0	0	0	2	6	6	11
A	1	2	3	4	5	6	K

Move 10: A4

K	6	5	4	3	2	1	B
5	0	1	0	1	4	0	
	0	0	0	0	7	7	11
A	1	2	3	4	5	6	K

Move 11: B3

K	6	5	4	3	2	1	B
5	0	1	1	0	4	0	
	0	0	0	0	7	7	11
A	1	2	3	4	5	6	K

Move 12: A6

K	6	5	4	3	2	1	B
5	1	2	2	1	5	1	
	0	0	0	0	7	0	12
A	1	2	3	4	5	6	K

Move 13: B2

K	6	5	4	3	2	1	B
6	2	3	3	2	0	1	
	0	0	0	0	7	0	12
A	1	2	3	4	5	6	K

Move 14: B1

PLAYER B CAPTURES 7 FROM A5

K	6	5	4	3	2	1	B
14	2	3	3	2	0	0	
	0	0	0	0	0	0	12
A	1	2	3	4	5	6	K

PLAYER B WINS. A 12 B 24

References

American National Standards Institute, "Correspondences of 8-bit Codes for Computer Environments—A Tutorial," *Communications of the Association for Computing Machinery*. 11: 783–789 (Nov 1968).

Cress, P., P. Dirksen, and J. W. Graham, *Fortran IV with Watfor and Watfiv*. Englewood Cliffs, N.J.: Prentice-Hall, 1970.

Haggerty, J. B., "Kalah—An Ancient Game of Mathematical Skill," *The Arithmetic Teacher* (May 1964).

Kemeny, J. G., and T. E. Kurtz, *Basic Programming*, second ed. New York: Wiley, 1971.

Knuth, D. E., *The Art of Computer Programming*: Vol. 1, *Fundamental Algorithms*. Reading, Mass.: Addison-Wesley, 1968.

Knuth, D. E., *The Art of Computer Programming*: Vol. 3, *Sorting and Searching*. Reading, Mass.: Addison-Wesley, 1973.

Martin, W. A., "Sorting." *Computing Surveys* 3: 147–174 (Dec 1971).

"Pits and Pebbles." *Time* (14 June 1963).

Price, C. E., "Table Lookup Techniques." *Computing Surveys* 3: 49–66 (Jun 1971).

Webster's New Collegiate Dictionary, seventh ed. Springfield Mass.: E. C. Merriam Co., 1958.

6 Compound Data Structures

6.1 How Can Tables be Represented?

A surveyor makes a sequence of measurements of elevation (height above sea level) along a proposed highway route. He makes these measurements at unequally spaced intervals, farther apart when the terrain is smooth and closer together near hills or canyons (Figure 6.1a). The measurements will be stored in a computer, using two lists with *corresponding* subscripts. For each measured point we store the horizontal distance (measured along the route) in the first list, Dist, using a subscript value, J, that increases by one at each measured point. In the other list, Elev, using the *same* subscript value, J, we store the corresponding elevation measurement value.

Table 6.1a *Example of Survey Profile*

Station No.	*Distance (feet)*	*Elevation (feet)*	*Station No.*	*Distance (feet)*	*Elevation (feet)*
1	0	571	13	5,903	623
2	617	631	14	6,236	687
3	1,685	692	15	7,092	650
4	2,138	763	16	7,313	617
5	2,819	928	17	7,794	537
6	2,909	855	18	8,237	434
7	3,098	981	19	8,507	391
8	3,275	1,003	20	9,241	375
9	4,427	924	21	9,756	416
10	4,796	815	22	10,217	451
11	4,851	699	23	10,499	507
12	5,267	594	24	10,830	568

Two or more lists related in this way form a *table* (as in Table 6.1a). If we use the same subscript value, we can find a set of related facts by looking in the related lists.

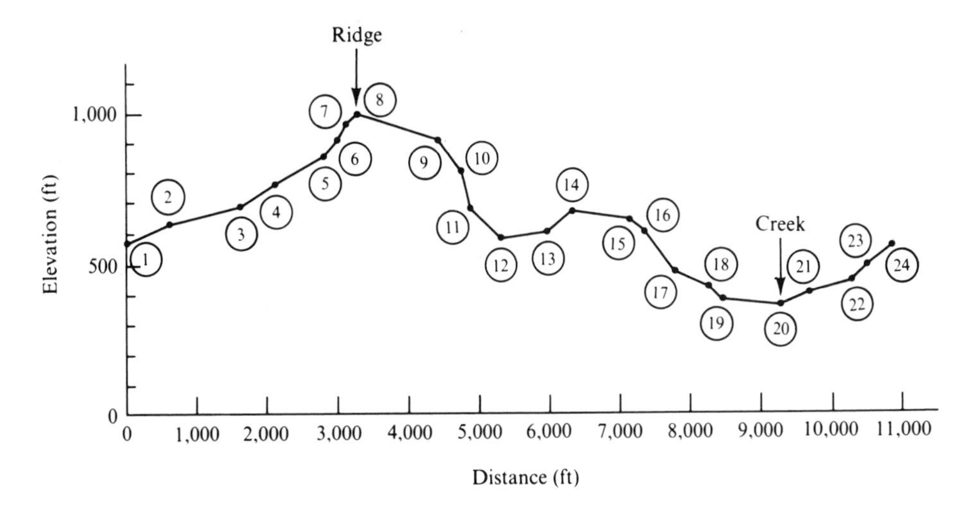

6.1a. *Example of a survey profile.*

Another table was used in the payroll example of Section 5.4. The subscripted variable BaseRate[EmpNo] was used to determine the basic pay rate of an employee, and then his gross pay was stored at GrossPay[EmpNo], keeping the same value for the subscript.

In this payroll example, we known in advance what subscript value to use to obtain a specified employee's basic pay rate or to store his gross pay. Much more commonly, as in the survey profile example, the entries in one list *identify* the entries in one or more other lists. This list may be searched to determine a *subscript value*, specifying the consecutive number of a desired item within the identifying list. This subscript becomes the value of a *pointer* variable, which in turn can be used as a subscript value for retrieving the related information from any of the other lists in the table.

In using the survey profile table, for example, we would most likely not know the station number in advance. Instead, we would search the list of horizontal distances to find the nearest entry to a desired value—say, a horizontal distance of 5,600 feet. As a result of this search, we find that the specified point lies between stations 12 and 13, and we use these subscript values to determine that the elevation at this point is about 600 feet.

Interpolation In practical applications, we often *interpolate* to find a value between two table entries. If we assume that the elevation varies *linearly* between measured stations (the way tables are often set up), we can find the elevation corresponding to a distance of 5,600 feet from the following formula (see also Figure 6.1b):

$$E = 594 + \frac{(5600 - 5276)}{(5903 - 5276)} \times (623 - 594)$$

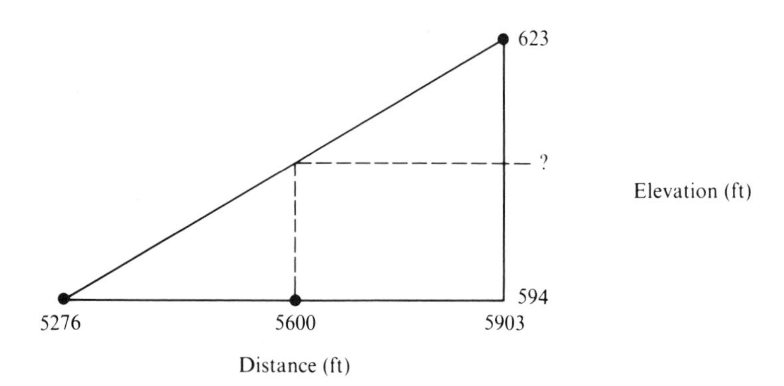

6.1b. Example of linear interpolation.

In general, using the tabulated values D1, D2, E1, and E2, we find the elevation E corresponding to the nontabulated distance D as follows:

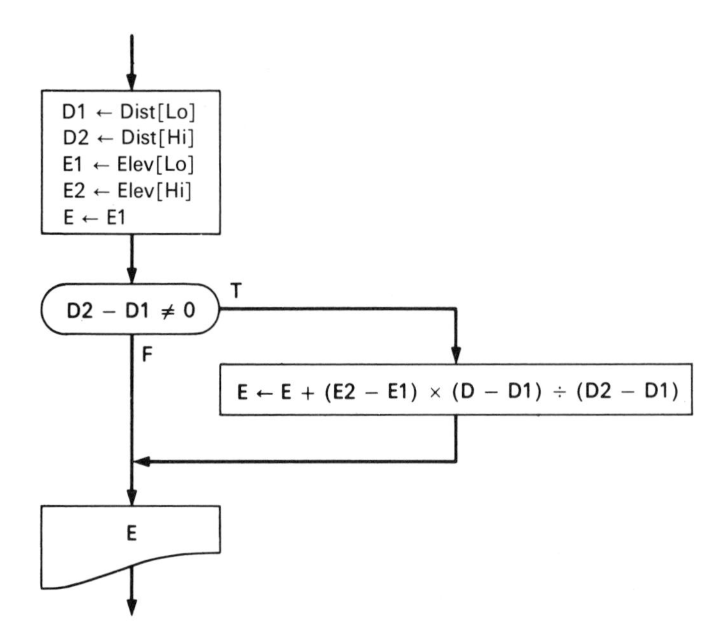

Here we assume that the values of Lo and Hi have been determined by the binary search procedure described in Section 5.4.

In the early days of computers, the use of tables for mathematical functions (such as exponential and trigonometric functions) was recommended. As the economics of computing have developed, the trend has changed from extensive tabular work to polynomial (or other) approximations for the standard mathematical functions. Tables are now used mainly to express relations not conveniently handled by analytical means—the outstanding examples being drawn from empirical or measured relationships, as in the previous examples.

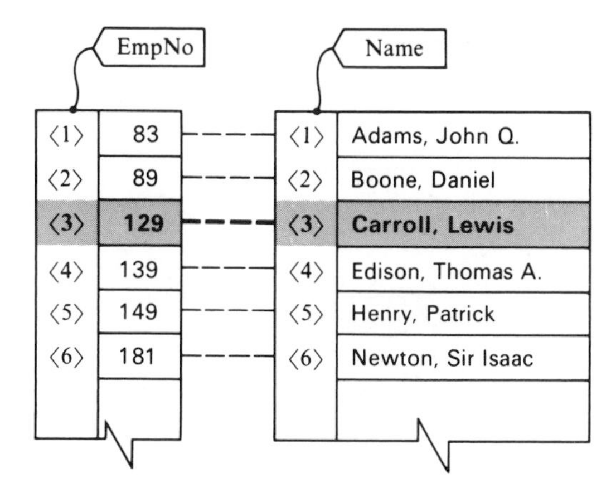

6.1c. *A* table *of correlated lists. Lewis Carroll's employee number is 129. His employee number is stored in the cell named* EmpNo [3] *and his name is the string referred to as* Name [3, *].

A review of the applications of lists indicates that most problems involve a group of correlated lists that form a table (often with one list carrying identifying information for the others). (See Figure 6.1c.) In a payroll computation, for example, there may be several lists, for storing various items of data concerning each employee —number of dependents, gross earnings for year to date, tax withheld for year to date, special payroll deduction codes, and the like. As an employee's payroll is processed, the same subscript value can be used with all these lists to fetch and store various items of information relating to the same employee.

Lists of Strings One column in a table may consist of *names* or other alphabetical information. This information can be stored in a list of strings. Such a list is declared as follows:

---- string *name*[*length, limit*] *name*

The first name is the name of the list of strings. The length value inside the bracket tells how many strings are in the list, and the limit value is the *maximum* length of each string in the list. The name at the end of the declaration refers to a corresponding list containing the *current* lengths of the strings. Thus the maximum length must be the same for all the strings in the list, but the current lengths can all be different so long as none exceeds the maximum.

To refer to a specified *character* in a string, we use *two* subscripts—the first designates one of the strings in the list, and the second designates the particular character in that string. For instance, if we have the following names:

'Jernegan, John D.'
'Martin, Edwin M.'
'Clark, Edward A.'

we can store them in a list of strings declared as follows:

```
string People[3, 20] LPeople
```

The lower-case letter a appears at the locations referred to as

People[1, 7] (in Jernegan)
People[2, 2] (in Martin)
People[3, 3] (in Clark)
People[3, 11] (in Edward)

To refer to an entire string from a list of strings, we use an asterisk in place of the second subscript. For instance,

People[2, *]

refers to the string

'Martin, Edwin M.'

The list of lengths is referred to by the name LPeople. Thus in this case, the value of LPeople[2] is the length of the string

'Martin, Edwin M.'

or 16. Note that this list of lengths is not used explicitly except in problems involving the individual characters of the strings.

The sorting program of Section 5.3 can be used to arrange a list of strings in order, provided that the string comparison operations give the proper ranking. In particular, both ASCII and EBCDIC codes will rank correctly, if nothing but capital letters, simple punctuation, and blanks are included in the string. For an application of this simple type, such as a telephone book *without* lower-case letters, we rewrite the sorting program as follows:

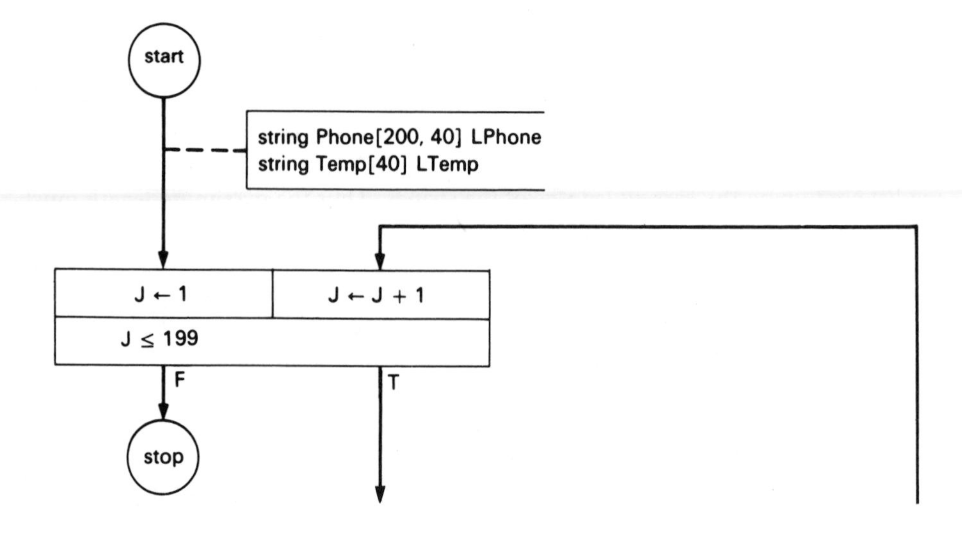

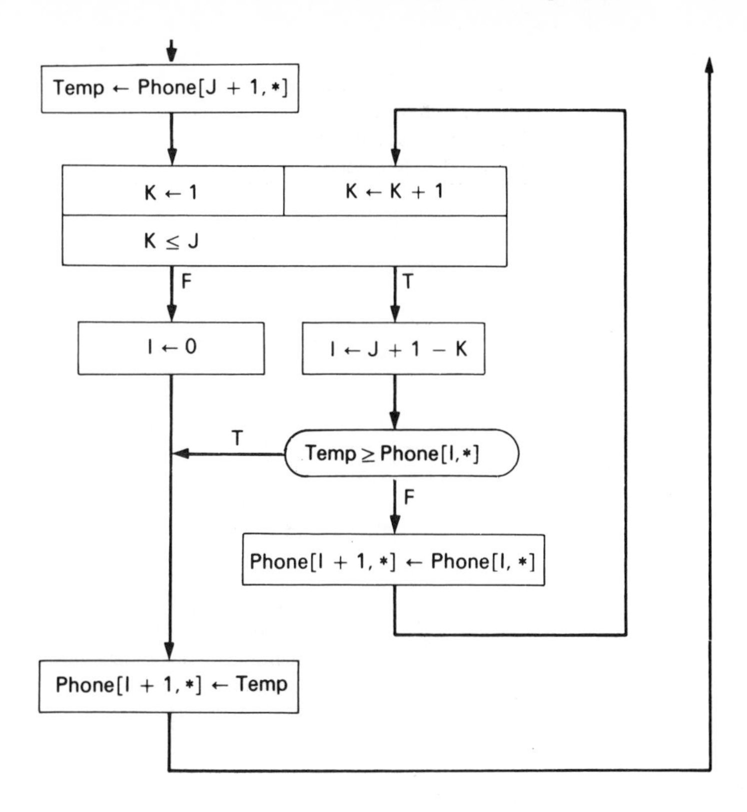

Indirect Sorting We may wish to arrange the entries of a table in order. The table consists of a group of correlated lists, some of which may be lists of strings. The entries in one list are to be used as *keys* to identify the items in the corresponding positions in the other lists. The problem, then, is to arrange the table so that the keys will be in the proper order. For instance, suppose we have a table composed of three lists, EmpNo, BaseRate, and NDepend, the latter list containing the number of dependents for tax withholding purposes. If the list needs to be arranged in order, we can write the following program:

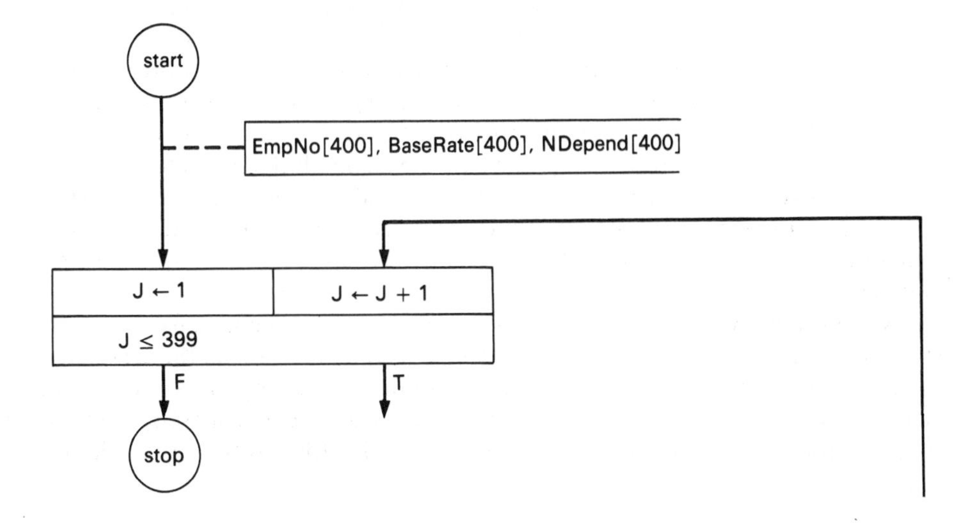

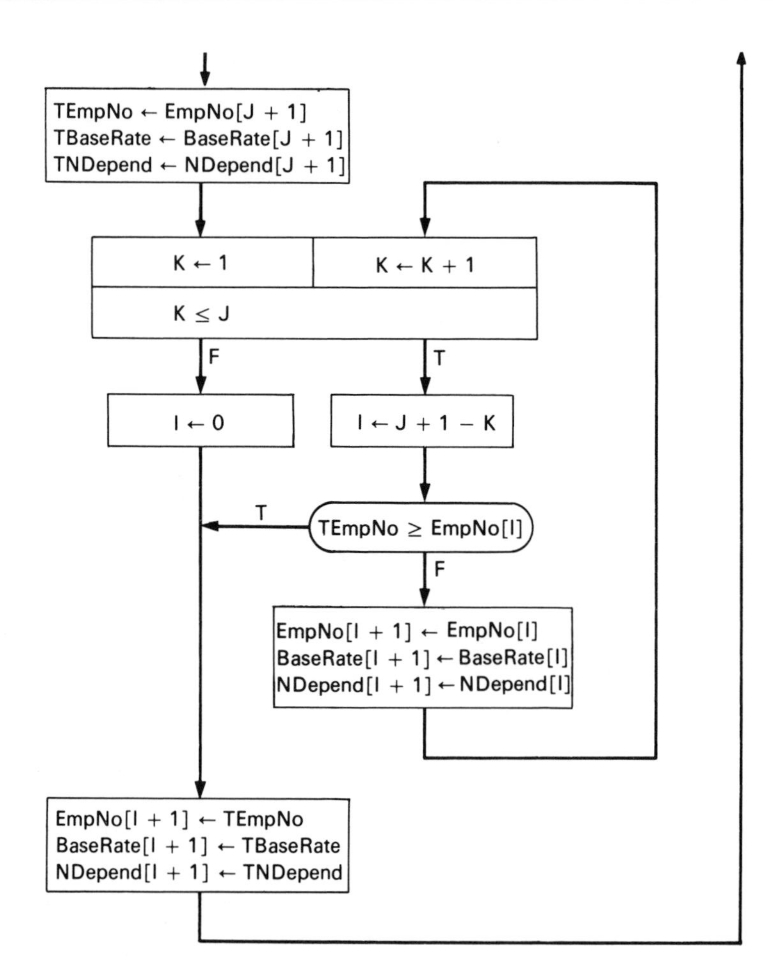

Notice that, in this example, all the *assignment* instructions involving the table entries must occur in groups of three, so that all three of the lists that compose the table will be arranged properly. On the other hand, the *comparison* instruction involves only one of the lists—that is, the list whose entries are being used as the sorting key.

If the table consists of a larger number of lists, say 20, a fair number of assignment instructions will be required to move the items on all the lists into the proper order. Also, a good deal of time will be required to do all this moving. Even for strings, where the multiple assignments of all the characters are handled "automatically" without a great increase in the number of instructions, the time required is still considerable.

We can simplify the sorting of a large table, then, by adjoining a list of *pointers*. This list, which we call Point, is the same length as the other lists in the table. Now, instead of moving the corresponding items of all the lists during the sort operation, we move only the pointers. The value of Point[I] can be used as a subscript to locate the list entries that would have been moved into the *I*th position in all the lists (Figures 6.1d, 6.1e).

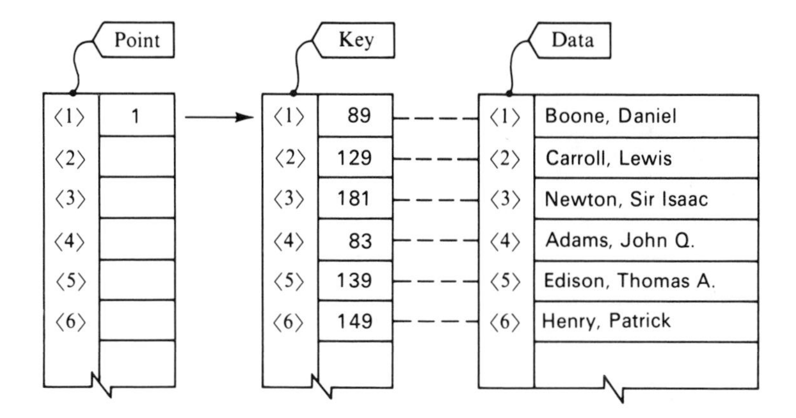

6.1d. *This figure shows the condition of the three lists* Point, Key, *and* Data *at the first stage of the indirect sort. All the data has been read in, and* Point [1] *has been provisionally set to* 1, *thus pointing to* Key[1].

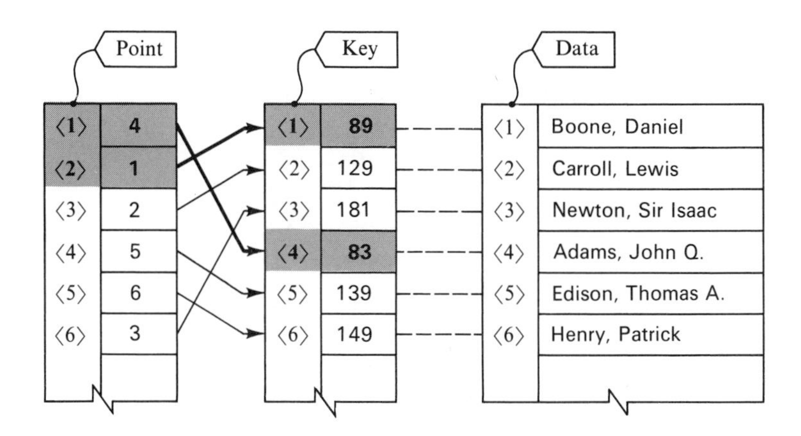

6.1e. *After the sixth stage of indirect sorting, the first six variables in the list* Point *have been given values that can be used as subscripts to variables in the list* Key. *Thus* Key[Point[1]] *is the smallest key,* Key[Point[2]] *is the second smallest, and so on.*

To make the comparison work, we must locate the items that would have been in positions I and J + 1 in the list of keys. This is accomplished with an *indirect reference*, using positions I and J + 1 in the pointer list.

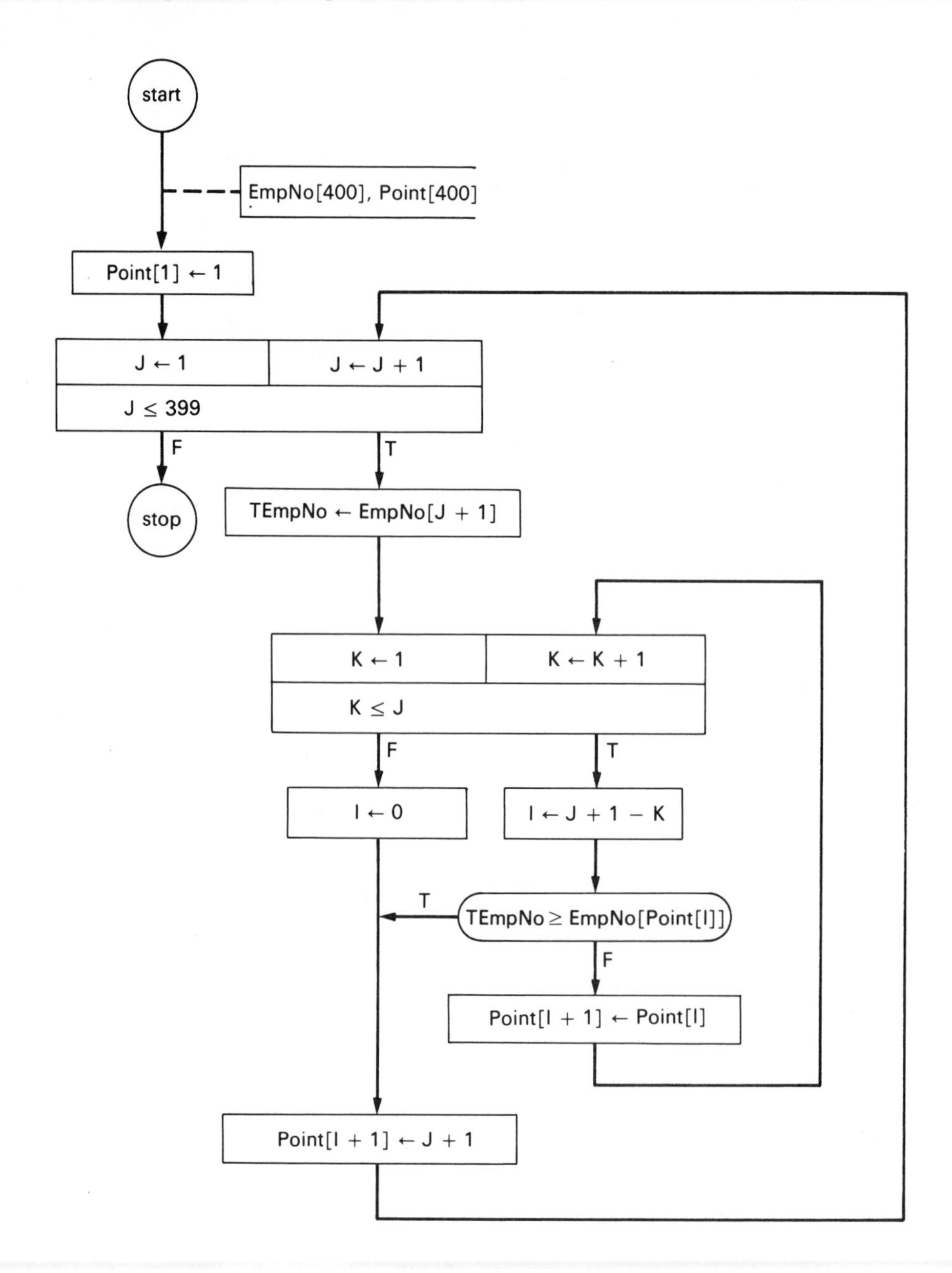

Notice especially that only the list of *keys* and the list of *pointers* are involved in the foregoing instructions. If we wish to continue by printing the values of EmpNo, BaseRate, and NDepend, in order of employee number, we can include the instructions

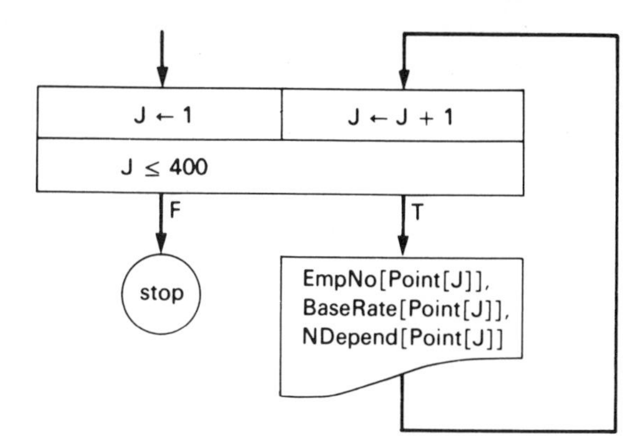

(Declarations must be included for the other lists as well.)

The same strategy can, of course, be used when the keys are in the form of strings. Thus, the data for a group of employees could be indirectly sorted into alphabetical order, for example.

It often happens that the same table must be ordered differently for different purposes. For example, a large company may keep a mailing list of all its employees, or an insurance company may have a file containing the names and addresses of all its policy holders. For most uses, these files must be arranged according to some numeric key, such as employee number or policy number. However, the company occasionally mails a general notice to all the people listed in the file, and because of postal regulations the file must be rearranged by Zip code. Again, it might be necessary for some purposes to ignore the numeric key and arrange the same file alphabetically, using the employee's name as a key.

Indirect sorting permits us to keep several lists of pointers simultaneously, each one arranged in a different order on the basis of a different key. The basic file of data may be arranged in some arbitrary order—perhaps chronologically, with new entries added at the end. The data itself is never rearranged during an indirect sort—only the pertinent keys are examined (by indirect reference) and the appropriate set of pointers is rearranged.

Flowchart

Indirect sorting:

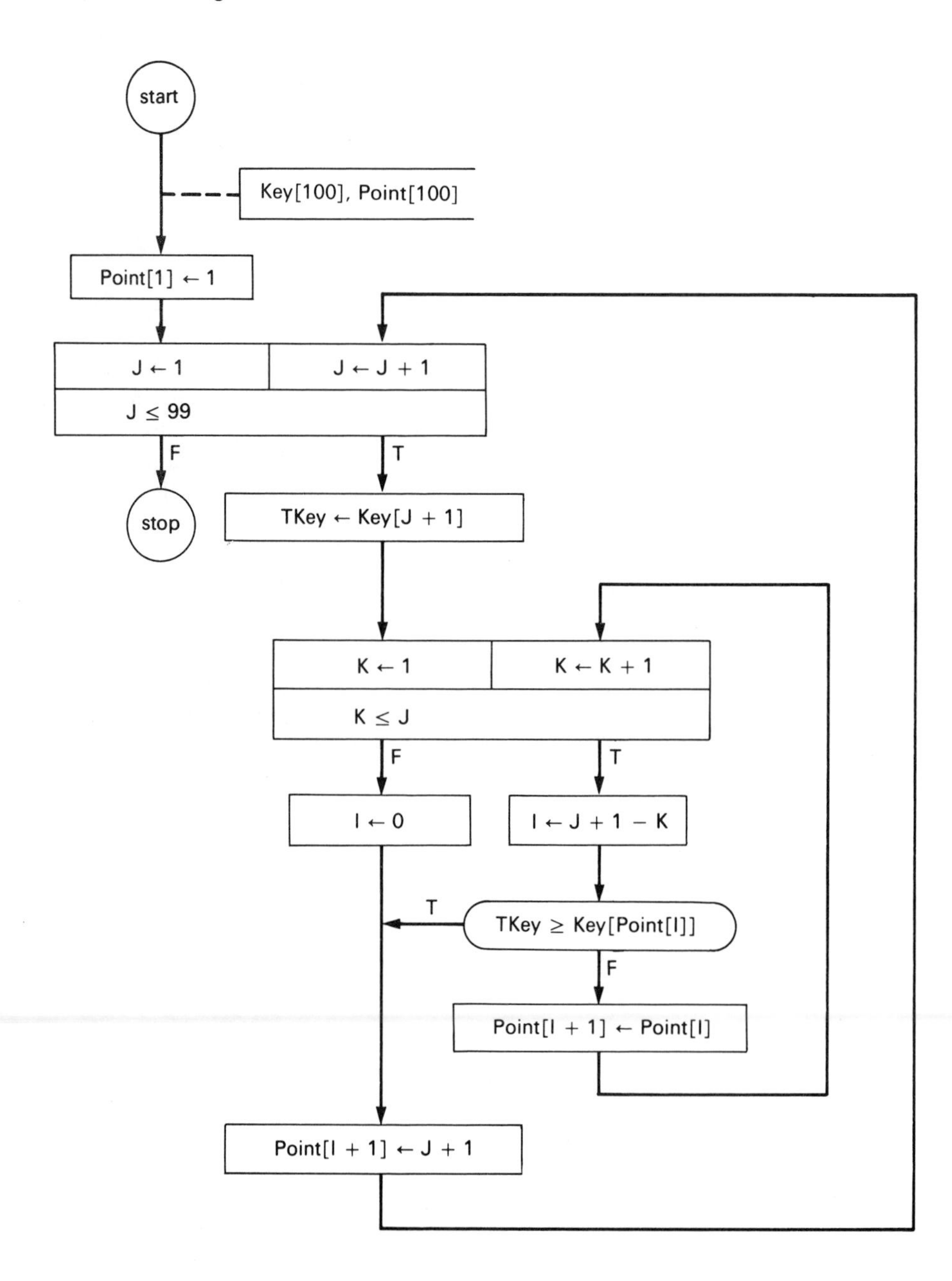

Basic

Indirect sorting:

```
100 DIM K[100], P[100]
105 LET P[1] = 1
110 FOR J = 1 TO 99
120    LET T = K[J + 1]
130    FOR I = J TO 1 STEP -1
140       IF T >= K[P[I]] THEN 180
150       LET P[I + 1] = P[I]
160       NEXT I
170    LET I = 0
180    LET P[I + 1] = J + 1
190    NEXT J
900 END
```

Fortran The subscripted variable POINT(J) cannot be used directly as a subscript; instead, the pointer value must be assigned to another variable.

Indirect sorting:

```
      INTEGER KEY(100), POINT(100), TKEY
      POINT(1) = 1
      DO 190 J = 1, 99
         TKEY = KEY(J + 1)
         DO 160 K = 1, J
            I = J + 1 - K
            MI = POINT(I)
            IF (TKEY GE. KEY(MI)) GO TO 180
            POINT(I + 1) = POINT(I)
160         CONTINUE
         I = 0
180      POINT(I + 1) = J + 1
190      CONTINUE
      STOP
```

PL/1

Indirect sorting:

```
RANK: procedure options (main);
  declare (KEY(1:100), TKEY) float;
  declare (POINT(1:100), I, J) fixed;
  POINT(1) = 1;
  do J = 1 to 99;
    TKEY = KEY(J + 1);
    do I = J to 1 by -1;
      if TKEY >= KEY(POINT(I)) then
        go to CONT;
    POINT(I + 1) = POINT(I);
    end;
  I = 0;
CONT: POINT(I + 1) = J + 1;
  end;
end RANK;
```

Words to Remember

table	interpolate	indirect sorting
key		

Exercises

1. Write a program to combine the *binary search* algorithm of Section 5.4 with the interpolation formula in this section. Run the program, using Table 6.1a and reading the nontabulated distance D. If the value of D is outside the table, print an appropriate message; otherwise, print the corresponding value of E. Values of D are: (a) 3,000; (b) 10,000; (c) 7,000; (d) 0; (e) 10,830; (f) 10,831; (g) 6,236.
2. As a part of a computer-based dialog, the student's response is compared with all the strings on a *list* of expected answers. Write a program that will read the student's response and then perform a sequence of *pattern matching* operations (see Section 5.2), using the student's string as the long string and each of the short strings in a stored list as the pattern. Set up a numeric list corresponding to the list of strings, and store in each position of this list, as the pattern matches are performed, the *location* within the student's response at which the match occurred (or zero if no match occurred). The dialog program might count the nonzero match locations to evaluate the student's achievement. The student's response is 'I do not know but I think it was either Abraham Lincoln or George Washington.' Patterns are 'Ab', 'Lin', 'do not know', 'Wash', 'Honest', 'Emancip', 'Johnson', 'Grant', 'log ca', 'rail'.
3. Write a program using *indirect* sorting to arrange the survey profile data of Table 6.1a in order by elevation. Use the pointers so obtained to print the data in sorted sequence.
4. Write a program to construct a symbol table, as required by a compiler. Read a numeric "next address"; then read a sequence of strings representing the variables and constants as they are encountered by the compiler during its processing of the source program. Create a table, which will include for each symbol the *string* that represents it, the *address* assigned by the compiler as the corresponding machine-oriented name, and the *initial value* to be stored in the cell before execution of the compiled instructions. As each string is read, compare it with all the strings (if any) already in the table. If the string is not already there, enter it in the table. Also allocate the "next address" to it and increase the value of the "next address" by one. If the string represents a constant, store the string as the corresponding initial value; if it is a variable, store the string 'undefined'. Print the symbol, the address, and the string defining the initial value. The next address is 4730. Symbols are 'PI', '3.141592', 'CI', 'DI', 'PI', 'CI'.

6.2 Rectangular Arrays

In many problems, a group of items may be thought of as a rectangular array. Most tables, in fact, have this form. However, we distinguish the more general case of the table, consisting of a group of lists each of which may contain a different kind of information, from the special case where all the items in the rectangular array are of the same nature. Instead of using separate list names with single subscripts, we use one name for the entire homogeneous array, with *two* subscripts to identify each individual item. The subscripts are the row number and the column number.

For instance, several different salesmen work in a shoe store, and each salesman sells various categories of shoes:

Salesman's number	(1) *Men's shoes*	(2) *Ladies' shoes*	(3) *Children's shoes*	(4) *Sports shoes*
(1)	14	6	4	8
(2)	5	3	20	4
(3)	10	9	6	12
(4)	4	2	8	5
(5)	7	9	6	4

These numbers might be stored in a rectangular array named Shoes, and referenced by using the array name and a pair of subscripts. The first subscript identifies the salesman; the second identifies the shoe category. For example,

```
Shoes[2, 3]
```

is the name of the cell containing the value 20, telling how many shoes of category 3 were sold by salesman number 2. The array appears in a declaration:

```
-----| Shoes[5, 4]
```

in which the size of the array is specified by means of the maximum size of each subscript. The first subscript in this case must not exceed 5, and the second must not exceed 4.

An Application: Row and Column Totals A typical problem, once the entries have been stored in an array, is to determine the total of all the amounts in each row and in each column, as well as the grand total for the entire array. This problem requires, besides the array, two lists for the row totals and the column totals.

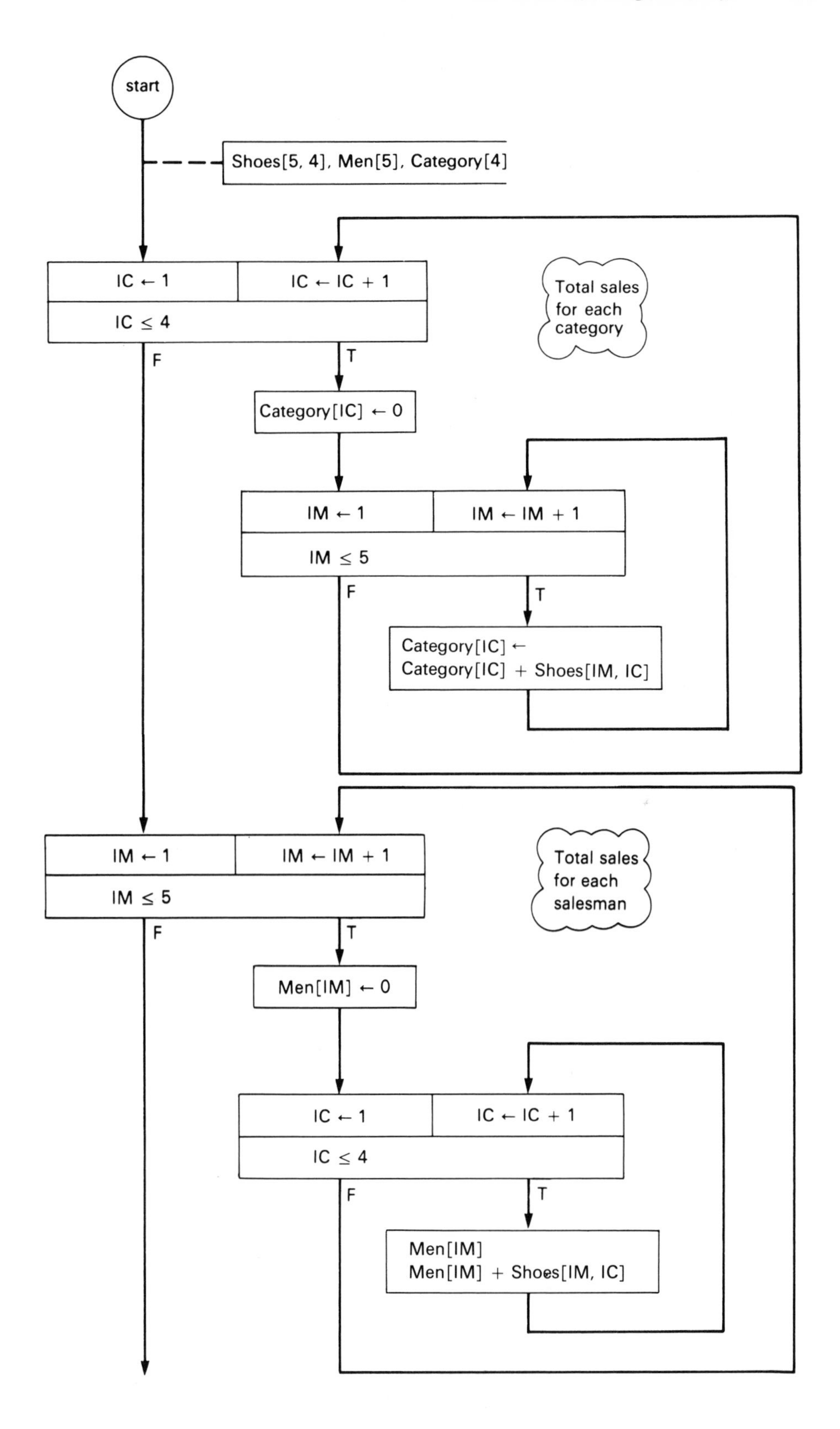
start
Shoes[5, 4], Men[5], Category[4]
IC ← 1
IC ← IC + 1
IC ≤ 4
F
T
Total sales for each category
Category[IC] ← 0
IM ← 1
IM ← IM + 1
IM ≤ 5
F
T
Category[IC] ←
Category[IC] + Shoes[IM, IC]
IM ← 1
IM ← IM + 1
IM ≤ 5
F
T
Total sales for each salesman
Men[IM] ← 0
IC ← 1
IC ← IC + 1
IC ≤ 4
F
T
Men[IM]
Men[IM] + Shoes[IM, IC]

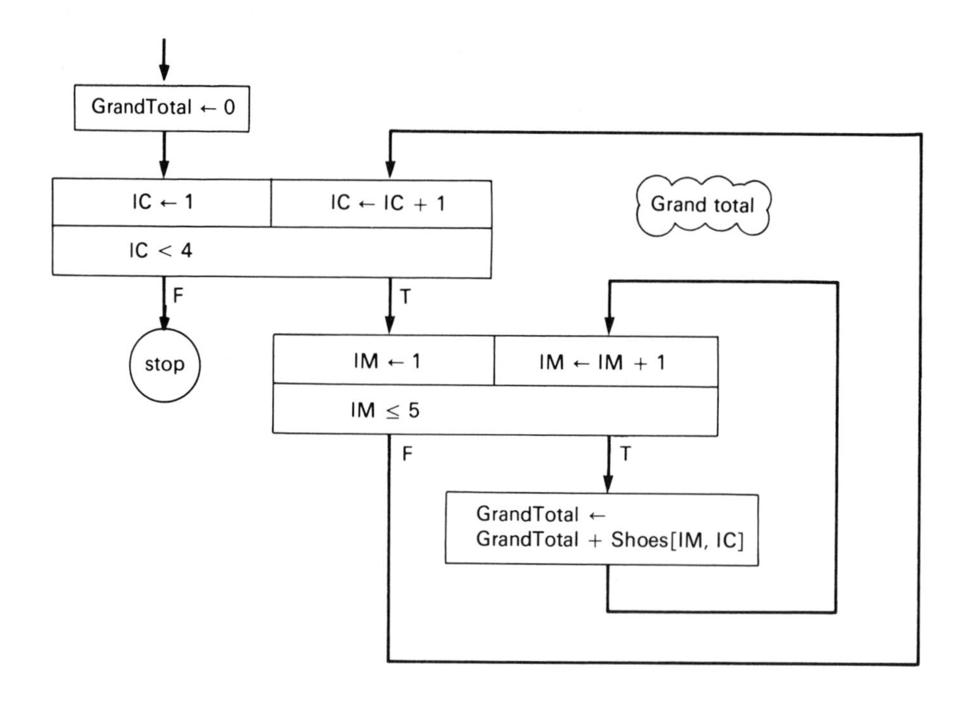

Mathematical Applications: Matrix Operations Adding a pair of rectangular arrays or multiplying all the elements of an array by a constant are easily programmed in a manner similar to the foregoing example, using a pair of nested loops. Such operations occur frequently in mathematical problems. In this context, a rectangular array is usually called a *matrix*.

Another operation that is commonly required in matrix applications is formation of the *scalar product* of a row of one matrix by a column of another. This operation makes sense mathematically only if the row length of the first matrix is equal to the column length of the second.

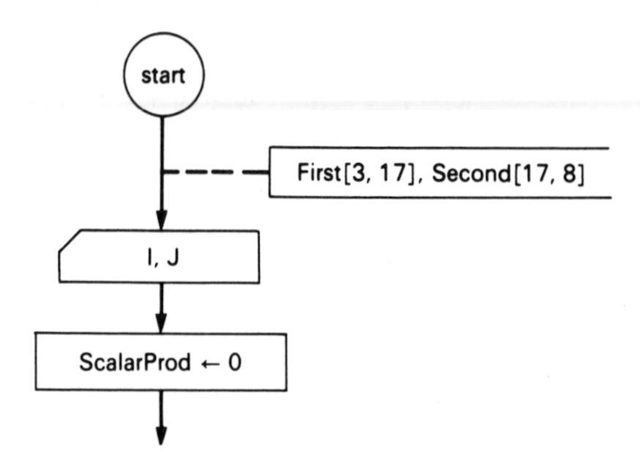

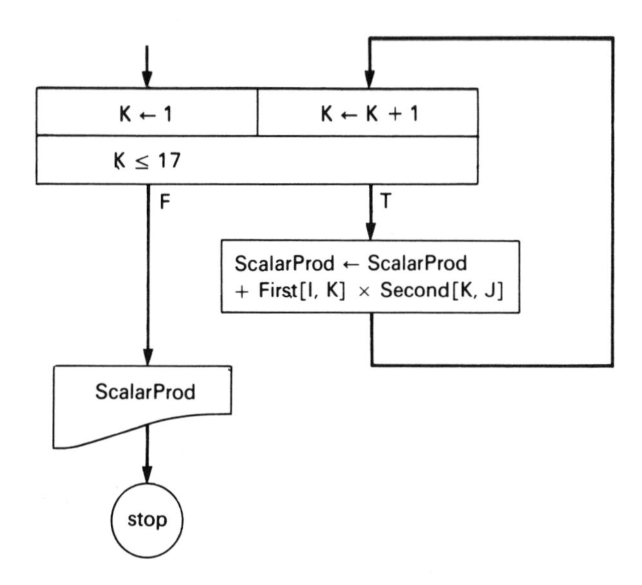

Matrix Representation of Chess A rectangular array can be used to represent a chess board. The legal moves of each kind of chessman may then be described as changes of the subscripts. For example, one possible move for a knight (from the middle of the board) would increase the first subscript by two and decrease the second subscript by one. Horizontal and vertical moves of the rook cause one subscript to change while the other remains unchanged. A diagonal move of the bishop changes both subscripts by the same absolute amount; however, both subscripts may increase or decrease together or one may increase while the other decreases.

Matrix Representation of a Graph A *graph* is a set of *nodes* with connections between some of them. Traffic patterns, electrical networks, and similar systems may be represented by a graph. In a *directed* graph, some of the connections are "one way"—making it possible to go directly from node *A* to node *B* but not from *B* to *A*. We can use a square matrix to store the connection matrix of a directed graph. There will be one row and one column for each node in the graph. The entry in row I column J will have a value indicating the number of paths from node I to node J (Figure 6.2a).

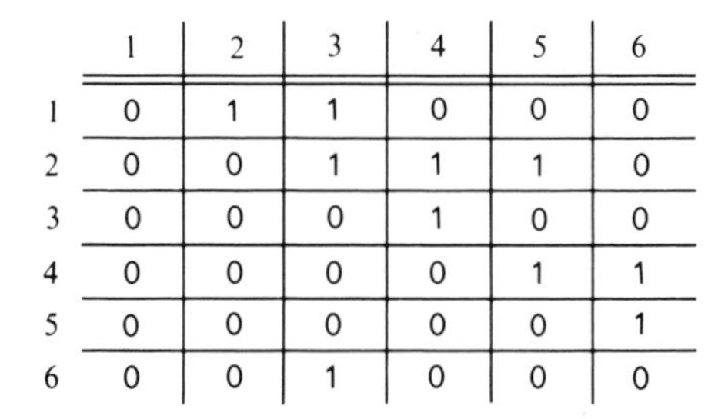

	1	2	3	4	5	6
1	0	1	1	0	0	0
2	0	0	1	1	1	0
3	0	0	0	1	0	0
4	0	0	0	0	1	1
5	0	0	0	0	0	1
6	0	0	1	0	0	0

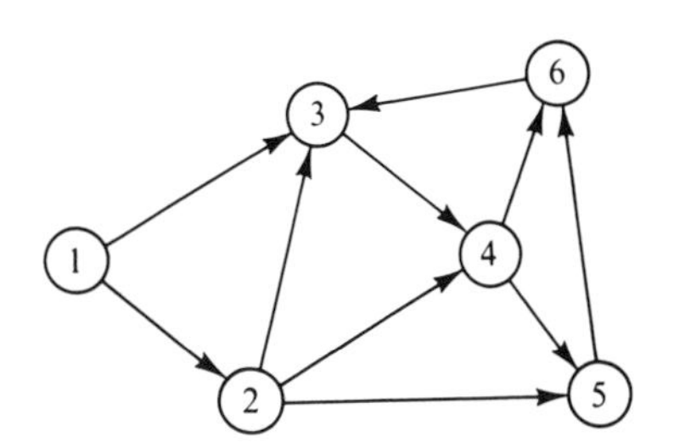

6.2a. *A graph and its connection matrix. The entry in row I and column J of the matrix indicates the number of paths from node I to node J of the graph.*

Flowchart

Row and column totals:

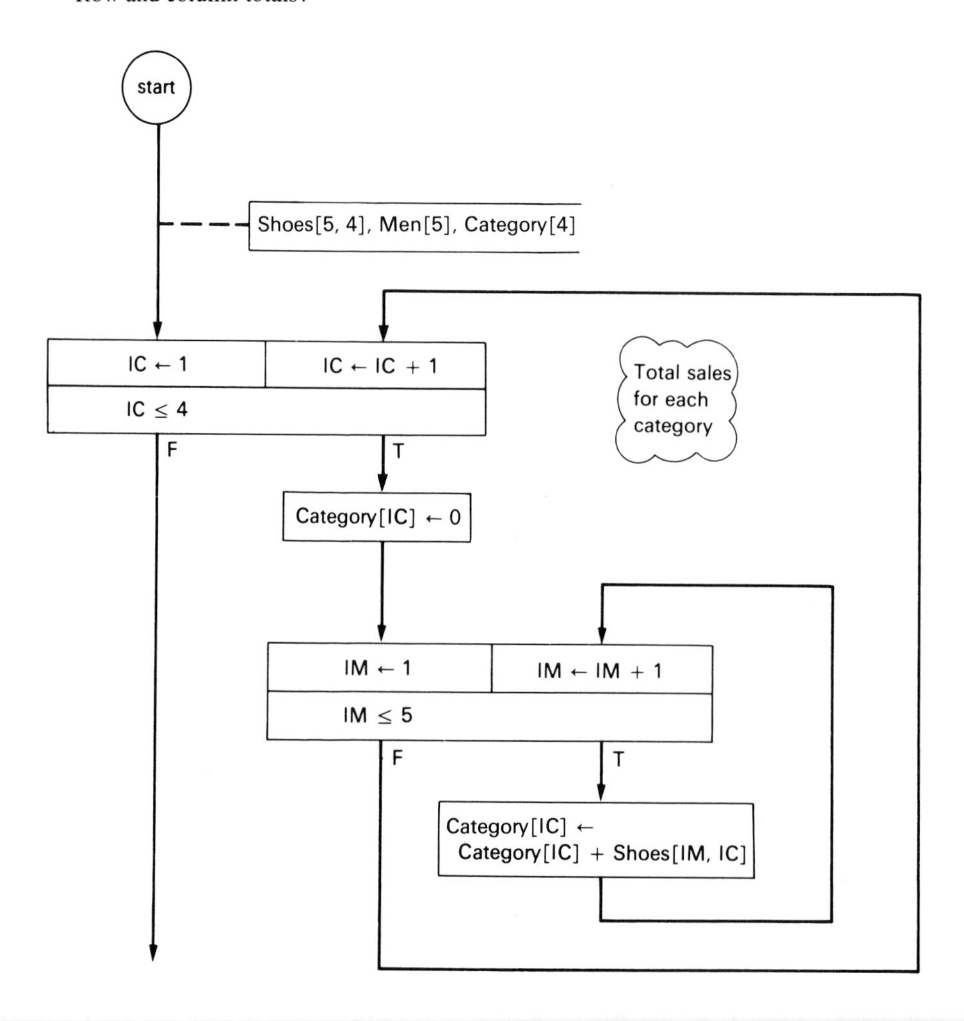

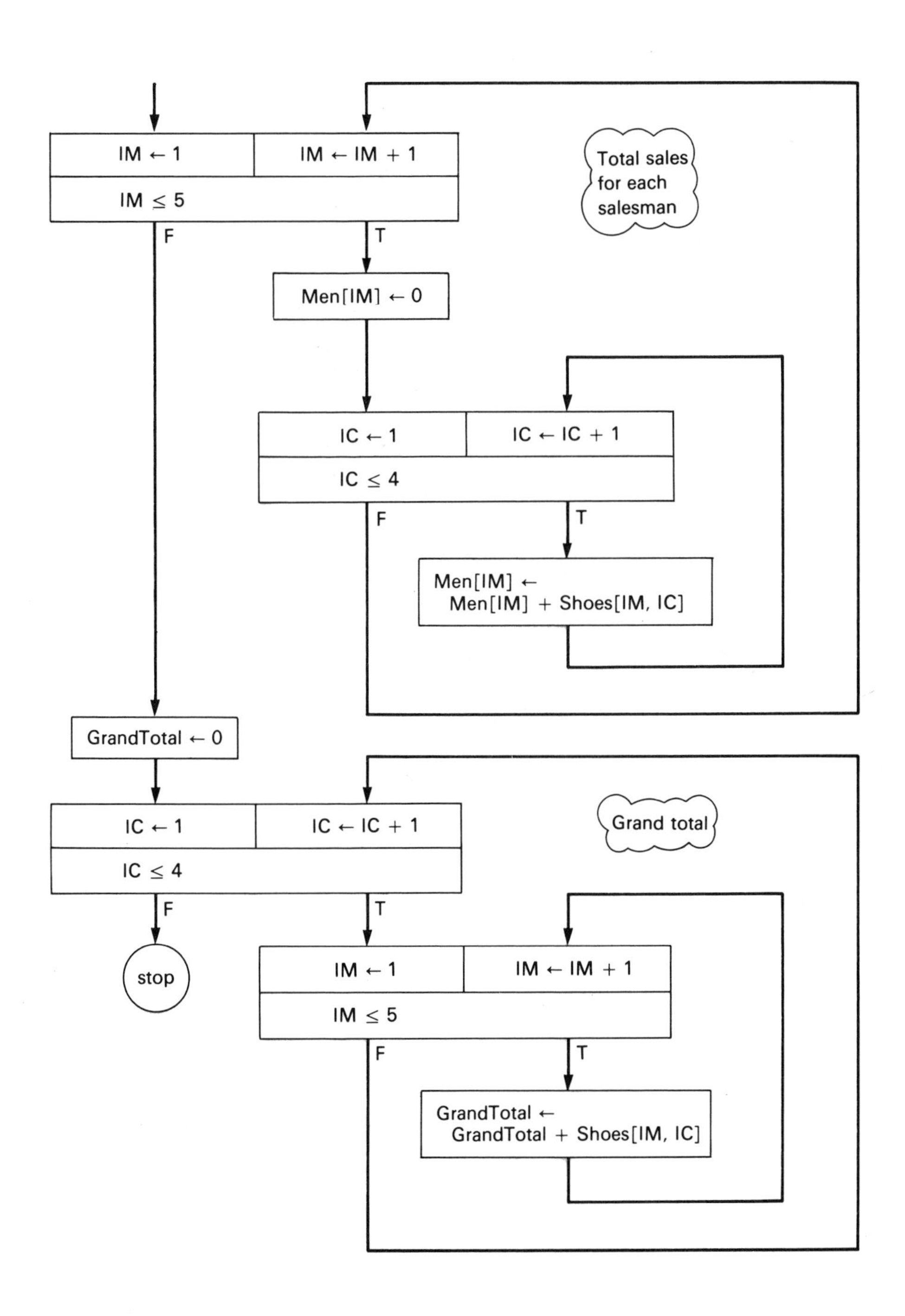
IM ← 1
IM ← IM + 1
IM ≤ 5
F
T
Total sales for each salesman
Men[IM] ← 0
IC ← 1
IC ← IC + 1
IC ≤ 4
F
T
Men[IM] ←
Men[IM] + Shoes[IM, IC]
GrandTotal ← 0
IC ← 1
IC ← IC + 1
IC ≤ 4
F
T
Grand total
stop
IM ← 1
IM ← IM + 1
IM ≤ 5
F
T
GrandTotal ←
GrandTotal + Shoes[IM, IC]

Scalar product:

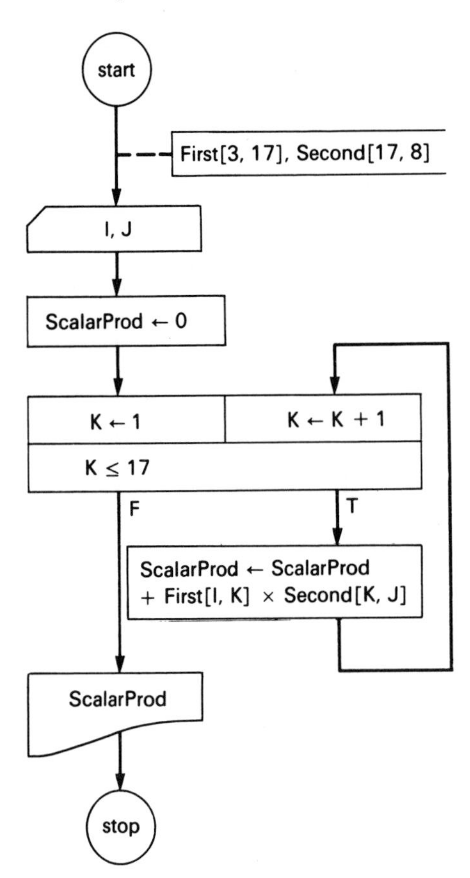

Basic

Row and column totals:

```
100 DIM S[5, 4], M[5], C[4]
200 FOR C1 = 1 TO 4
205    REM: TOTAL EACH CATEGORY
210    LET C[C1] = 0
220    FOR M1 = 1 TO 5
230       LET C[C1] = C[C1] + S[M1, C1]
240       NEXT M1
250    NEXT C1
300 FOR M1 = 1 TO 5
305    REM: TOTAL EACH SALESMAN
310    LET M[M1] = 0
320    FOR C1 = 1 TO 4
330       LET M[M1] = M[M1] + S[M1, C1]
340       NEXT C1
350    NEXT M1
399 REM: GRAND TOTAL
400 LET G = 0
410 FOR C1 = 1 TO 4
420    FOR M1 = 1 TO 5
430       LET G = G + S[M1, C1]
440       NEXT M1
450    NEXT C1
```

Scalar product:

```
100 DIM F[3, 17], S[17, 8]
110 INPUT I, J
120 LET P = 0
130 FOR K = 1 TO 17
140    LET P = P + F[I, K] × S[K, J]
150    NEXT K
160 PRINT P
```

Fortran

Row and column totals:

```
      DIMENSION SHOES(5, 4)
      DIMENSION SMEN(5), CATEG(4)
      DO 250 IC = 1, 4
C        TOTAL FOR EACH CATEGORY
         CATEG(IC) = 0.0
         DO 240 IM = 1, 5
            CATEG(IC) = CATEG(IC) +
     1         SHOES(IM, IC)
240      CONTINUE
250   CONTINUE
      DO 350 IM = 1, 5
C        TOTAL FOR EACH SALESMAN
         SMEN(IM) = 0.0
         DO 340 IC = 1, 4
            SMEN(IM) = SMEN(IM) +
     1         SHOES(IM, IC)
340      CONTINUE
350   CONTINUE
C     GRAND TOTAL.
      G = 0.0
      DO 450 IC = 1, 4
         DO 440 IM = 1, 5
            G = G + SHOES(IM, IC)
440      CONTINUE
450   CONTINUE
```

Scalar product:

```
         DIMENSION FIRST(3, 17),
         DIMENSION SECOND(17, 8)
      READ(5, 81) I, J
81    FORMAT (8 I 10)
      SCPR = 0.0
      DO 150 K = 1, 17
         SCPR = SCPR + FIRST(I, K) *
     1      SECOND(K, J)
150   CONTINUE
      WRITE (6, 83) SCPR
83    FORMAT (8 F 15.3)
```

PL/1

Row and column totals:

```
declare (SHOES(1:5, 1:4), MEN(1:5),
   CATEGORY(1:4), GRANDTOTAL)
   float;
declare (IC, IM) fixed;
do IC = 1 to 4;
   /* Total sales for each category */
   CATEGORY(IC) = 0;
   do IM = 1 to 5;
      CATEGORY(IC) = CATEGORY(IC)
         + SHOES (IM, IC);
      end;
   end;
do IM = 1 to 5;
   /* Total sales for each salesman */
   MEN(IM) = 0;
   do IC = 1 to 4;
      MEN(IM) = MEN(IM) +
         SHOES(IM, IC);
      end;
   end;
/* Grand total */
GRANDTOTAL = 0;
do IC = 1 to 4;
   do IM = 1 to 5;
      GRANDTOTAL = GRANDTOTAL
         + SHOES(IM, IC);
      end;
   end;
```

Scalar product:

```
declare (FIRST(1:3, 1:17),
   SECOND(1:17, 1:8),
   SCALARPROD) float;
declare (I, J, K) fixed;
get list I, J;
SCALARPROD = 0;
do K = 1 to 17;
   SCALARPROD = SCALARPROD +
      FIRST(I, K) × SECOND(K, J);
   end;
put skip list (SCALARPROD);
```

Words to Remember

rectangular array	row	column
matrix	graph	

Exercises

1. Write a program to read the row and column numbers of two rooks on a chess-board, and determine whether either one can capture the other.
   a. (2, 5) and (5, 2)
   b. (4, 7) and (4, 2)
   c. (3, 3) and (5, 1)
2. Determine which of the following can be captured by any of the others:
   a. a queen (2, 5) and a knight (5, 2);
   b. a rook (4, 7) and a bishop (4, 2);
   c. three queens (1, 1), (8, 8), (5, 7);
   d. eight queens (1, 3), (2, 6), (3, 4), (4, 1), (5, 8), (6, 5), (7, 7), (8, 2).
3. Write a program to compute *all* the elements of a 3 by 8 matrix C, which is the product of the 3 by 17 matrix A and the 17 by 8 matrix B. To find the element in row I column J of C, form the scalar product of row I of the matrix A by row J of the matrix B.
4. Write a program to read the connection matrix of a graph, and then read two node numbers and construct a path leading from the first to the second, if this is possible. If there is no such path, the program should determine this fact and print an appropriate comment. (This exercise may prove rather difficult.) Using Figure 6.2a, try to construct a path from (a) 3 to 6; (b) 6 to 5; (c) 5 to 3; (d) 4 to 1.

6.3 *How Can Space Be Allocated and Released During Execution of a Program?*

With the increase in nonnumeric information handling tasks that became evident in the early 1960s, computer users realized the need for data structures more flexible than tables and arrays. The most evident limitation of those structures is the difficulty of deleting and especially of adding items in the middle. With these inflexible structures, the usual way to add an item is to reconstruct the entire structure, copying the portion that precedes the new item, inserting the new item, and then copying the remainder of the file.

The shift to more flexible data structures was concurrent with an important development in the external or auxiliary storage hardware that was used to store files too large to be held in the internal storage cells. Until about 1960, auxiliary files were almost invariably kept on magnetic tape. During the ensuing decade, however, "random access" storage devices, the most important being the magnetic disc, came into much more widespread use.

Files on magnetic tape are inherently *sequential*—the only way to reach a particular point on the file is to physically move the tape past all the items on the tape that precede the desired item. Thus, the usual procedure for modifying a magnetic tape file is to generate a revised copy of the entire file, processing all the entries sequentially. As each item on the "old" file is reached, it is copied to the "new" file or deleted from it, and new items are inserted when the proper position on the new file is reached. With a magnetic disc, on the other hand, each item can be accessed directly. A magnetic storage device normally contains a group of individual discs, arranged in somewhat the same manner as the records in a juke box, so that the active reading mechanism can be positioned at a particular disc without scanning past all the information on the other discs. Furthermore, each disc contains a number of separate tracks of recorded information, analogous to the bands on a phonograph record, and the reader can be positioned at the beginning of any one of these tracks. Thus an item can be inserted or replaced in the file without affecting the remaining items.

The availability of this newer equipment, on which files can be stored in a more flexible manner, gave impetus to the development of flexible data structure concepts. It was then soon realized that these new concepts could be applied advantageously to internal storage, as well as to external random access storage.

A later development is the concept of "virtual memory" (Denning, 1970, 1971), in which the programmer writes instructions as though the amount of storage is unlimited. The computer's operating system assigns internal storage as long as it lasts and then allocates blocks of storage on an external auxiliary medium on demand. Sophisticated algorithms are provided in modern operating systems to keep the allocation of storage as efficient as possible, so that the data items used most often occupy the available internal storage cells, while the less accessible storage is used for items that are used less frequently. All of this requires no extra effort by the programmer, who writes instructions in a human-oriented language that is concerned with the inherent relationships among the items of data while ignoring the actual location of the data in internal or external storage.

Doubly Linked Lists Among the flexible, nonsequential ways of organizing data into structures are such arrangements as stacks, queues, trees, and linked lists (Knuth, 1968; Berztiss, 1970; Maurer, 1972, Chap. 4; Dodd, 1969; Sedelow, 1970). We shall not attempt to discuss all of these or explain how each is used—such a discussion is beyond the subject matter of this book. However, we shall consider one of the most important examples, the organization of data in a doubly linked list.

In a doubly linked list, each item consists of a *datum* and two *pointers*, a forward pointer and a backward pointer. The datum contains the basic information of the item—it is comparable to one row of a table. For instance, it might contain several kinds of information about an employee, including his name and address, his basic pay rate, the number of his dependents, his total wages for the year to date. The structure of the datum will vary according to the application. It may include strings as well as numeric data. However, the structure of a datum remains fixed throughout any particular application. For purposes of discussion, we may think of the datum as consisting of some fixed number of cells.

Space for the datum and the pointers is allocated during execution from a *pool* of available cells (or possibly "virtual" cells, some of which may actually be located on an external auxiliary storage medium). The cells in this pool are numbered consecutively. The forward and backward pointers have integer values, giving the location of the next items in the list in the forward and backward directions.

As entries are added to and deleted from the list, space is allocated from and released to the pool of available cells. The allocation of space for a new item is not under control of the user; an allocation algorithm is made available as part of the list-processing system, and the user merely "requests" that a set of cells be allocated or released.

Example Suppose that we want to establish a doubly linked list, containing information about the employees of the Blank Manufacturing Company. We first give a declaration, describing the arrangement of the datum along with its pointers:

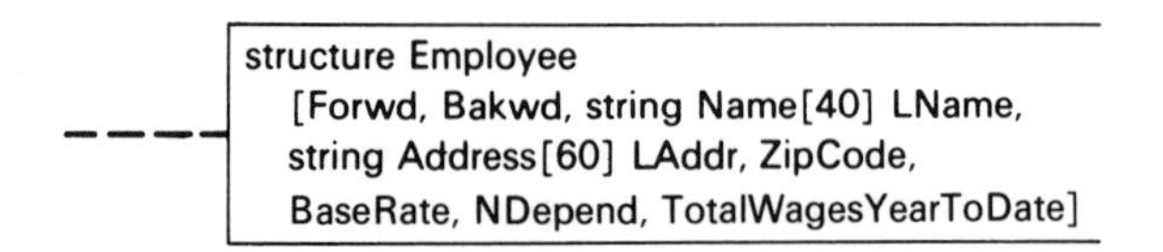

structure Employee
[Forwd, Bakwd, string Name[40] LName,
string Address[60] LAddr, ZipCode,
BaseRate, NDepend, TotalWagesYearToDate]

This declaration does *not* relate to a particular area of storage; instead, it describes a way of *arranging* a datum and its pointers. The name Employee now refers to this *structure*—that is, to an arrangement consisting of two pointers, two strings of a certain maximum length, and four numerical variables.

To assign some storage cells to be used for the list, we give an *allocate* instruction. In this instruction, we include the structure name Employee, indicating the amount of storage and its manner of arrangement, and we include the name of a *pointer variable.*

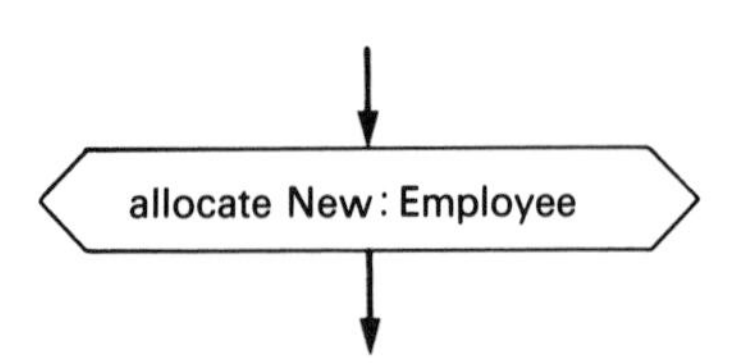

The compiler will translate this instruction, using the structural description previously declared, and will take the necessary steps to make available enough space for one item of the form Employee. A pointer to this space will be stored as the value of the pointer variable New, when the space is allocated. Execution of the *allocate* instruction does not store any values in the cells that are allocated, however.

We now give *read* instructions, to bring data from the reader into the newly allocated Employee area.

New: Employee[Name, Address, ZipCode,
BaseRate, NDepend, TotalWagesYearToDate]

This instruction specifies a read operation, which assigns values to variables in the area just allocated. Notice that we reference the elements of the structure by giving first the name of the cell containing the *pointer* to the specified instance of the structure; then the *structure name*, to indicate the arrangement of the data in the item; and finally the identification of the particular *variables* and strings within the structure.

So far, we have created one item of the linked list, along with a pointer to it, and we have provided values for the two strings and the four numerical variables in this item by means of a read operation. The two pointers in the item will eventually be used to indicate the preceding and following items in the linked list, but so far we have created no such items. Thus we need to give these pointers a value indicating that they are "null"—pointers with nothing to point to. We arbitrarily assign the numerical value 0 to indicate such a null pointer.

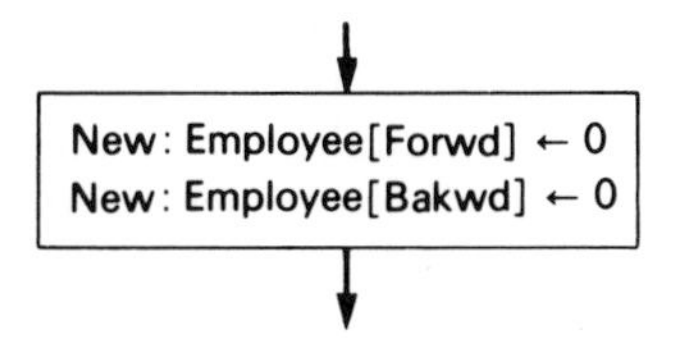

Now we establish pointers from "outside" the list to this newly created list item. We create a pointer variable Alpha, pointing "forward" to the first list item, and a pointer variable Omega, pointing "backward" to the last list item. Because there is only one item on the list at the moment, this item is both Alpha and Omega, the first and the last. We therefore give each of these "outside" pointer variables the value of New, so that they point to the newly allocated item:

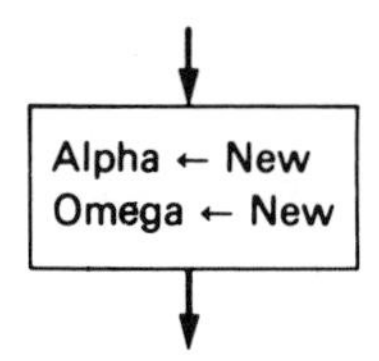

Figure 6.3a illustrates the situation so far. We have one item in the linked list, and two pointers outside the list pointing to the beginning and the end of it. The initial setup of the list is now complete:

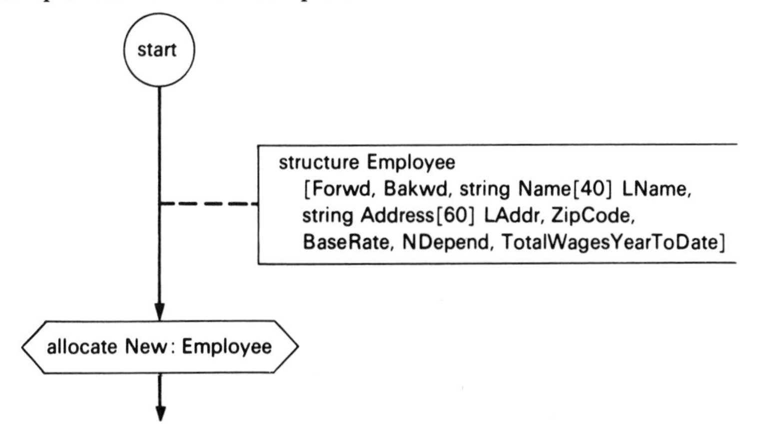

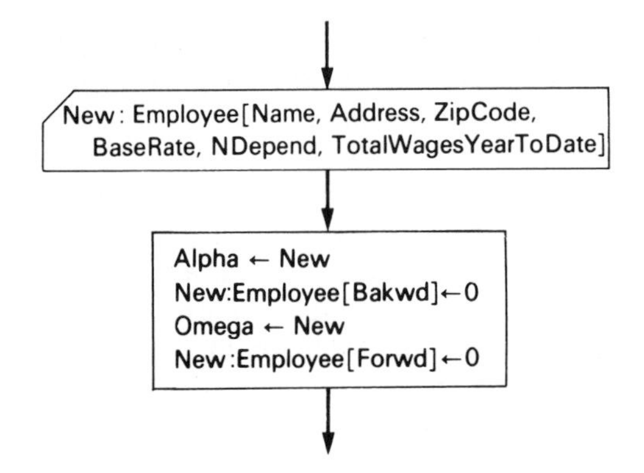

We have grouped the pointer handling instructions so as to emphasize the fact that Alpha and New : Employee[Bakwd] are related and that Omega and New: Employee[Forwd] are related. The (null) backward pointer from the first item of the list is related to the external pointer Alpha, which points from a fixed cell outside the list to the first item of the list. Similarly, the (null) forward pointer from the last item of the list corresponds to the external pointer Omega, which points to the last item from outside the list.

Now we are ready to add an item to the list. We allocate space for the new item, and read data into it.

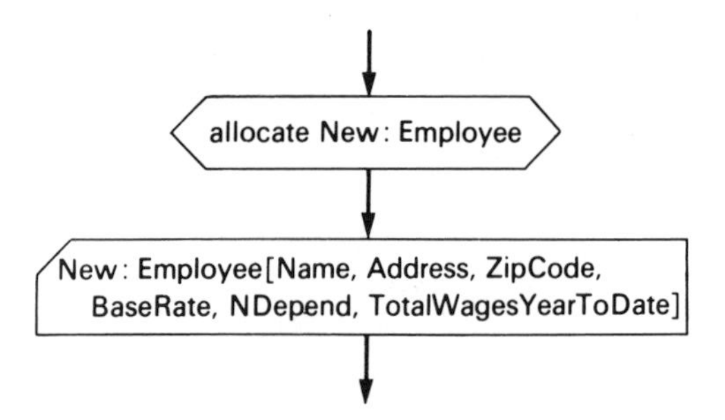

The new item can now be linked to the list merely by adjusting the pointers (see Figure 6.3b).

If we want to add the new item to the *end* of the list, we must do the following:

(1) make the "old" last item point forward to the new item;
(2) make the new item point backward to the "old" last item;
(3) make the Omega pointer point (backward) to the new item;
(4) make the forward pointer of the new item a null pointer, since the new item is now the last item.

The following four instructions accomplish this:

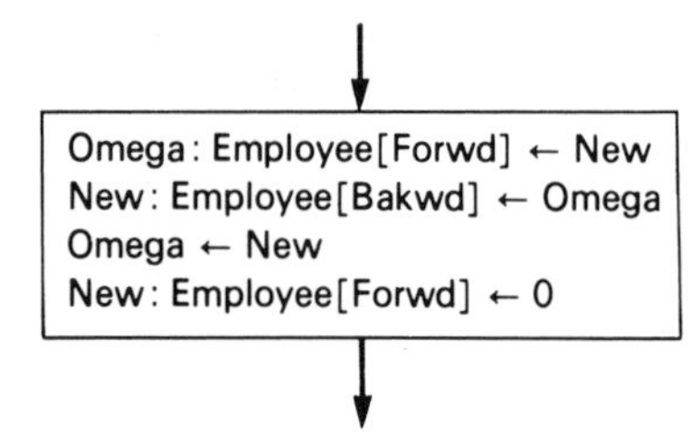

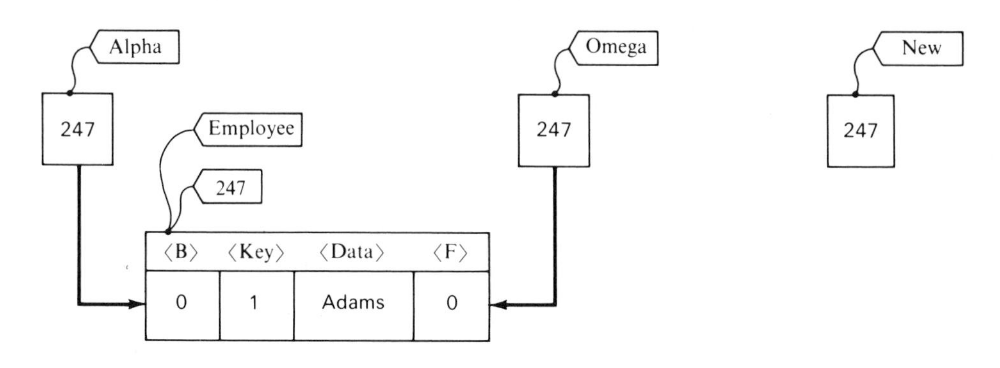

6.3a. *Using a doubly linked list.* Part 1: *The list has been created with one entry.*

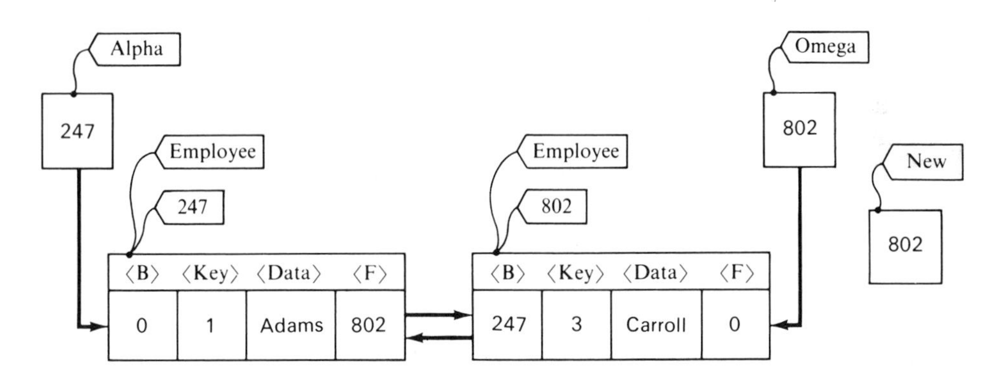

6.3b. *Using a doubly linked list.* Part 2: *A second item has been added after the first one.*

We can repeat this sequence of instructions, each time allocating space for another item, reading the data, and rearranging the pointers to keep the items properly linked together. We might agree to stop when the data for an input item is encountered having a null string as the employee's name. Immediately following the read instruction, we insert:

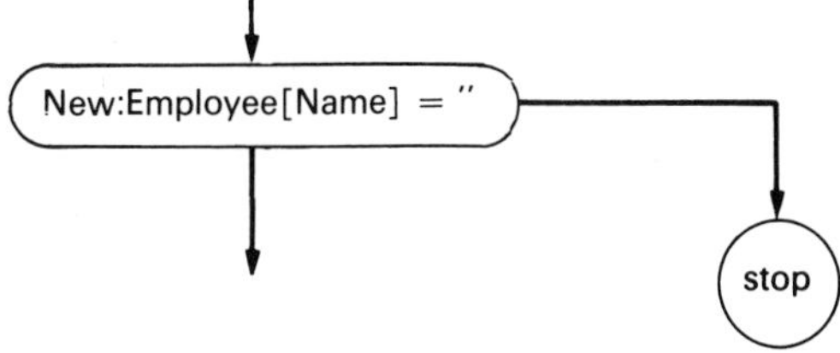

Now suppose that we want to insert some new items in a doubly linked list at the correct position in sequence rather than at the end of the list. The desired sequence is established by using one portion of the datum (say the employee name) as a *key*. We must scan the list, following the forward links and comparing the keys of the items already in the list with the key of the new item.

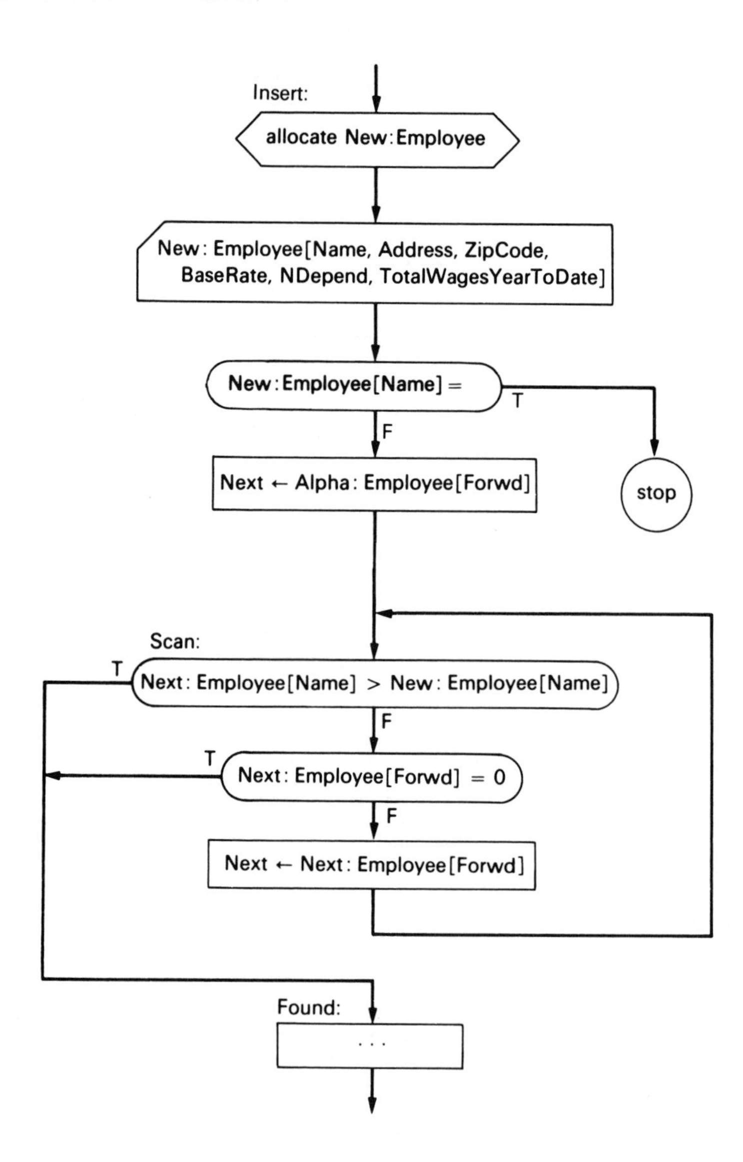

At this point, we know where to insert the new item. The pointer variable Next is pointing to the item that is to follow the new item (see Figure 6.3c). If we write the instruction

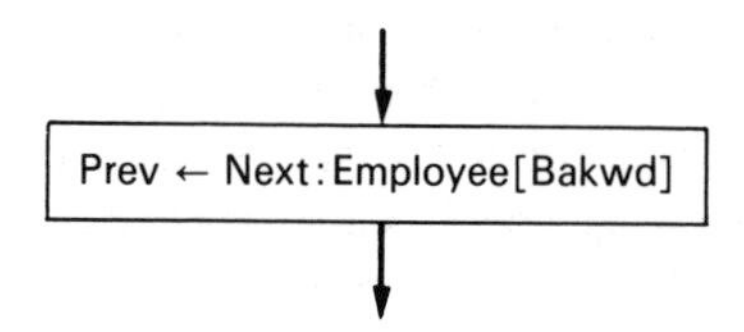

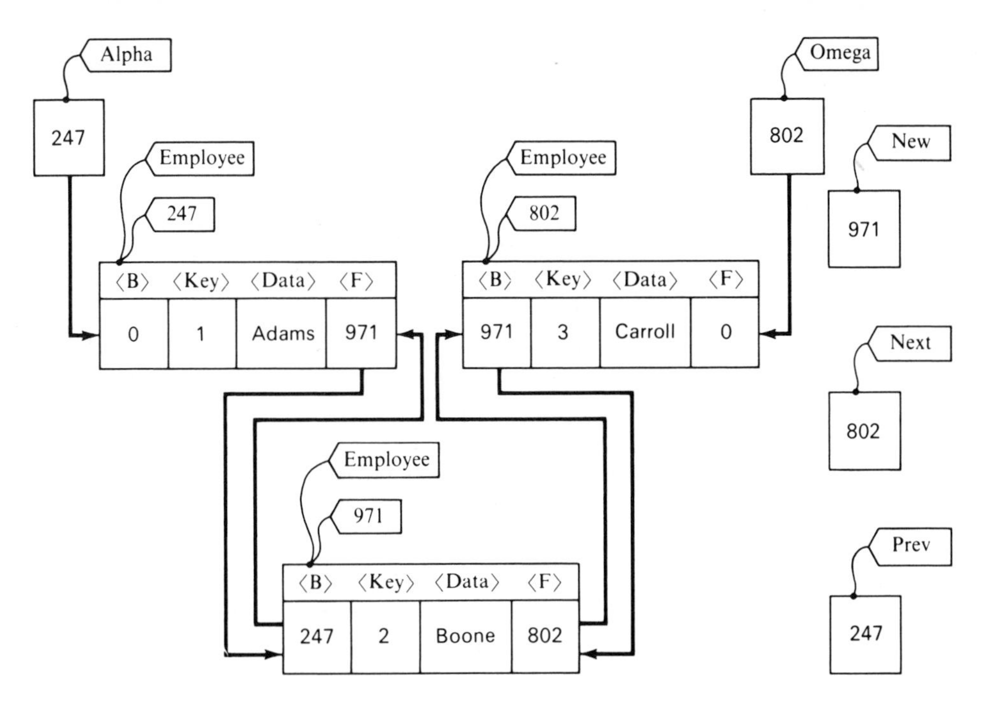

6.3c. *Using a doubly linked list.* Part 3: *An item has been inserted in the middle of the list. The only parts of the previous items that had to be changed were the forward and backward pointers.*

the pointer variable Prev will be set to point to the item that should *precede* the new item. Again we must take four steps[1] to rearrange the pointers so as to insert the new item:

(1) make the previous item point forward to the new item;
(2) make the new item point backward to the previous item;
(3) make the following item point backward to the new item;
(4) make the new item point forward to the following item.

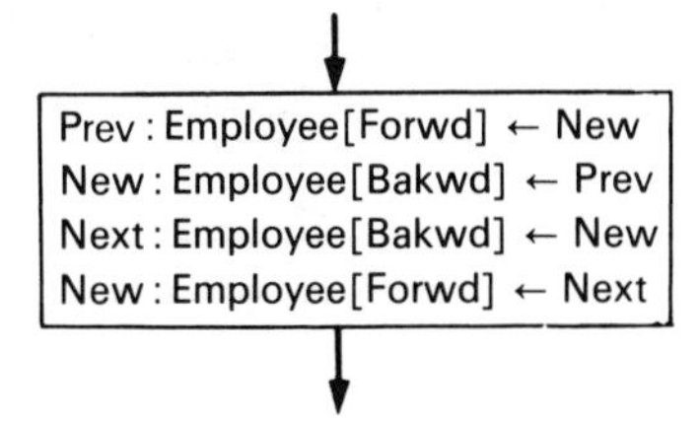

To delete items, we must first search as before and, when we find a matching item, rearrange the pointers and delete the item no longer wanted.

---

[1] I have taken one liberty here—the instructions as written will work only when the new item is not inserted at the beginning or the end of the linked list. Some tests for null pointer values will remove this limitation. A more elegant solution, which avoids this difficulty, is to include in every doubly linked list a *head*, which is an item with pointers but no meaningful data values. The forward and backward pointers of this item then play the roles of our Alpha and Omega, and only one pointer from outside the list, to this head item, is required.

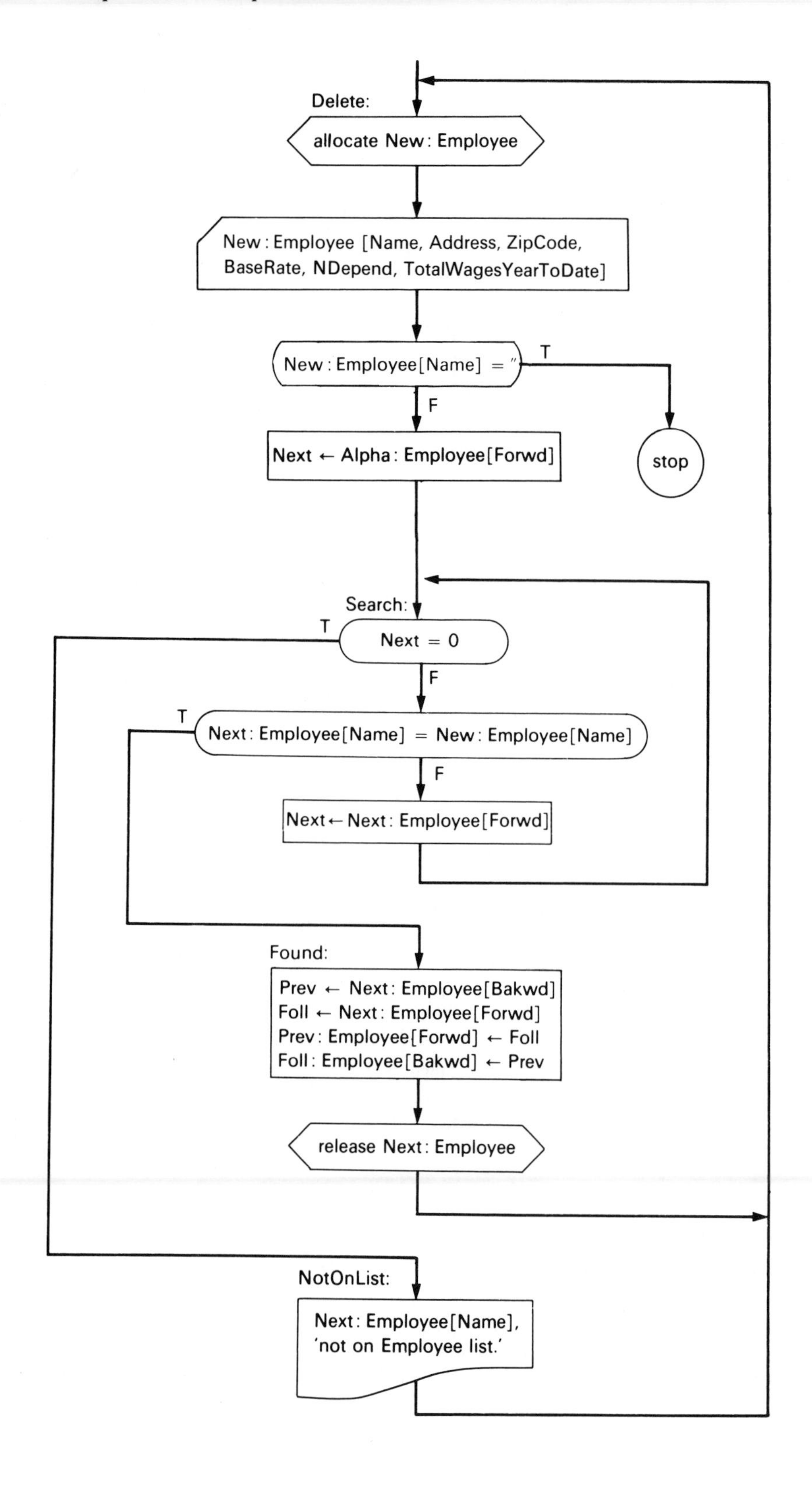
Delete:
allocate New: Employee
New: Employee [Name, Address, ZipCode, BaseRate, NDepend, TotalWagesYearToDate]
New: Employee[Name] = ''
T
F
stop
Next ← Alpha: Employee[Forwd]
Search:
Next = 0
T
F
Next: Employee[Name] = New: Employee[Name]
T
F
Next ← Next: Employee[Forwd]
Found:
Prev ← Next: Employee[Bakwd]
Foll ← Next: Employee[Forwd]
Prev: Employee[Forwd] ← Foll
Foll: Employee[Bakwd] ← Prev
release Next: Employee
NotOnList:
Next: Employee[Name], 'not on Employee list.'

Flowchart

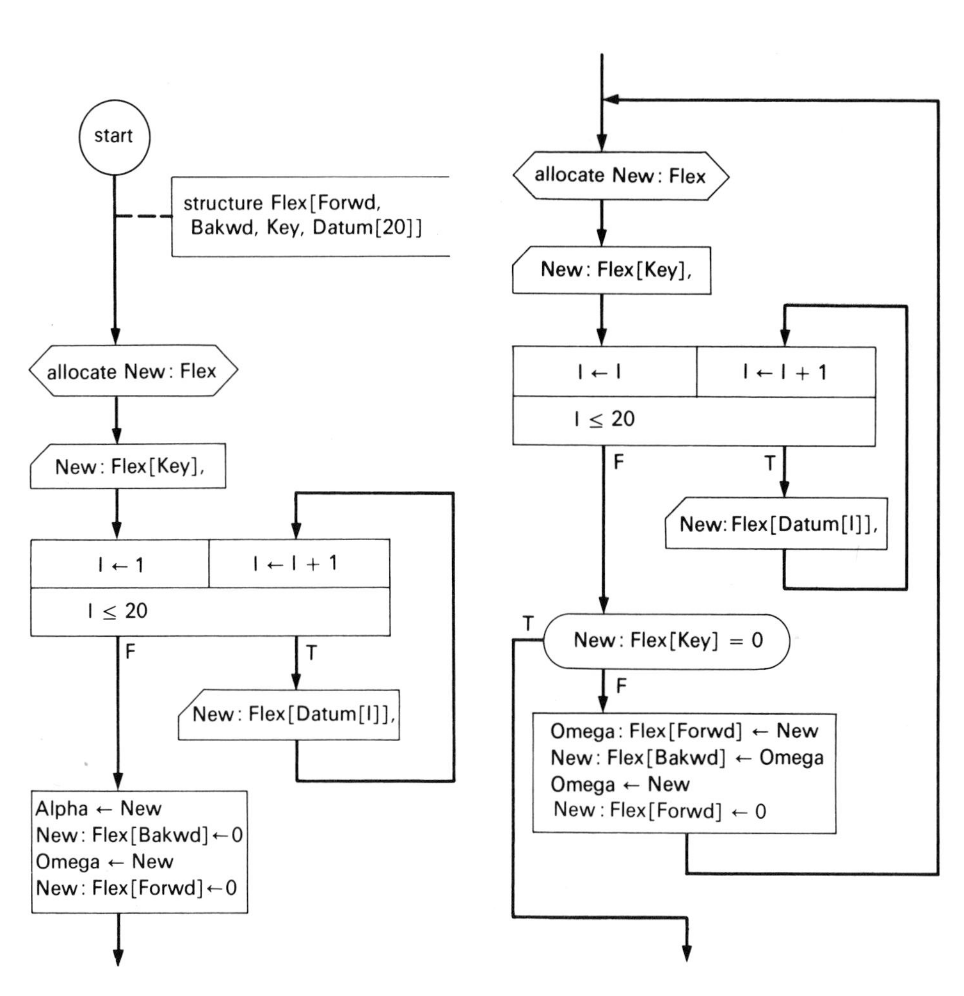
Initialize:
Add:
start
structure Flex[Forwd, Bakwd, Key, Datum[20]]
allocate New: Flex
New: Flex[Key],
I ← 1
I ← I + 1
I ≤ 20
F
T
New: Flex[Datum[I]],
Alpha ← New
New: Flex[Bakwd] ← 0
Omega ← New
New: Flex[Forwd] ← 0
allocate New: Flex
New: Flex[Key],
I ← I
I ← I + 1
I ≤ 20
F
T
New: Flex[Datum[I]],
T
New: Flex[Key] = 0
F
Omega: Flex[Forwd] ← New
New: Flex[Bakwd] ← Omega
Omega ← New
New: Flex[Forwd] ← 0

Insert:

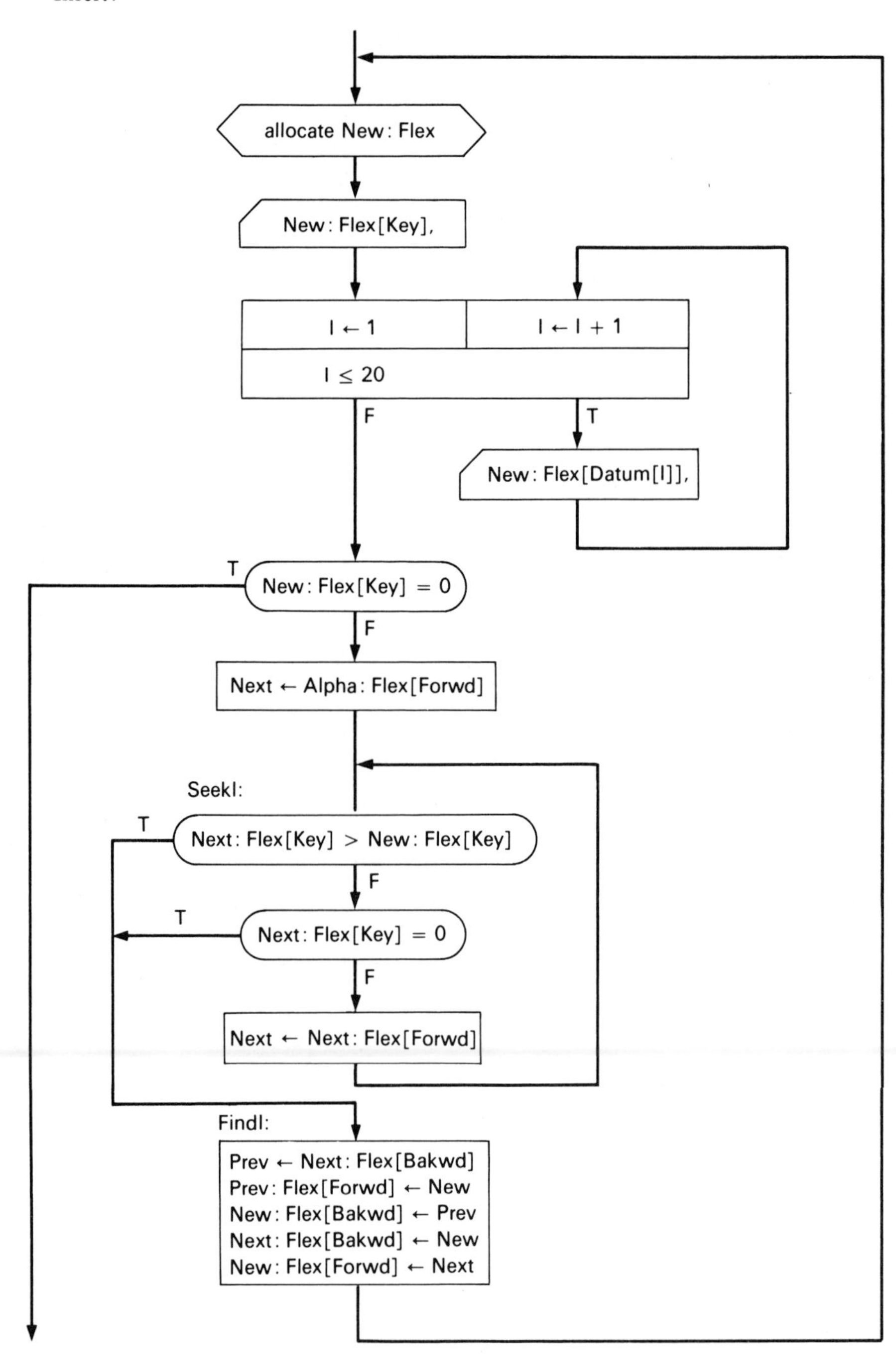
allocate New: Flex
New: Flex[Key],
I ← 1
I ← I + 1
I ≤ 20
F
T
New: Flex[Datum[I]],
T
New: Flex[Key] = 0
F
Next ← Alpha: Flex[Forwd]
SeekI:
T
Next: Flex[Key] > New: Flex[Key]
F
T
Next: Flex[Key] = 0
F
Next ← Next: Flex[Forwd]
FindI:
Prev ← Next: Flex[Bakwd]
Prev: Flex[Forwd] ← New
New: Flex[Bakwd] ← Prev
Next: Flex[Bakwd] ← New
New: Flex[Forwd] ← Next

Delete:

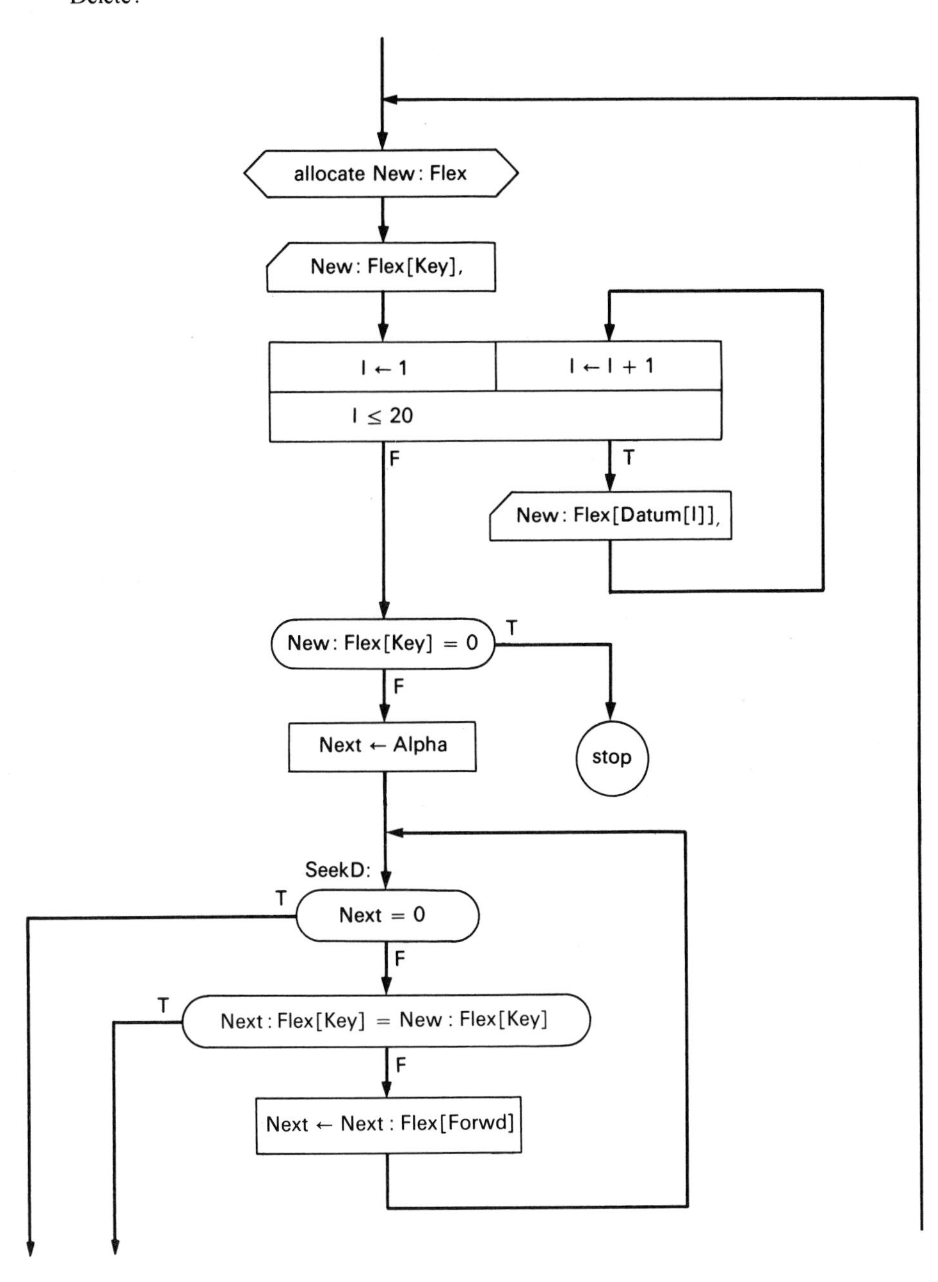

allocate New: Flex
New: Flex[Key],
I ← 1
I ← I + 1
I ≤ 20
F
T
New: Flex[Datum[I]],
New: Flex[Key] = 0
T
F
stop
Next ← Alpha
SeekD:
T
Next = 0
F
T
Next: Flex[Key] = New: Flex[Key]
F
Next ← Next: Flex[Forwd]

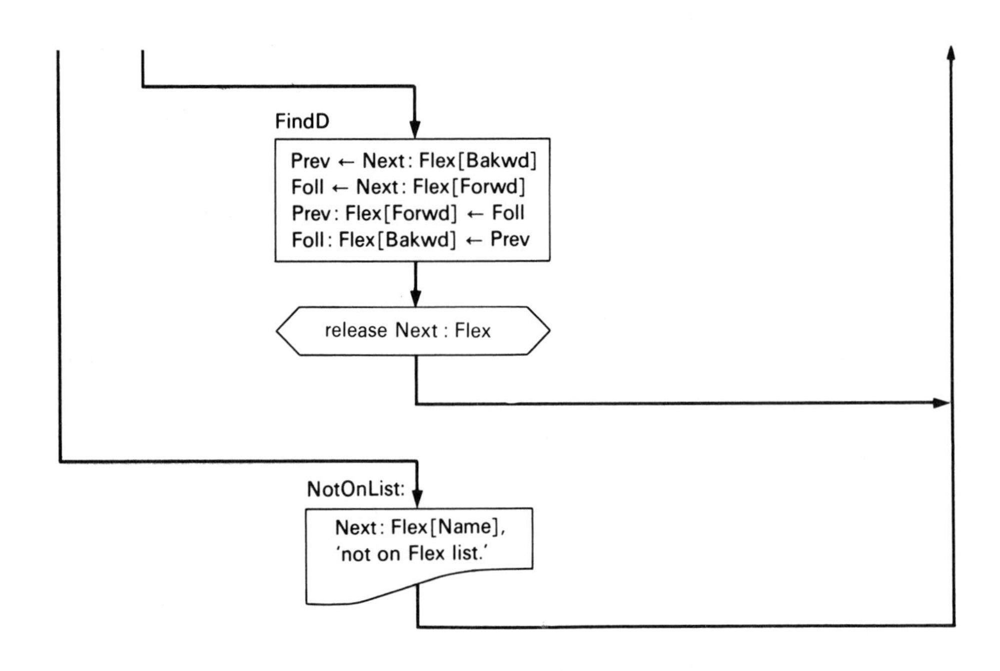
FindD
Prev ← Next: Flex[Bakwd]
Foll ← Next: Flex[Forwd]
Prev: Flex[Forwd] ← Foll
Foll: Flex[Bakwd] ← Prev
release Next : Flex
NotOnList:
Next: Flex[Name],
'not on Flex list.'

Basic Basic has no facilities for dynamic storage allocation during program execution.

Fortran Fortran has no facilities for dynamic storage allocation during program execution. However, such features may be simulated using arrays (see Berztiss, 1970). Furthermore, extensions to Fortran can be made by providing list processing functions; the best known extension of this type is SLIP (Weizenbaum, 1963).

PL/1*

```
LINKLIST: procedure options (main);
   declare 1 FLEX based,
            2 FORWD pointer,
            2 BAKWD pointer,
            2 KEY float,
            2 DATUM(1 : 20);
   declare (NEW, FIRST, LAST, NEXT,
      PREV, FOLL) pointer;
INITIAL: allocate FLEX set (NEW);
   get list (NEW -> FLEX);
   FIRST = NEW;
   NEW -> BAKWD = NULL;
   LAST = NEW;
   NEW -> FORWD = NULL;
ADD: allocate FLEX set (NEW);
   get list (NEW -> FLEX);
   if NEW -> KEY = 0 then go to
      INSERT;
   LAST -> FORWD = NEW;
   NEW -> BAKWD = LAST;
   LAST = NEW;
   NEW -> FORWD = NULL;
   go to ADD;

INSERT: allocate FLEX set (NEW);
   get list (NEW -> FLEX);
   if NEW -> KEY = 0 then go to
      DELETE;
   NEXT = FIRST -> FORWD;
SEEKI: if NEXT = NULL then go to
      FINDI;
   if NEXT -> KEY > NEW -> KEY
      then go to FINDI;
   NEXT = NEXT -> FORWD;
   go to SEEKI;
FIND: PREV = NEXT -> BAKWD;
   PREV -> FORWD = NEW;
   NEW -> BAKWD = PREV;
   NEXT -> BAKWD = NEW;
   NEW -> FORWD = NEXT;


DELETE: allocate FLEX set (NEW);
   get list (NEW -> FLEX);
   if NEW -> KEY = 0 then go to QUIT;
   NEXT = FIRST;
SEEKD: if NEXT = NULL
      then go to NOTONLIST;
   if NEXT -> KEY = NEW -> KEY
      then go to FINDD;
   NEXT = NEXT -> FORWD;
   go to SEEKD;
FINDD: PREV = NEXT -> BAKWD;
   FOLL = NEXT -> FORWD;
   PREV -> FORWD = FOLL;
   FOLL -> BAKWD = PREV;
   free NEXT -> FLEX;
   go to DELETE;
NOTONLIST: put skip list (NEW ->
   KEY, 'not on Flex list.');
   go to DELETE;
QUIT : free NEW → FLEX
   end LINKLIST;
```

* See Fike (1970), Chapter 10.

Words to Remember

file	datum	release
item	allocate	structure
doubly linked list	dynamic allocation	insert

Exercise

(If you are learning PL/1, write the program; otherwise, draw a flowchart.) Modify the program for adding, inserting, and deleting the items of a doubly linked list, by giving the list a head. The forward pointer of the head item then performs the role of the external pointer Alpha, and the backward pointer of the head item performs the role of the external pointer Omega. A single external pointer to the head item is now required. Note that searches in such a list need not be restricted to the case where the item sought does not precede the first item nor follow the last item of those already in the list.

6.4 *How Are Data Structures Organized Inside a Computer?*

As we have seen, a numeric variable (or constant) refers to a cell. The cell performs a central role in manipulating information, especially in floating binary form. Numeric operations that involve a portion of a cell, or a group of cells, are more complicated for the computer's internal mechanism to perform than those that use just the amount of information contained in a single cell.

Lists The next simplest data structure, after the single cell, is a (sequential linear) list called a vector in mathematical applications. In the internal arrangement of storage, a list is just a group of adjacent cells. As we have seen, to designate one variable in a list, we must give the array name and an integer subscript value. The list address, which is the internal machine-oriented form of the list name, is the address of the *first* variable in the list. To evaluate any expression containing a subscripted variable, it is necessary to use the current value of the subscript, along with this list address, to determine the address of the particular cell.

Modern computers have *index registers* to facilitate this process. The machine-oriented instructions produced by the compiler include array addresses, but these are "flagged" to designate an associated index register. The compiler generates extra instructions to evaluate the subscript expression and store it in the appropriate index register before the flagged instruction is executed. The computer automatically uses the contents of the index register, along with the flagged array address, to determine the actual "effective address" of the desired cell.

Strings Alphabetical (and other) characters do not, as we have seen, require as much space as is usually provided for numbers in computer storage. However, computers vary greatly in the facilities they provide for compact storage and efficient processing of strings. Until fairly recently, they were designed with little attention to nonnumeric applications. On many of these systems, strings can be efficiently handled only if each character is stored in a separate cell, and this is wasteful of storage. A practical solution on such systems is to use an entire cell for each character in those strings (or portions of strings) that are being actively manipulated, and to store inactive characters in packed form. The extra effort of packing and unpacking the characters is required only when they are moved to or from inactive status, and this effort can be minimized by careful planning.

Many computers that were designed after the importance of string manipulation became evident incorporate in their machine-oriented instruction repertories the means for addressing the individual characters of a string, even when several characters are stored in the same cell. One method, supposing that four 8-bit characters are stored in a 32-bit cell, for example, is to give consecutive address numbers to each 8-bit group (or *byte*); then the address used to reference a cell for numeric purposes will always be a multiple of four. Other machines still give consecutive addresses to the cells but provide for additional information (either in the machine instruction along with the address or in auxiliary registers) to specify a portion of the cell.

Matrix Storage When items are structured in a rectangular array, all the items in a *row* have a common property, while all the items in a *column* have another common property. For instance, we studied an example in which the entries in a row were sales of different kinds of shoes by the same salesman, while the entries in a column were sales of the same kind of shoes by different salesmen. How should the array be stored in the cells of a computer? If all the information about a given salesman is stored compactly in a group of adjacent cells, then the information about a given category of shoes will necessarily be spread out into nonadjacent cells. If each salesman's items of data are stored as a list, in four adjacent cells, and followed immediately by the next salesman's data, then it is easy to see that the cells pertaining to a given category of shoes will be *equally spaced*—in fact, every fourth cell will contain data belonging to the same category.

Thus, to find a particular cell of the array, we need to know the salesman's number (first subscript), the category number (second subscript), and one other number. If each row is stored compactly (as with PL/1, Figure 6.4a), then the extra information we need is the *row length*. For instance, we need this information to locate the item having subscripts (2, 1), that is, the first item in the second row. On the other hand, if each column is stored compactly (as with Fortran, Figure 6.4b), then we need to know the *column length* to locate the item having subscripts (1, 2), that is, the first item in the second column. The compiler uses the values specified in the array declaration to compute the cell address from the array address, the subscript values, and the row or column length specified in the declaration.

Matrix

⟨1,1⟩	⟨1,2⟩	⟨1,3⟩	⟨2,1⟩	⟨2,2⟩	⟨2,3⟩	⟨3,1⟩	⟨3,2⟩	⟨3,3⟩
8	1	6	3	5	7	4	9	2

⟨1,1⟩	⟨1,2⟩	⟨1,3⟩
8	1	6

⟨2,1⟩	⟨2,2⟩	⟨2,3⟩
3	5	7

⟨3,1⟩	⟨3,2⟩	⟨3,3⟩
4	9	2

6.4a. *Matrix storage in PL/1.*

⟨1,1⟩	8
⟨2,1⟩	3
⟨3,1⟩	4
⟨1,2⟩	1
⟨2,2⟩	5
⟨3,2⟩	9
⟨1,3⟩	6
⟨2,3⟩	7
⟨3,3⟩	2

⟨1,1⟩	8
⟨2,1⟩	3
⟨3,1⟩	4

⟨1,2⟩	1
⟨2,2⟩	5
⟨3,2⟩	9

⟨1,3⟩	6
⟨2,3⟩	7
⟨3,3⟩	2

6.4b. *Matrix storage in Fortran.*

Dynamic Structures Depending on the application, dynamic structures may contain lists, strings, and arrays. In fact, flexibility is a key characteristic of these structures.

In the examples of Section 6.3, we saw how to assign cells as they are required from a storage pool. In those examples, however, no reason was given for not assigning all the cells of the pool in advance, so that they would be ready to use when needed. When a single flexible list is included in a program, there is in fact no advantage in allocating and releasing the storage dynamically, since the entire pool must be available all the time and no other use can be made of storage that has not yet been allocated or has already been released. More complicated examples, including several flexible structures, would show that one linked list might be increasing while another was decreasing; in this case, the several structures would share the pool, and storage would be more efficient.

The structures used with dynamic storage can be *simulated* with tables or arrays. An item, with its pointers, keys, and other data, occupies a row of the table (or array). Pointer values are used as subscripts to refer to the various portions of the structure. With some extra care, we can even arrange the program to permit several different structures to occupy the same cells at different times, thus simulating the allocation and release of cells in a dynamic manner (see Berztiss, 1970).

We note, furthermore, that in many operations with linked structures only the *pointers* and the *keys* are actually referred to. If these occupy a small amount of storage, relative to the rest of the data in the structure, it may be possible to segregate the other data in "inactive" storage until a true need for it arises, meanwhile keeping the pointers and keys in a more accessible portion of the available storage space. For instance a *directory* may be established, consisting of a set of keys that can be searched to locate an item, along with pointers to the less active data (which may be on an auxiliary device such as a magnetic disc).

Word to Remember

directory

Exercise

The array

8	1	6
3	5	7
4	9	2

which appears in the illustration of computer storage for matrix elements in PL/1 and in Fortran (Figures 6.4a and 6.4b) is a magic square. The sum of the three numbers in any row or column, or on either of the two main diagonals, is fifteen. Write a program that will read the elements of a square matrix of any size, up to 10 by 10, and will determine whether these numbers form a magic square—that is, whether all the row sums, all the column sums, and the two diagonal sums have the same value.

References

Berztiss, A. T., *Data Structures: Theory and Practice*. New York: Academic Press, 1970.

Denning, P. J., "Virtual Memory." *Computing Surveys* 2: 153–190 (Sep 1970).

Denning, P. J., "Third Generation Computer Systems." *Computing Surveys* 3: 175–216 (Dec 1971).

Dodd, G. G., "Elements of Data Management Systems." *Computing Surveys* 1: 117–133 (Jun 1969).

Fike, C. T., *PL/1 for Scientific Programmers*. Englewood Cliffs, N.J.: Prentice-Hall, 1970.

Knuth, D. E., *The Art of Computer Programming*: Vol. 1, Fundamental Algorithms. Reading, Mass.: Addison-Wesley, 1968.

Maurer, W. D., *Programming: An Introduction to Computer Techniques*. San Francisco: Holden-Day, 1972.

Sedelow, S. Y., "The Computer in the Humanities and Fine Arts." *Computing Surveys* 2: 89–110 (June 1970).

Weizenbaum, J., "Symmetric List Processor." *Communications of the Association for Computing Machinery* 6: 524–544 (Sep 1963).

7 Program Structures

7.1 Why and How Are Programs Subdivided?

As we saw in Chapter 3, a principal motive for using the special iteration pattern is that it helps to keep the control instructions of a loop *separate* from the instructions being repeated.

Similarly, in a program for adding fractions, we might wish to separate the instructions that calculate the greatest common divisor from the remainder of the program.

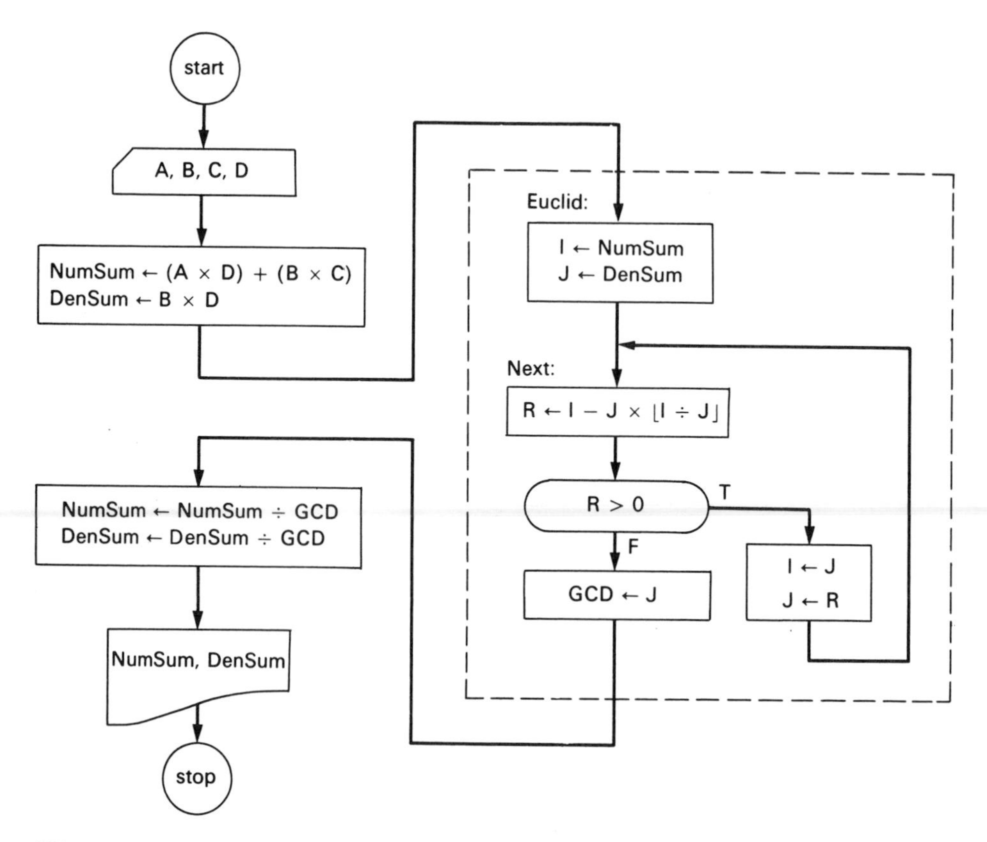

The separate group of instructions at the right applies the Euclidean algorithm to NumSum and DenSum to find their greatest common divisor, GCD.

In this example, little is gained by separating the instructions into two groups. Yet it permits the programmer to consider each group of instructions as a unit, separate from the rest of the program. Thus, when he looks at the instructions on the right he can concentrate on the Euclidean algorithm, ignoring the parts of the program that do not interact with those instructions. The most important reason for introducing *structure* into a program, then, is comprehension. A program that is too complicated to be comprehended all at once may be subdivided into manageable chunks. A segment can again be subdivided if this seems desirable.

For example, a complete program for preparing manuscripts would first be subdivided into three sections, for input, editing, and output. The input section would include blocks for initializing the file, for accepting the individual characters from the keyboard, for detecting control characters in the input stream, and for storing away character strings in the text file. The editing section would include blocks for reading control information from the keyboard, for locating strings in the text file, for displaying strings in visible form to the editor, for replacing characters in a string (including additions and deletions), and for inserting new strings at arbitrary places in the text file (with proper linkages). The output section would include blocks for dividing the text into units such as paragraphs, lines, and sentences, for justifying and hyphenating lines (if this is desired), for inserting headings, footnotes, page numbers, and other special elements at the proper places, and for generating the ultimate coded output for the device on which the text is to be printed. Each block is a nontrivial program, worth separate consideration, and in many cases requiring still further segmentation; yet all the blocks are unified by the need to process text in a common coded form, stored on a single basic text file.

As another example (of intermediate complexity), we might consider a program for computing the grade-point averages of a number of students. Again we might have separate input and output sections; the input section, for example, would need to include steps to detect a change from one student's data to the next. A block for performing the actual averaging might refer to sub-blocks for converting back and forth between letter grades and their associated numerical equivalents. The output block should probably use a sub-block to sort the final results in order by student's name, by grade-point average, or both.

Besides comprehension, however, the programmer may also have other motives for isolating parts of a program. One reason might be to allow a part of the program to be modified with little effect on the remainder of the program. In a payroll computation, for instance, a group of instructions which computes tax withholding might have to be rewritten when a new income tax law is passed. Isolating this part of the program minimizes the chance of unwanted repercussions elsewhere in the program when an instruction in the isolated section is changed. Or suppose it is suspected that the program contains a mistake somewhere. The programmer could make copies of various sections of the program and test each in a simple environment, free from possible interactions with other parts of the program.

Sometimes, when a programmer writes a group of instructions for a particular problem, he suspects that they might also be useful in other applications. The Euclidean algorithm is an example of this kind. Groups of instructions for calculating the cube root or the factorial of a number or for searching a string or sorting a list, might also be designed for use in many programs.

Procedures Let us now see how we might use a group of instructions at more than one point in a program. An example relates to the calculation of the elements of Pascal's triangle (see Section 3.7, exercise 6, Section 2.4, exercise 3, and Section 2.7, exercise 9) according to the formula based on the *factorials* of n, r, and $n - r$:

$$C[n, r] = \frac{n!}{r! \cdot (n - r)!}$$

A program to implement this calculation, say for values of n up to 10 and for r up to n, might include three copies of the instructions for finding a factorial.

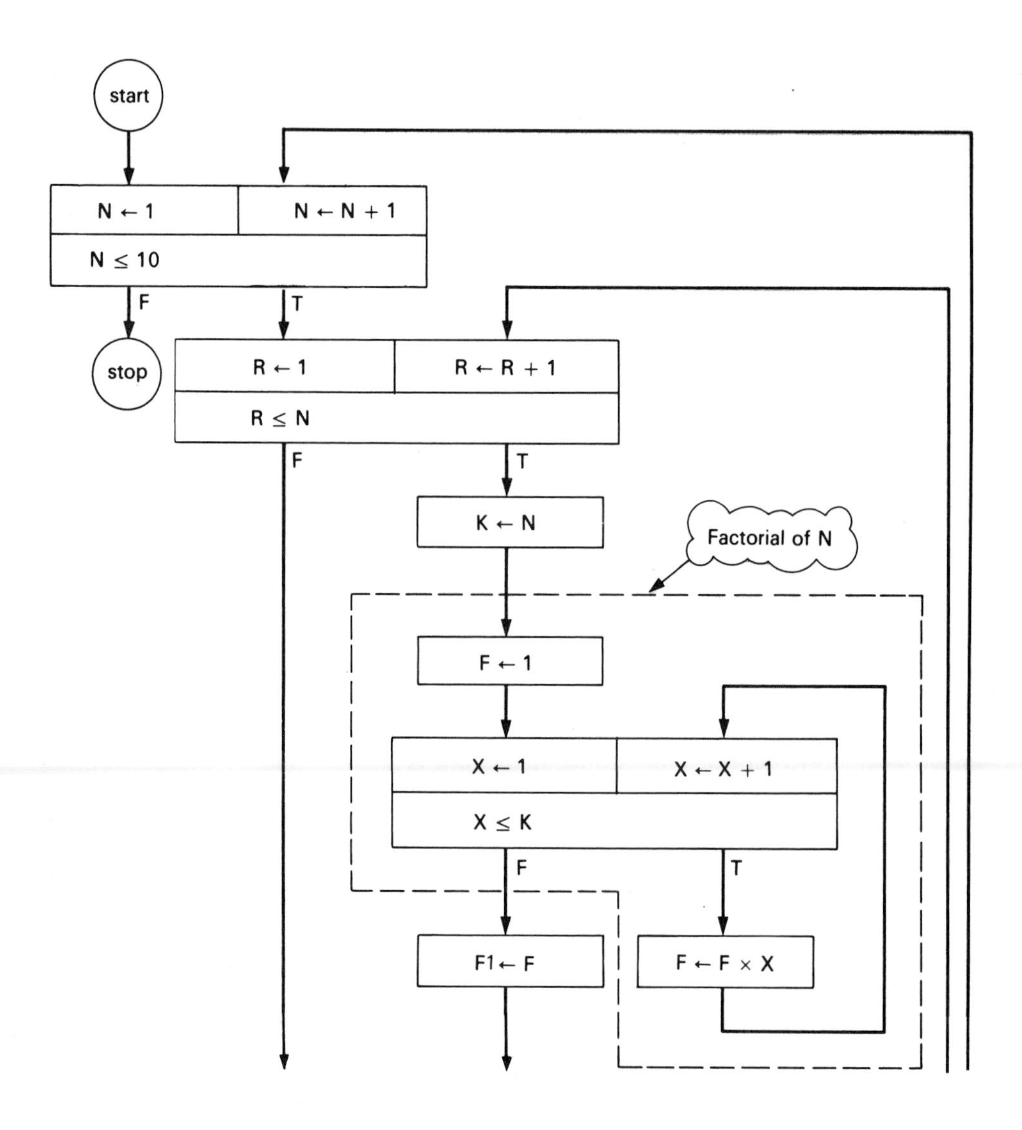

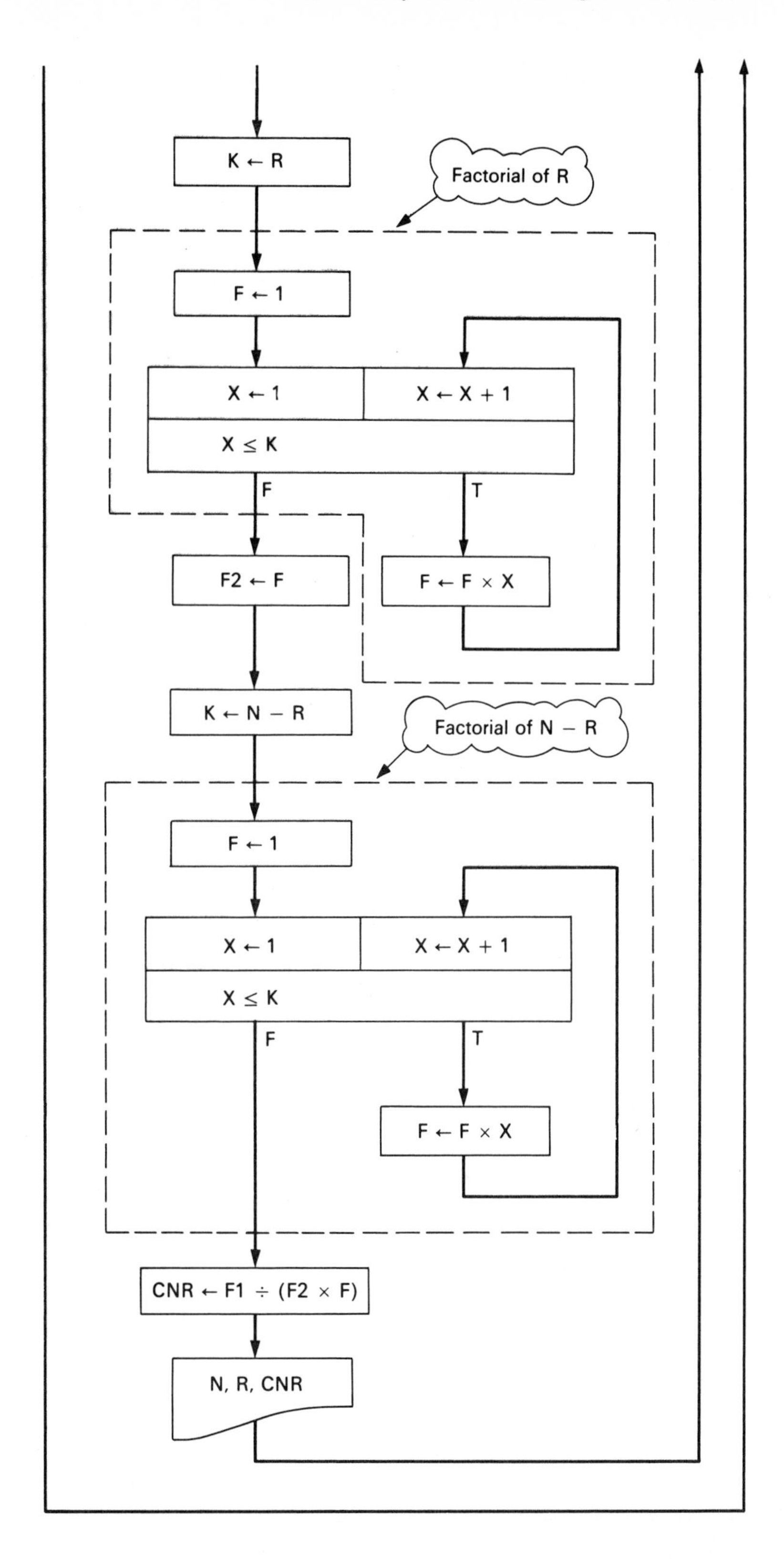
K ← R
Factorial of R
F ← 1
X ← 1
X ← X + 1
X ≤ K
F
T
F2 ← F
F ← F × X
K ← N − R
Factorial of N − R
F ← 1
X ← 1
X ← X + 1
X ≤ K
F
T
F ← F × X
CNR ← F1 ÷ (F2 × F)
N, R, CNR

Because the group of instructions for calculating the factorials is repeated identically three times, we are motivated to rewrite the program with only one copy of these instructions, using control instructions to relate them to the remaining instructions of the program.

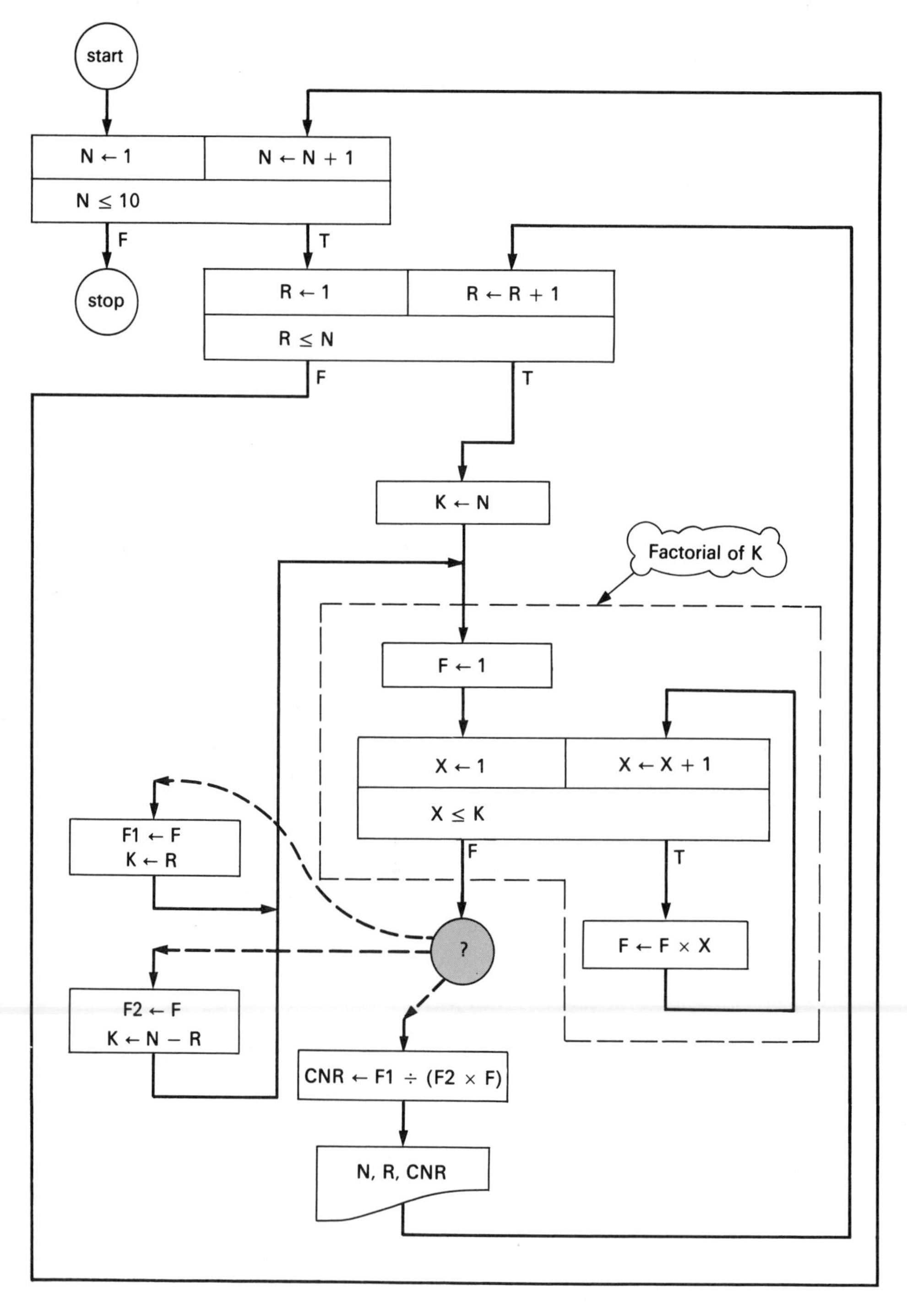

Some sort of multiple branching instruction seems to be needed at the end of the factorial calculation. We could, for example, introduce an integer variable, and give it a distinctive integer value before each execution of the group of instructions. At the end of the group, we could then make a sequence of tests, and return to the proper point in the program after completing the factorial calculation.

The awkwardness and inflexibility of such a scheme is evident. Early in the development of programming languages, it was realized that an *automatic return* feature could be provided, if the isolated group of instructions were programmed as a *procedure*. Instead of using a go to instruction to begin executing a procedure, we use a call instruction. Executing such an instruction not only transfers to the block (as a go to instruction would) but also causes control information to be stored, so that it is possible to resume execution at the instruction following the call. The last instruction to be executed in the procedure is a return instruction, which interprets the stored control information and acts as a go to instruction. Thus the return instruction will cause execution to resume at various places, depending on which call instruction was executed most recently.

The following version of the Pascal's triangle program shows how we indicate procedures in flowcharts, and how we represent the instructions which activate them. First we define the procedure for calculating the factorial. We name this procedure Fact. Notice the procedure declaration just ahead of the starting point of the procedure, and notice that execution ends with a return instruction rather than stop.

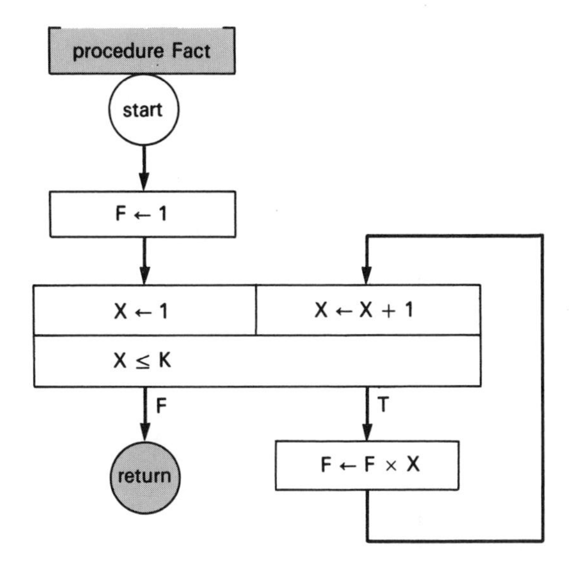

When a call instruction is used to activate this procedure, the proper control information will be stored so that, after the return step, the program that activated the procedure will continue from the point of activation. The call or procedure activation instruction in the main program (that is, in the portion of the program from which the activation occurs) is indicated by a flowchart box with a pair of vertical stripes enclosing the name of the procedure.

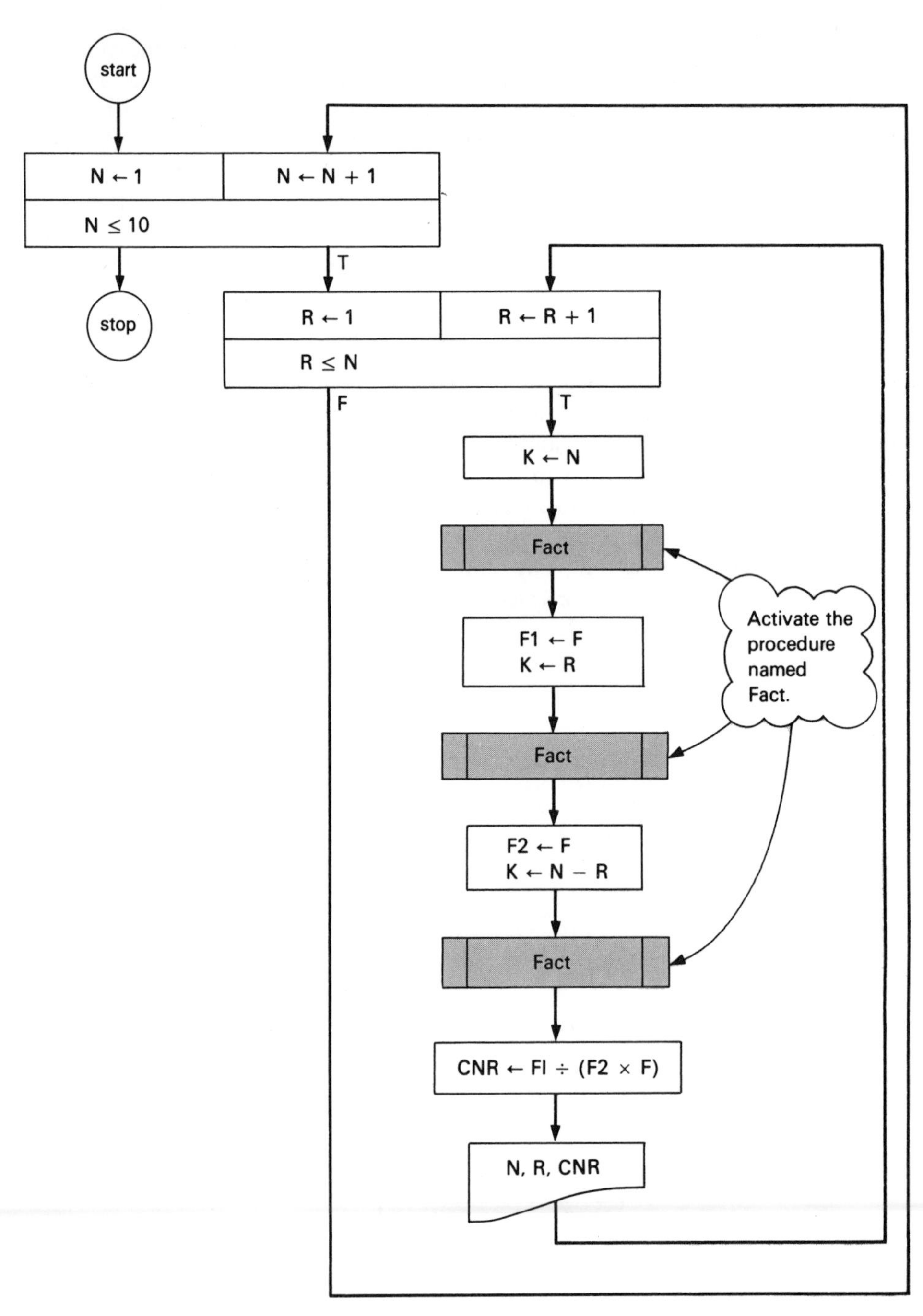

This is our first illustration of *linkage*. The general idea of program segmentation is to create a *separate* group of instructions, so that *interaction* between the separate group and the rest of the program is held to a minimum. Because the instructions in this group are more or less isolated, we must supply them with information and obtain the results from them. This information can be transmitted only in certain stylized ways.

A procedure must know what to do after it completes its task. This information is provided *automatically* by the "automatic return" feature, when we use the call instruction to activate a procedure.

The fact that a procedure can be referred to from a number of points in the same program provides the most important *practical* motive for grouping the instructions in this way. A programmer often finds himself writing the same instructions more than once in a program. He can avoid this redundant effort by organizing the repeated instructions into a procedure and using call instructions to activate the procedure.

Global Variables During execution of a procedure, the computer must fetch or store the values of the variables referred to in the procedure. There is no difficulty here, if the same variables appear in the procedure as elsewhere. If a variable in the procedure also occurs elsewhere in the program, it is called a *global* variable. That is, it is not just "local" within the procedure. For example, in Fact, the variables K and F are global. By means of these global variables, information (in the form of variable values) is transmitted to and from the rest of the program.

In the Pascal's triangle problem, the value for each of the three numbers for which the factorial is to be computed is assigned to the global variable K before the procedure Fact is activated. Similarly, the result of each factorial calculation must be fetched from the cell designated by the global variable F after the procedure is executed.

The global variables used in the procedure may or may not occur "naturally" in the remainder of the program. If the procedure is called from only one point in the program, there is no need to move values to and from special global variables. The variables already in use elsewhere in the program can be used also in the procedure. This strategy will also work when there are several calls to the procedure, provided that there are no conflicts between the global variables. For example, suppose we wish to read the value of K and print the factorials of K and of K^2.

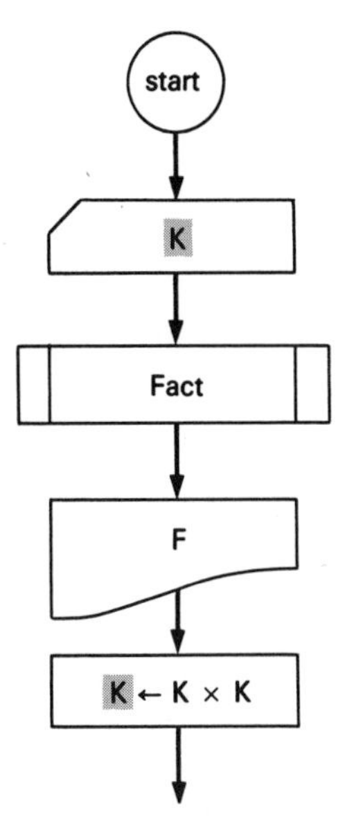

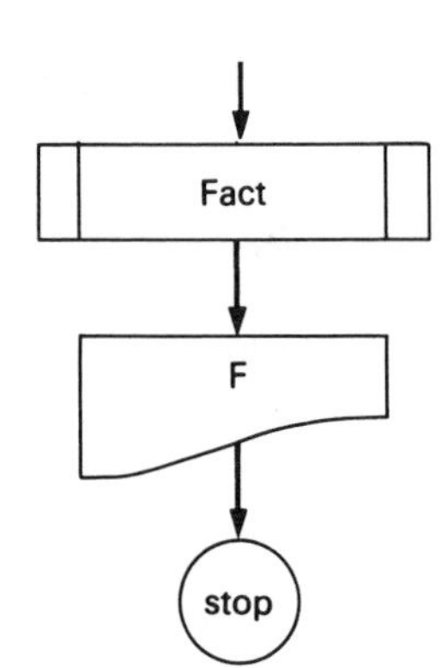

Here the global variables K and F are used during two references to the Fact procedure. There is no conflict in the two uses of K and F. The previous value of K is not needed after K ← K × K is executed, and the old value of F need not be saved after it is printed.

Contrast this with the Pascal's triangle program. In that example, the cell named K cannot be used throughout the program to hold the values of N, R, and N − R, since the first two must be saved until the end. Instead, these three values are assigned to cells having no relation to the procedure, and the extra global variable K is temporarily given the appropriate value each time the procedure is called. Similarly, the same variable F cannot hold all three results, which are needed for the final calculation of CNR, so the auxiliary variables F1 and F2 are used to store the intermediate results.

Another procedure that uses global variables is the cube root calculation presented in Section 3.6.

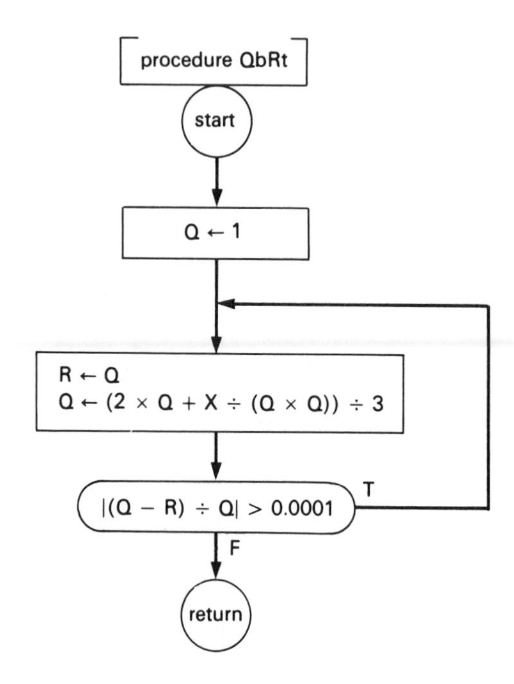

The global variables in this procedure are X, the number whose cube root is to be found, and Q, the result.

Global List and String Names The name of a list or of a string, as well as an individual variable, can be used in a global context when it occurs both in a procedure and in the program that activates the procedure. For example, we might want a procedure to replace each item in a list by its absolute value:

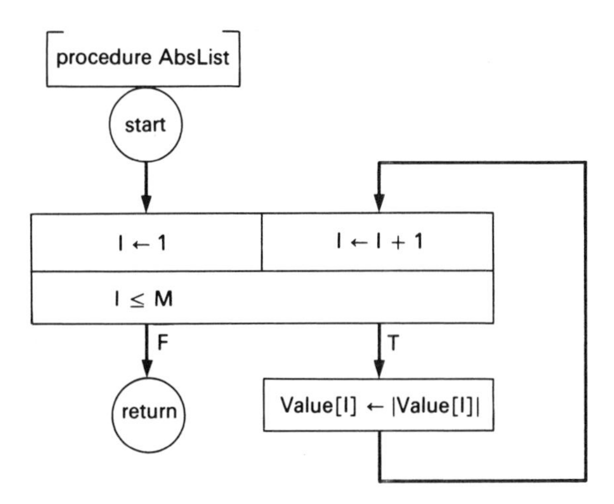

The main program using this procedure would assign a value to the global variable M, generate values for the subscripted variables related to the global list name Value, and call AbsList.

The sorting algorithm of Chapter 5 can also be written as a procedure using a global list name.

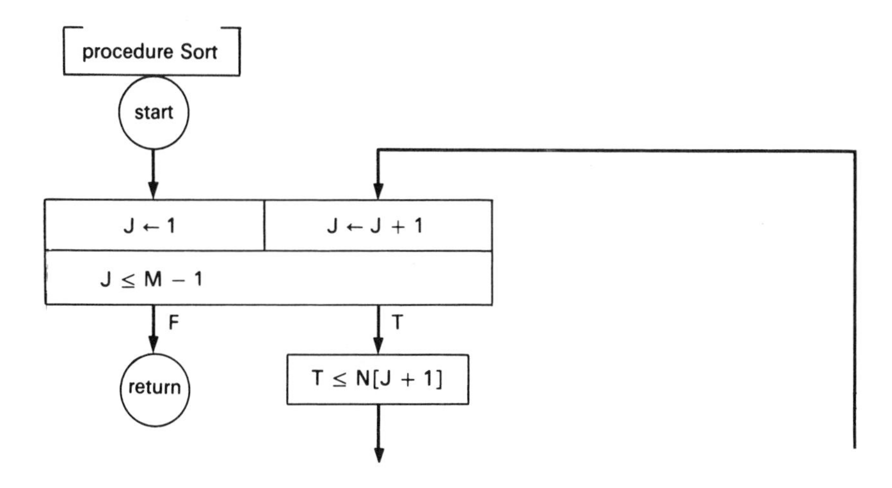

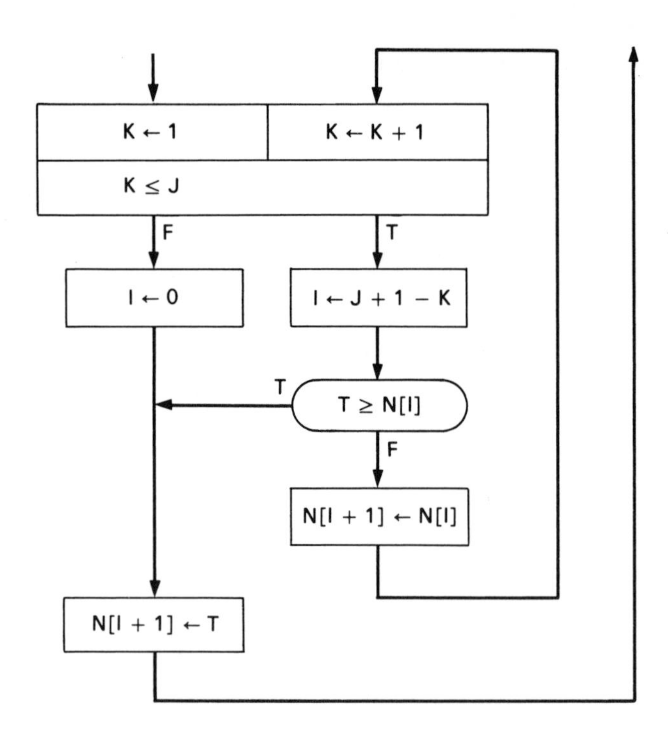

In this procedure, M is a global variable, as before. Now there is also the global list name, N. A program using Sort will store the data items to be sorted in the list N, and assign to M a value indicating how many data items there are. Activation of the procedure Sort will then result in the rearrangement of the data in the list N.

In another example, suppose we need to count the occurrences of a particular character in a string:

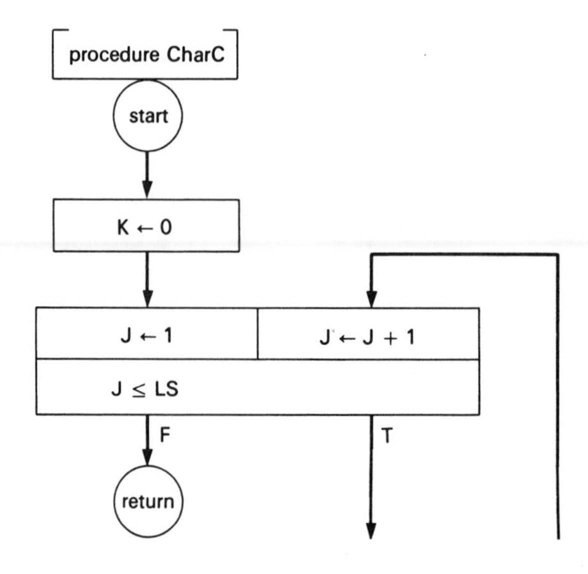

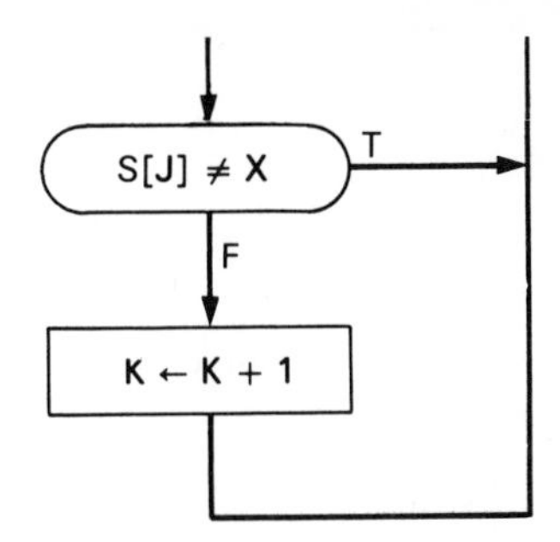

The string to be counted is named S; its length is the value of LS. The character to be searched for is in the cell named X, and the count will be returned as the value of the variable K. Here S is a global string name, and LS, X, and K are global variables.

Flowchart

Add fractions with Euclidean procedure:

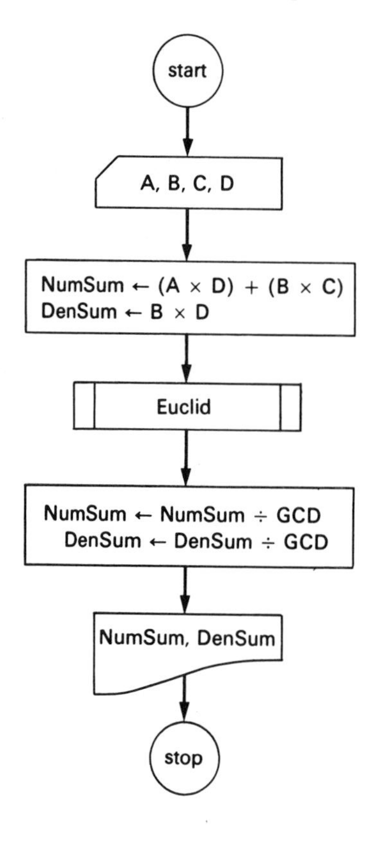

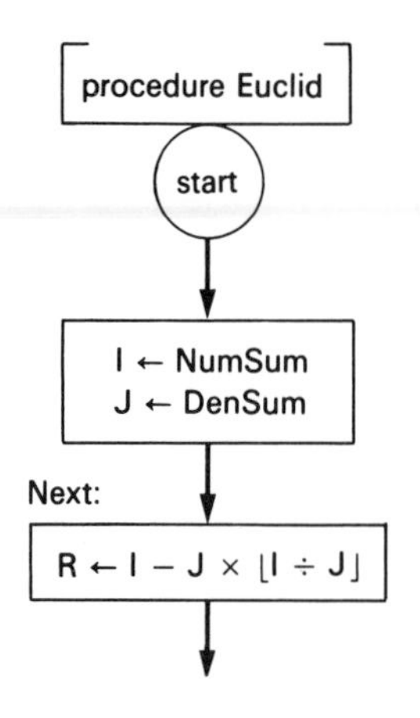

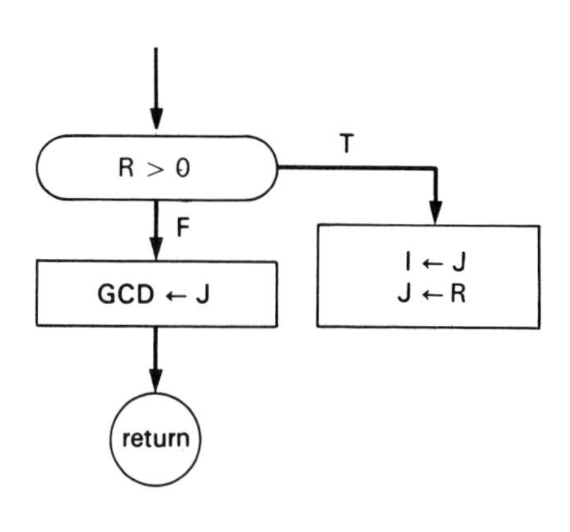

Cube root procedure:

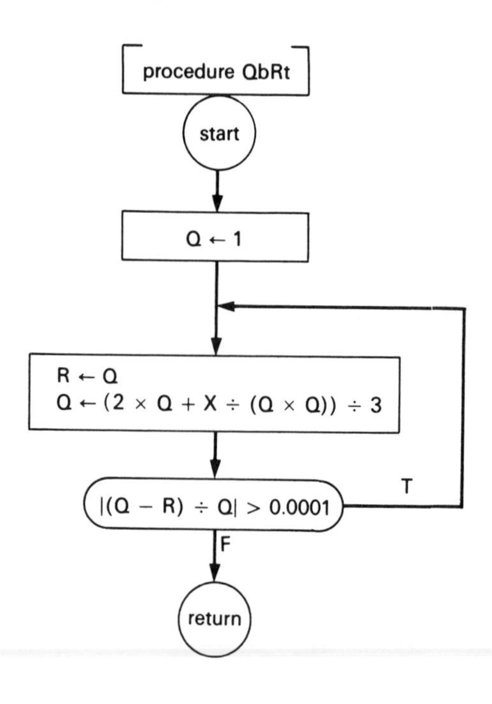

Pascal's triangle with factorial procedure:

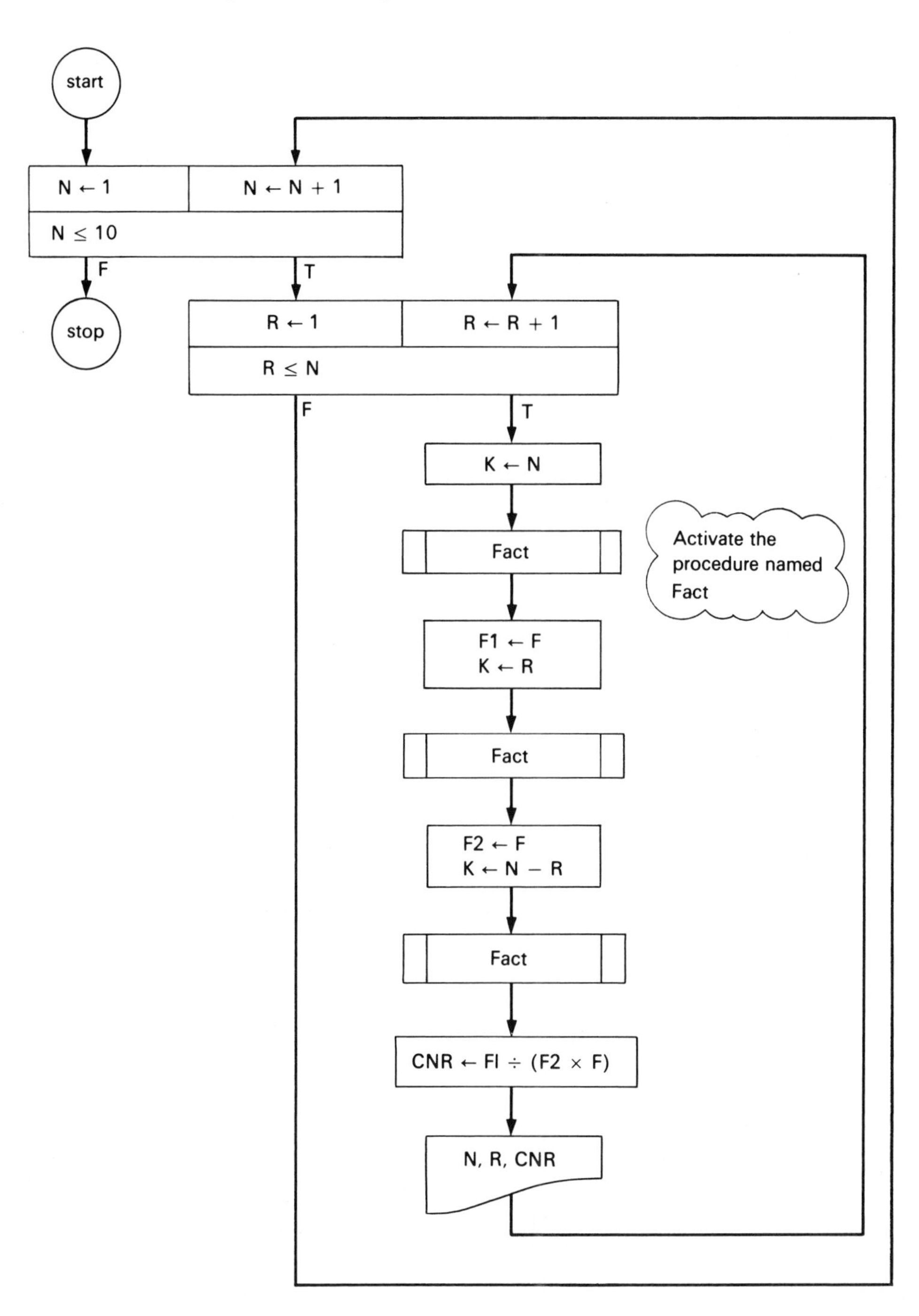

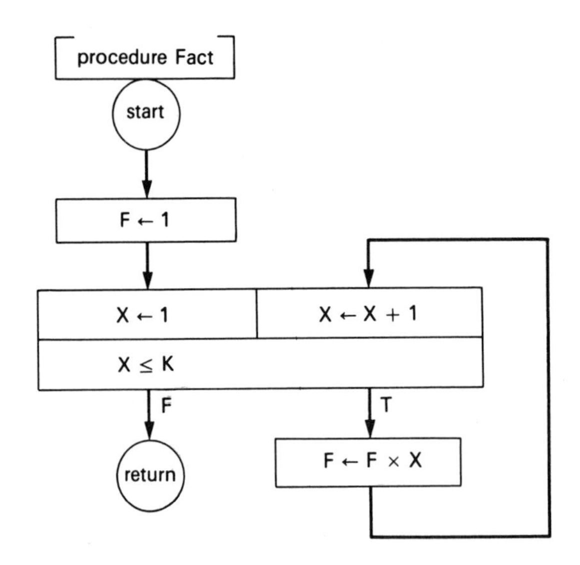

Sorting procedure:

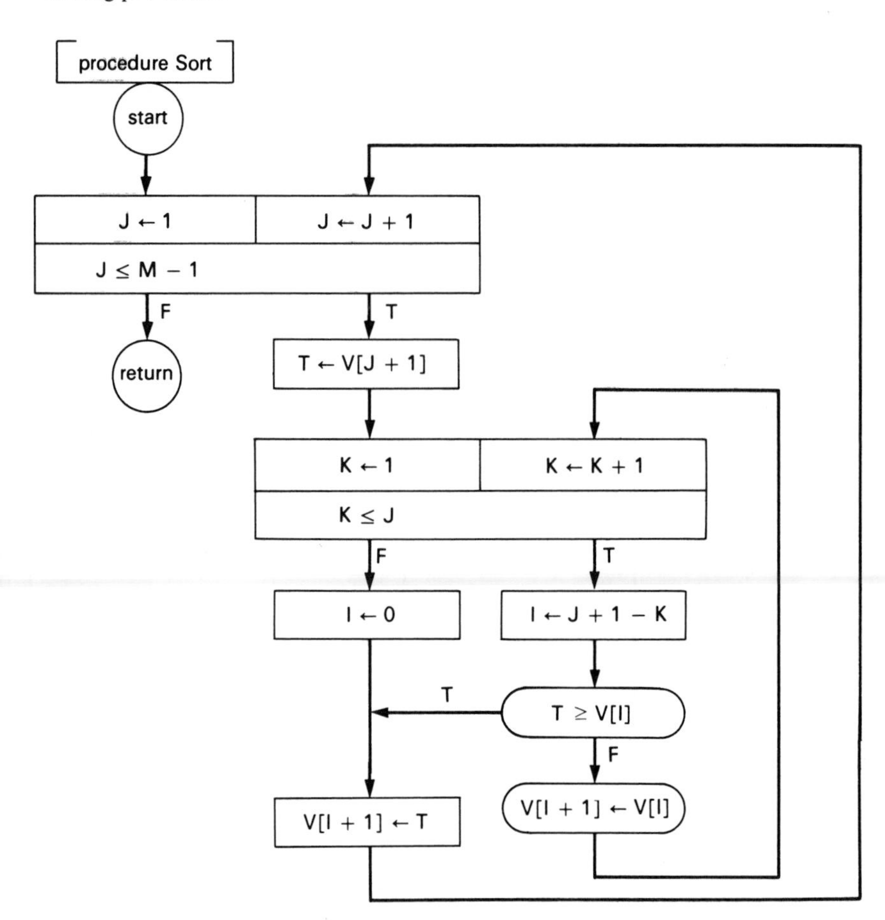

Basic A procedure in Basic does not have any special heading, so care must be taken not to enter it improperly. To call a Basic procedure, the GOSUB instruction is used. The RETURN instruction in the procedure causes execution to continue from the line following the GOSUB instruction.

Add fractions with Euclidean procedure:

```
110 READ A, B, C, D
120 LET N1 = (A * D) + (B * C)
130 LET D1 = B * D
135 REM: FIND GREATEST
136 REM: COMMON DIVISOR
140 GOSUB 210
150 LET N1 = N1 / G
160 LET D1 = D1 / G
170 PRINT N1, D1
180 GO TO 900

210 LET I = N1
220 LET J = D1
230 LET R = I - J * INT (I / J)
240 IF R = 0 THEN 280
250 LET I = J
260 LET J = R
270 GO TO 230
280 LET G = J
290 RETURN

900 END
```

Cube root procedure:

```
310 LET Q = 1
320 LET R = Q
330 LET Q = (2 * Q + X / (Q * Q)) / 3
340 LET E = ABS ((Q - R)/R)
345 IF E > .0001 THEN 320
350 RETURN
```

Pascal's triangle with factorial procedure:

```
105 FOR N = 1 TO 10
110   FOR R = 1 TO N
115     LET K = N
117     REM: FACTORIAL OF N.
120     GOSUB 410
125     LET F1 = F
130     LET K = R
132     REM: FACTORIAL OF R.
135     GOSUB 410
140     LET F2 = F
145     LET K = N - R
147     REM: FACTORIAL OF (N - R).
150     GOSUB 410
155     LET C = F1 / (F2 * F)
160     PRINT N, R, C
165     NEXT R
170   NEXT N
175 GO TO 900

410 LET F = 1
420 FOR X = 1 TO K
430   LET F = F * X
440   NEXT X
450 RETURN

900 END
```

Sorting procedure:

```
205 FOR J = 1 TO M - 1
210   LET T = V[J + 1]
215   FOR K = 1 TO J
220     LET I = J + 1 - K
225     IF T >= V[I] THEN 245
230     LET V[I + 1] = V[I]
235     NEXT K
240   LET I = 0
245   LET V[I + 1] = T
250   NEXT J
255 RETURN
```

Fortran Procedures in Fortran are called *subroutines.* Fortran has *common* cells that can be referred to both in the subroutine and in the main program. The name used in a Fortran subroutine to refer to common cells, however, does not have to be the same as the name used in the main routine; the correspondence is established by the lists of names in the COMMON declarations. Hence, Fortran does not have global variables, but the common cells serve much the same purpose.

Add fractions with Euclidean procedure:

```
      INTEGER A, B, C, D, DS, GCD
      COMMON NS, DS, GCD
      READ (5, 81) A, B, C, D
81    FORMAT (4 I 10)
      NS = (A * D) + (B * C)
      DS = B * D
C     FIND GREATEST COMMON DIVISOR.
      CALL EUCLID
      NS = NS / GCD
      DS = DS / GCD
      WRITE (6, 81) NS, DS
      STOP
      END

      SUBROUTINE EUCLID
      COMMON NS, DS, GCD
      INTEGER DS, GCD, R
      I = NS
      J = DS
230   R = MOD (I, J)
      IF (R .EQ. 0) GO TO 280
      I = J
      J = R
      GO TO 230
280   GCD = J
      RETURN
      END
```

Cube root procedure:

```
      SUBROUTINE QBRT
      COMMON X, Q
      Q = 1.0
320   R = Q
      Q = (2.0 * Q + X / (Q * Q)) / 3.0
      IF (ABS ((Q - R) / Q) .GT. .0001)
     1 GO TO 320
      RETURN
      END
```

Pascal's triangle with factorial procedure:

```
      COMMON F, K
      INTEGER F1, F2, F, R, CNR
      DO 170 N = 1, 10
        DO 165 R = 1, N
          K = N
C         FACTORIAL OF N
          CALL FACT
          F1 = F
          K = R
C         FACTORIAL OF R
          CALL FACT
          F2 = F
          K = N - R
C         FACTORIAL OF (N - R)
          CALL FACT
          CNR = F1 / (F2 * F)
          WRITE (6, 81) N, R, CNR
81        FORMAT (8 I 10)
165       CONTINUE
170     CONTINUE
      STOP
      END

      SUBROUTINE FACT
      COMMON F, K
      INTEGER X, F
      F = 1
      IF (K .LT. 1) GO TO 441
      DO 440 X = 1, K
        F = F * X
440     CONTINUE
441   CONTINUE
      RETURN
      END
```

Sorting procedure:

```
      SUBROUTINE SORT
      COMMON V, M
      REAL V(50)
      IF (M .LT. 2) GO TO 251
      DO 250 J = 1, M
        T = V(J + 1)
        DO 235 K = 1, J
          I = J + 1 - K
          IF (T .GE. V(I)) GO TO 245
          V(I + 1) = V(I)
235       CONTINUE
        I = 0
245     V(I + 1) = T
250     CONTINUE
251   CONTINUE
      RETURN
      END
```

PL/1

Add fractions with Euclidean procedure:

```
ADDFRAC: procedure options (main);
  declare (A, B, C, D, NUMSUM,
    DENSUM, GCD) fixed;
  get list (A, B, C, D);
  NUMSUM = (A * D) + (B * C);
  DENSUM = B * D;
  /* Find Greatest Common Divisor. */
  call EUCLID;
  NUMSUM = NUMSUM / GCD;
  DENSUM = DENSUM / GCD;
  put skip list (NUMSUM, DENSUM);
EUCLID: procedure;
  declare (I, J, R) fixed;
  I = NUMSUM;
  J = DENSUM;
NEXT: R = mod (I, J);
  if R = 0 then go to DONE;
  I = J;
  J = R;
  go to NEXT;
DONE: GCD = J;
  return;
  end EUCLID;
  end ADDFRAC;
```

Cube root procedure:

```
  QBRT: procedure;
  declare R float;
  Q = 1;
ITER: R = Q;
  Q = (2 * Q + X / (Q * Q)) / 3;
  if abs ((Q - R) / Q) > .0001
    then go to ITER;
  return;
  end QBRT;
```

Pascal's triangle with factorial procedure:

```
PASCAL: procedure options (main);
  declare (F1, F, F2, CNR) float;
  declare (N, R, K) fixed;
  do N = 1 to 10;
    do R = 1 to N;
      K = N;
      /* Factorial of N. */
      call FACT;
    R1: F1 = F;
      K = R;
      /* Factorial of R. */
      call FACT;
    R2: F2 = F;
      K = N - R;
      /* Factorial of (N - R). */
      call FACT;
    R3: CNR = F1 / (F2 * F);
      put skip list (N, R, CNR);
      end;
    end;
FACT: procedure;
  declare X fixed;
  F = 1;
  do X = 1 to K;
    F = F * X;
    end;
  return;
  end FACT;
  end PASCAL;
```

Sorting procedure:

```
SORT: procedure;
  declare (V(1:40), T) float;
  declare (M, I, J, K) fixed;
  do J = 1 to M - 1;
    T = V(J + 1);
    do I = J to 1 by -1;
      if T >= V(I) then go to CONT;
      V(I + 1) = V(I);
      end;
    I = 0;
  CONT: V(I + 1) = T;
    end;
  return;
  end SORT;
```

Note that global variables are declared in the main program; nonglobal variables are declared in the procedure.

Words to Remember

procedure
automatic return
call
return
linkage
global variable
global list name

Exercises

1. In each of the following short procedures, which variables are global? Which are not?
   (a) Procedure to interchange the values of two variables;

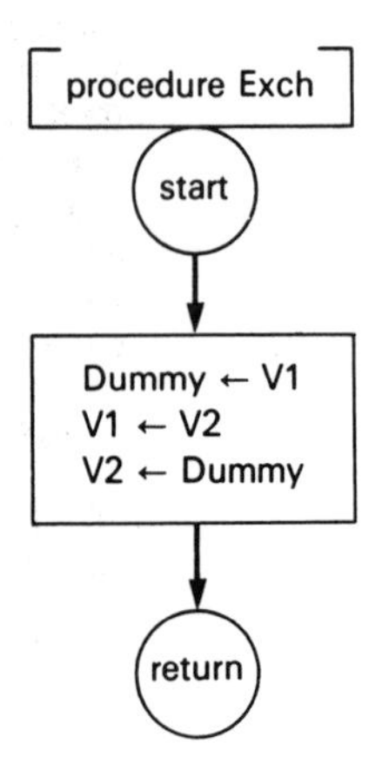
procedure Exch
start
Dummy ← V1
V1 ← V2
V2 ← Dummy
return

   (b) Procedure to find the absolute value of a complex number:

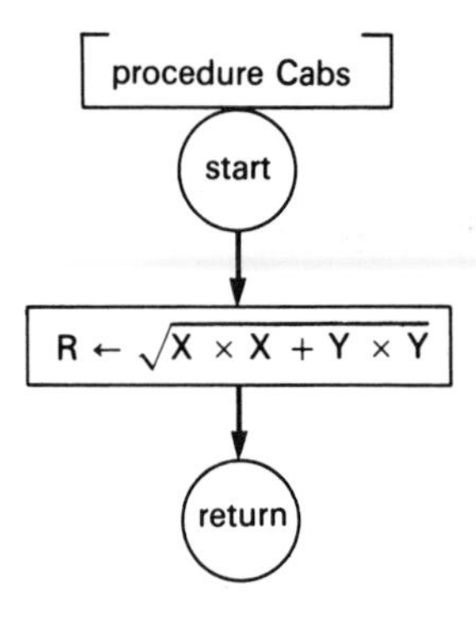
procedure Cabs
start
R ← √X × X + Y × Y
return

   (c) Procedure to find the largest and smallest of a pair of numbers:

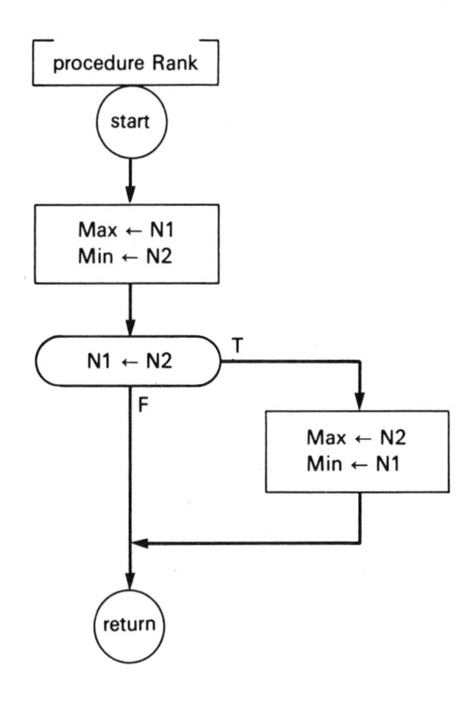

2. Identify the global variables in the program for adding fractions. Explain the need for the instructions I ← NumSum and J ← DenSum at the beginning of the procedure Euclid.
3. Write a main program that will use the cube root procedure given in this section. The main program should read the values of two variables, X and Y; then if the value of X is larger than that of Y, it should compute and print the cube root of X ÷ Y. On the other hand, if Y is larger, it should compute and print the cube root of Y ÷ X.

	Value of X	*Value of* Y
a.	16.3	41.2
b.	144	18
c.	−1	27

4. Write a program that will use the procedure Rank defined in exercise 1(c). The main program should do the following: Read the values of two variables, A and B, and use the procedure Rank to obtain Max and Min, the larger and smaller of A and B. Then compute Max + Min and Max − Min, and use Rank again to find the larger and smaller of these two values. Print A, B, the larger and smaller of A and B; also print the larger and smaller of the values of Max + Min and Max − Min.

	Value of A	*Value of* B
a.	89	144
b.	1	1.001
c.	55	34
d.	−89	144

5. Rewrite exercise 3, using both the Rank and QbRt procedures and the same data.

7.2 *How Can Pointers Be Used to Reference Procedure Arguments?*

The technique described in Section 7.1, for handling global list and string names is appropriate when the main program contains just one list or string that is to be processed by the subprogram. Values as they are generated or read are stored directly in the cells referred to by the global names. These cells may, of course, be used to store different values at different times.

However, if the procedure is to operate on two or more lists, one after the other, and if all the lists must be available at the same time, then the only way to use a global list name is to move values to the corresponding cells before calling the procedure, and to move them out again if necessary after the procedure has been executed. Suppose we have two different lists that are to be used extensively in a computation, and all the operations on the lists will involve the *absolute values*. So we use a procedure AbsList to replace the entries by their absolute values at the beginning of the computation. We must move the FirstList and SecondList entries to and from Value each time we call AbsList:

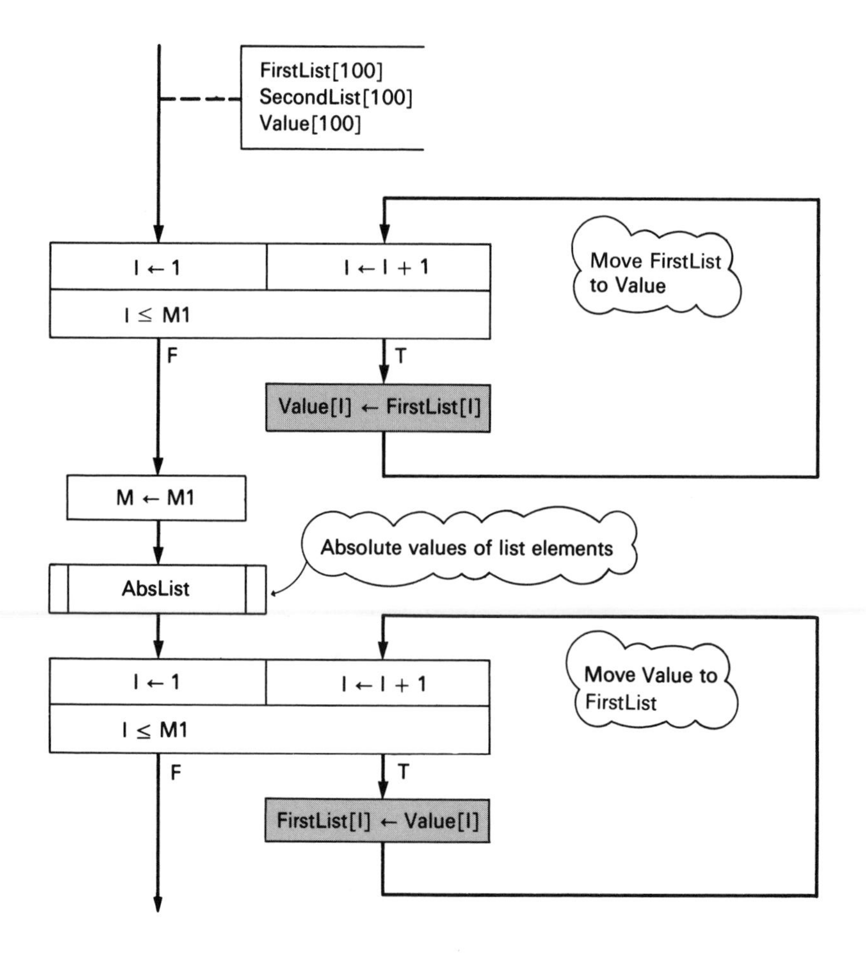

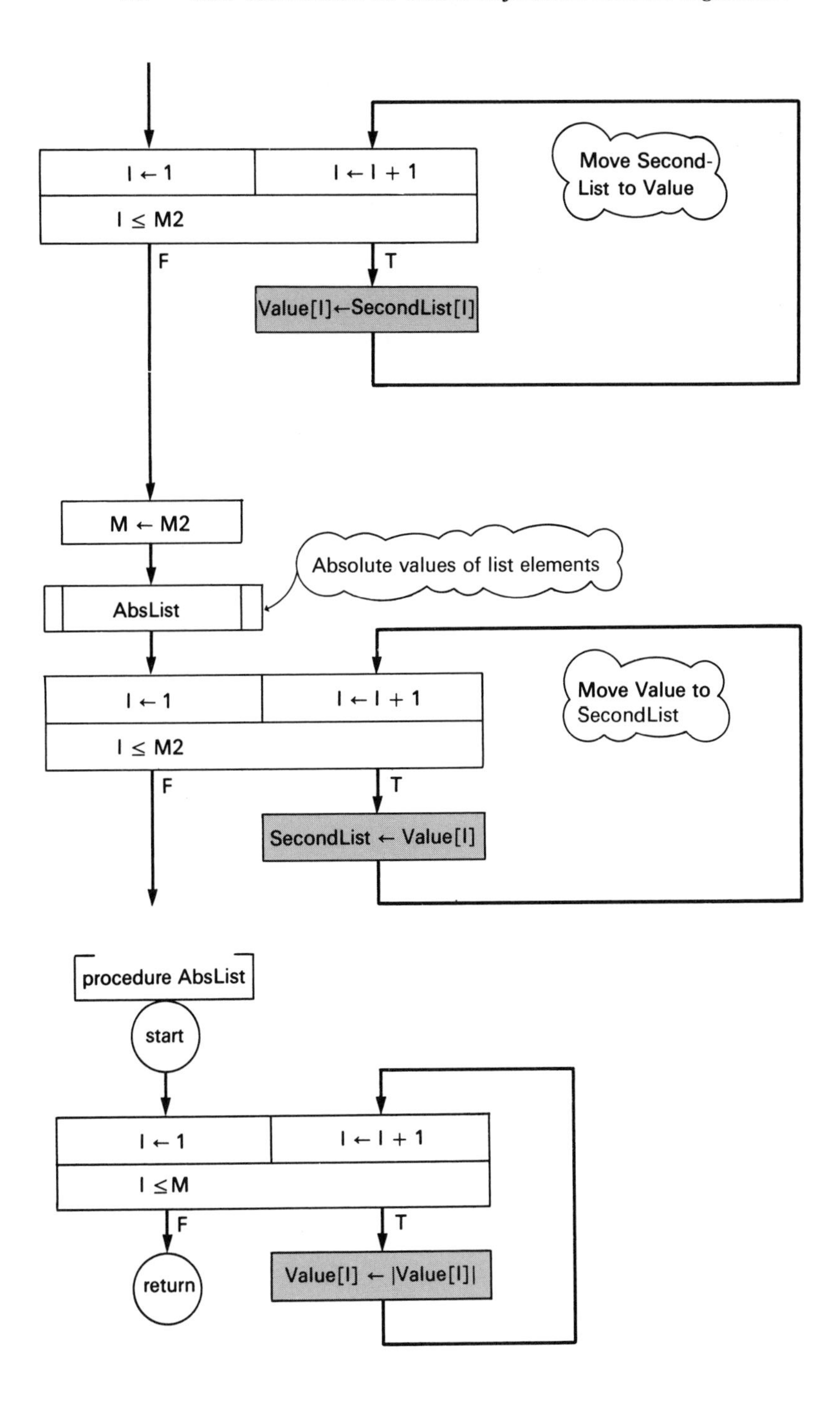

We soon realize that it is more work to move the values of the FirstList and SecondList entries to and from Value than to forget about the procedure and just compute the absolute values in the main program. Actually, when the data is voluminous and the procedure is simple, it is often better to eliminate the procedure and rewrite the instructions in the main program, several times if necessary.

An attractive alternative, in cases where the procedure is slightly less trivial than in the foregoing example, is to use a global variable whose value is a *pointer* to the list. When the main program contains several lists, we merely set the pointer to refer to the proper list before we call the procedure. Of course, the procedure must be written to refer to the list "indirectly" through the pointer.

We have already seen how to obtain a pointer to a string using a structure declaration and an allocate instruction. Here there is no need to increase or decrease the storage allocation during execution of the program, so we place the required number of allocate instructions at the beginning of the main program. Rewriting the previous example, we use the pointers P1 and P2, referring to two lists allocated according to the structure Arg. Thus P1:Arg[Value] and P2:Arg[Value] replace the former list names FirstList and SecondList. We set the global pointer P to the value of the pointer P1 or the pointer P2 before calling AbsList. Thus, P:Arg[Value], with the global pointer P, replaces the former global list name Value.

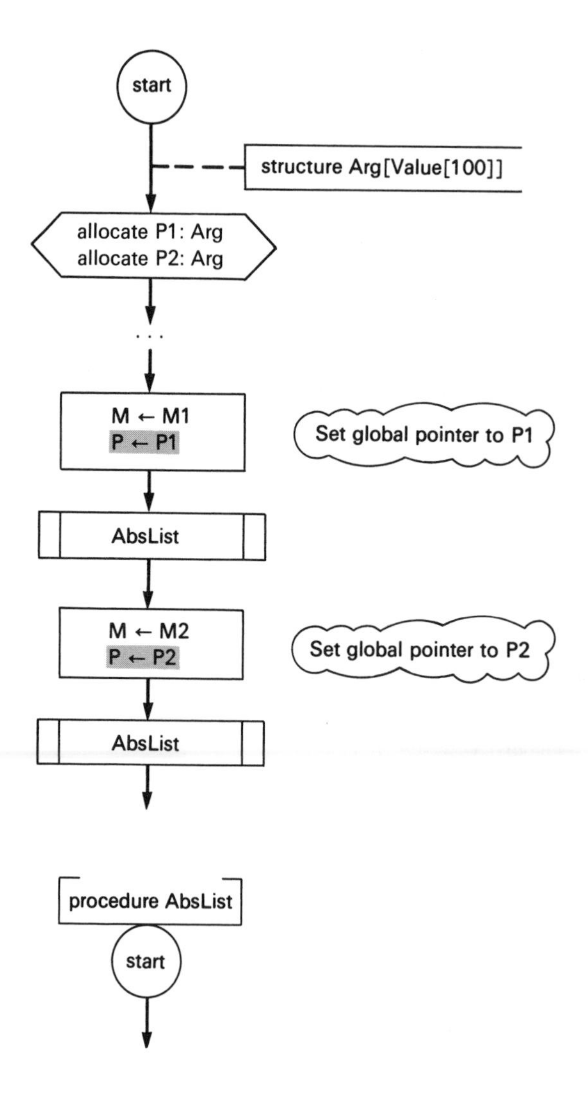

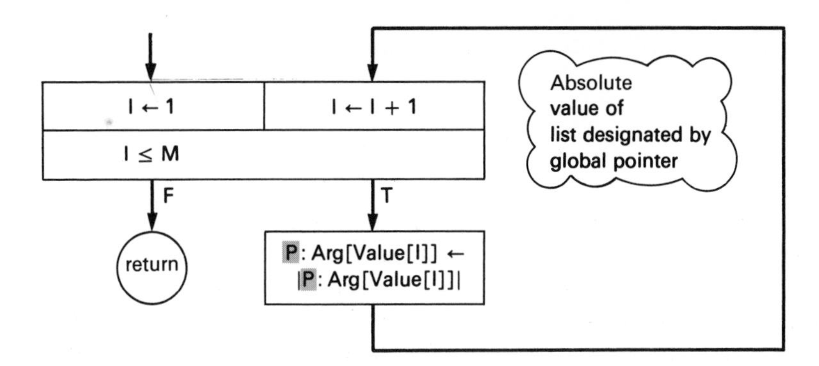

Note that there is a global variable name, M, besides the global pointer name, P.

As another example, suppose there are several different lists to be sorted in the same program. We can write a Sort procedure that uses a global pointer P to reference the list indirectly:

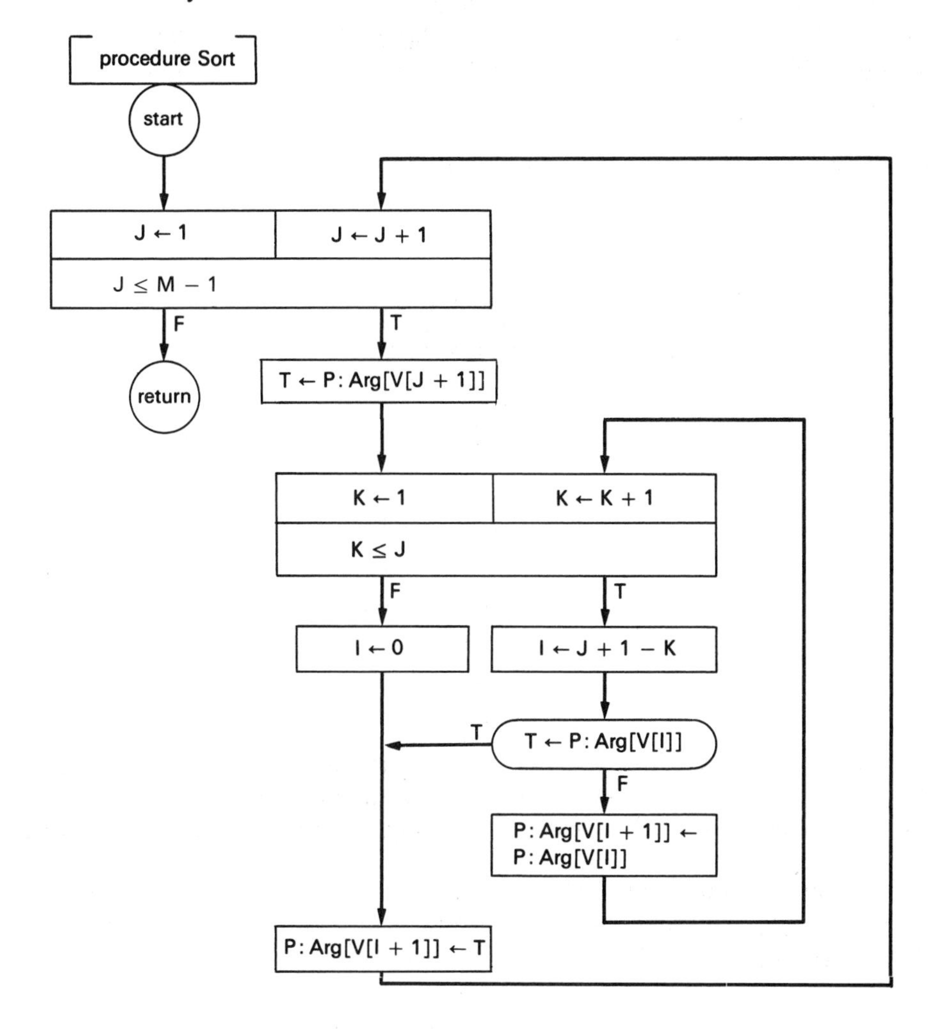

The main program will contain the declaration

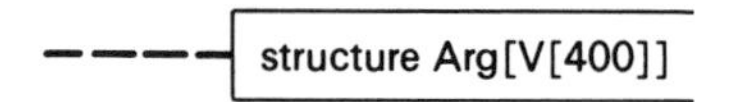

It will also contain allocate instructions for all the different lists to be sorted, if they all need to be kept available separately at the same time. Each allocate instruction will use a separate pointer. When a list is to be sorted, the pointer to it is fetched and assigned to the global pointer P, for use by the procedure. Of course the global variable M, as well as the variables in the list V in the area referenced by the global pointer, must be given a correct set of values before the procedure is called.

Parameters and Arguments Another automatic feature that is available for use with procedure calls consists of the automatic correspondence between *parameters* and *arguments.*

An *argument* is a name included in a call instruction, whose appearance causes a pointer to be created automatically by the compiler. This makes it unnecessary for the programmer to incorporate extra allocate instructions in his program just to obtain the necessary pointers.

Corresponding to an argument in a call instruction, the procedure declaration specifies a *parameter*, which is enclosed in parentheses and follows the procedure label:

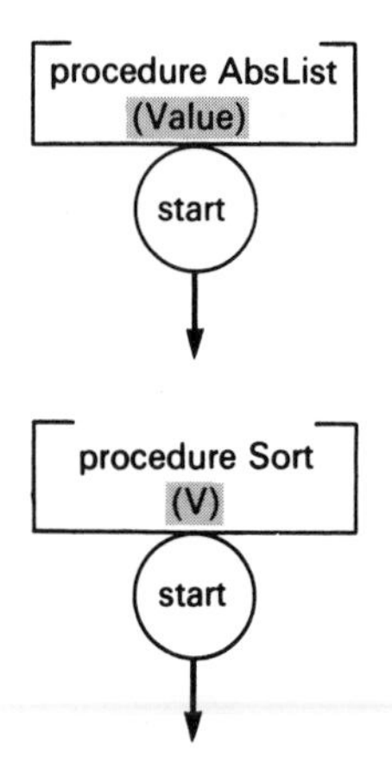

The parameter is used within the procedure as though it were an ordinary name, although it is actually a shorthand way of writing an indirect reference to the corresponding *argument* name, by means of the pointer that has been automatically created.

Rewriting the procedure AbsList to use the list Value as a parameter, we can now call the procedure from different points in the main program, using a different argument (such as FirstList or SecondList) each time.

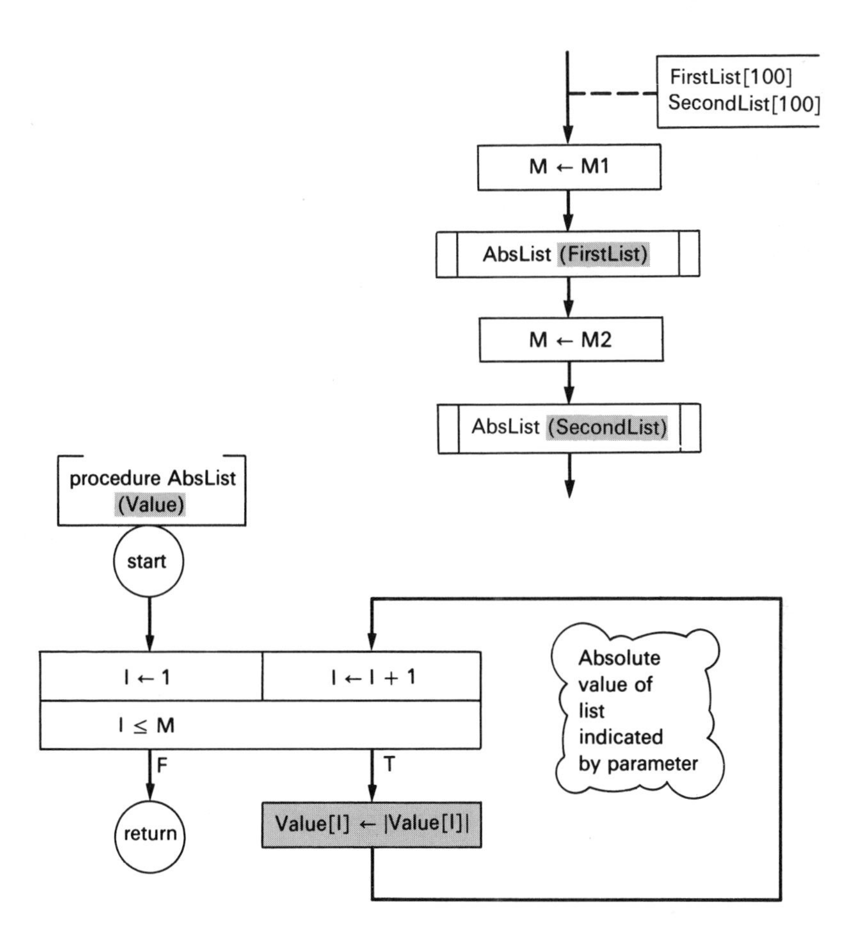

This shorthand notation almost completely obscures the understanding of what is going on when the procedure is called with an argument, although it considerably simplifies the writing of the program and completely eliminates the need for extra allocate instructions. The compiler creates a pointer, which is given a value at the time the call instruction is executed, so that it will point to the list specified as the argument in the call. As contrasted to the pointers created by allocate instructions, those created by a call are not accessible to the programmer. (Pointers to arguments are usually stored in registers rather than in cells.) However, the appearance of the parameter at the beginning of the procedure enables the compiler to set up indirect references to the list and to subscripted variables within the list, when these references occur in the procedure.

The instruction

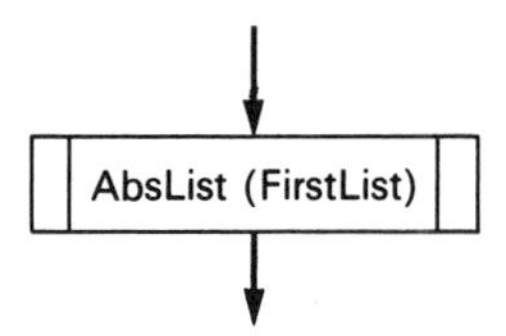

creates a pointer to the list named FirstList and stores it in a place (such as a register) where the procedure expects to find it. When the subscripted variable Value[I] appears in the procedure, the compiler takes into account the fact that the list name has appeared as a parameter and generates indirect references to the array FirstList by means of the pointer (Figure 7.2a).

A procedure may have more than one parameter. In this case, the call instruction will include corresponding arguments. The arguments, which indicate the *actual* lists to be used, correspond *in sequence* to the parameters named at the beginning of the procedures. A pointer is created for each argument, according to its position in the call. References to the various parameters inside the procedure are interpreted as indirect references, by means of the pointers, to the actual lists named as arguments.

The use of parameter lists antedated the provision of explicit pointers (such as those created by the allocate instruction) that are accessible to the programmer. Modern languages, such as PL/1, often provide both facilities. Basic and Fortran have parameter lists but not explicit pointers.

Variables, Expressions, and Constants as Arguments We see the advantage of using pointers when necessary to avoid moving large quantities of data to and from global lists. Once the parameter–argument shorthand has been provided, however, we are tempted by its convenience to use parameters to reference all the values to be processed by the procedure. We note that the program using AbsList with the parameter Value still contained the global variable M. This arrangement is usually more efficient than an extra pointer, but it involves a few extra instructions to move the values (M1 or M2, as the case may be) to the global cell named M.

Even in the earlier example, where we created the pointer explicitly with an allocate instruction, we could have included the variable M in the structure along with the list Value:

---- structure Arg[Value[100], M]

Then in the procedure we could refer to M indirectly with the pointer P.

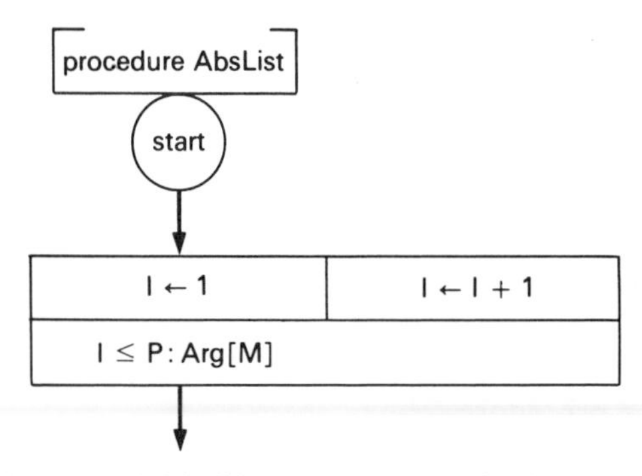

In the same way, we can include the variable M as a parameter in the procedure:

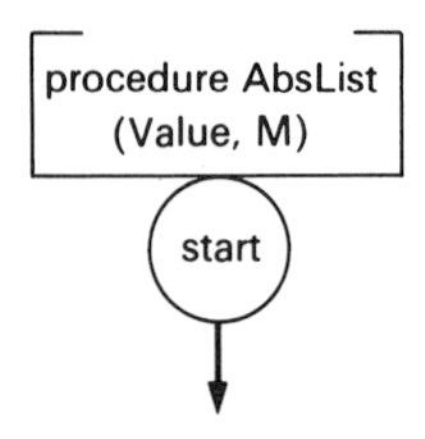

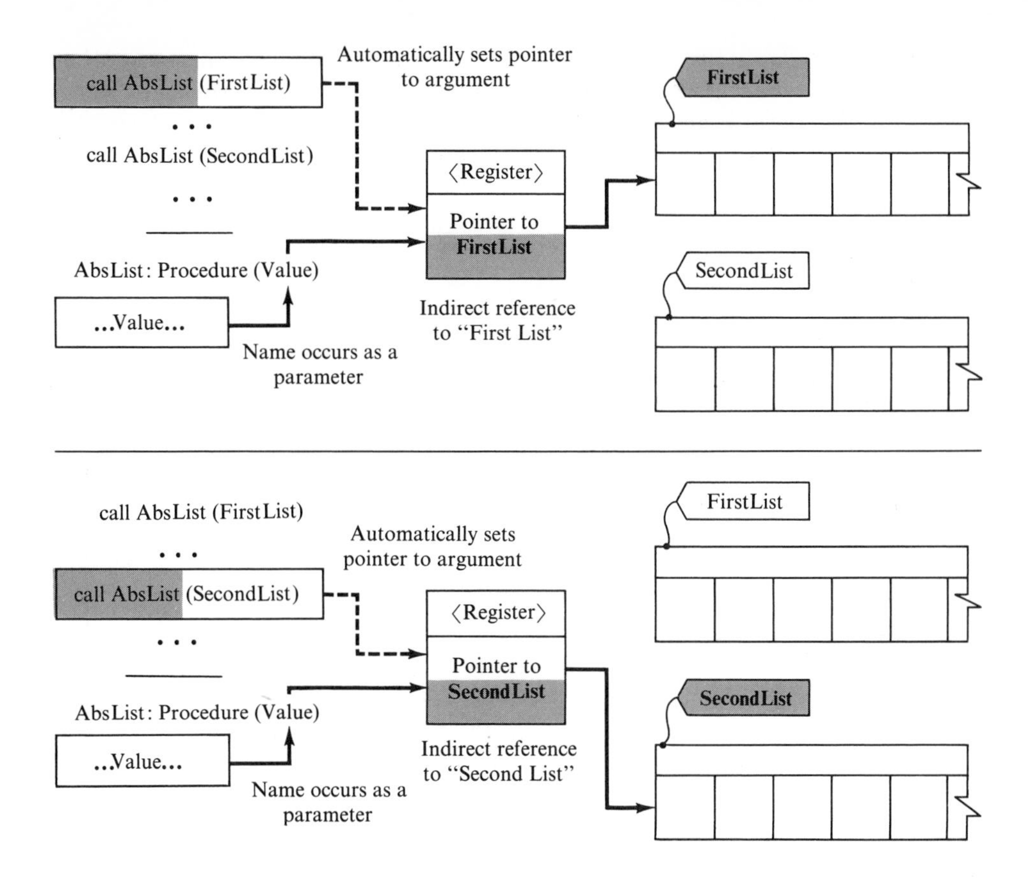

7.2a. Automatic parameter–argument correspondence. A register *holds a* pointer, *thus permitting a* parameter *reference to be interpreted as an indirect reference to the corresponding* argument.

The two calls

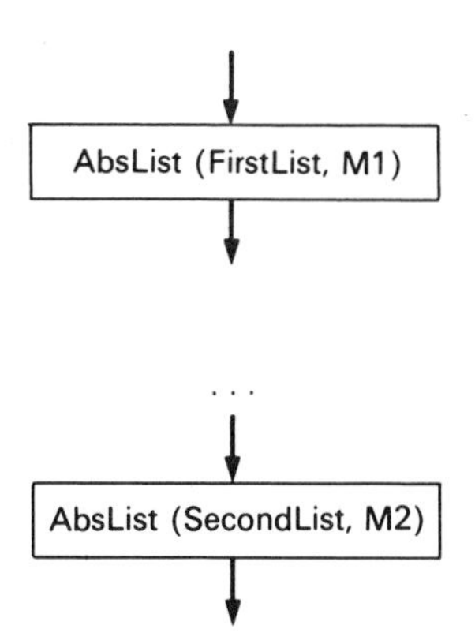

will create a pointer to the variable M1 or M2, besides the pointer to the list FirstList or SecondList. These pointers will be used for indirect references when the names Value and M appear in the procedure.

As another example, using variables but no lists as arguments, we rewrite the Pascal's triangle problem:

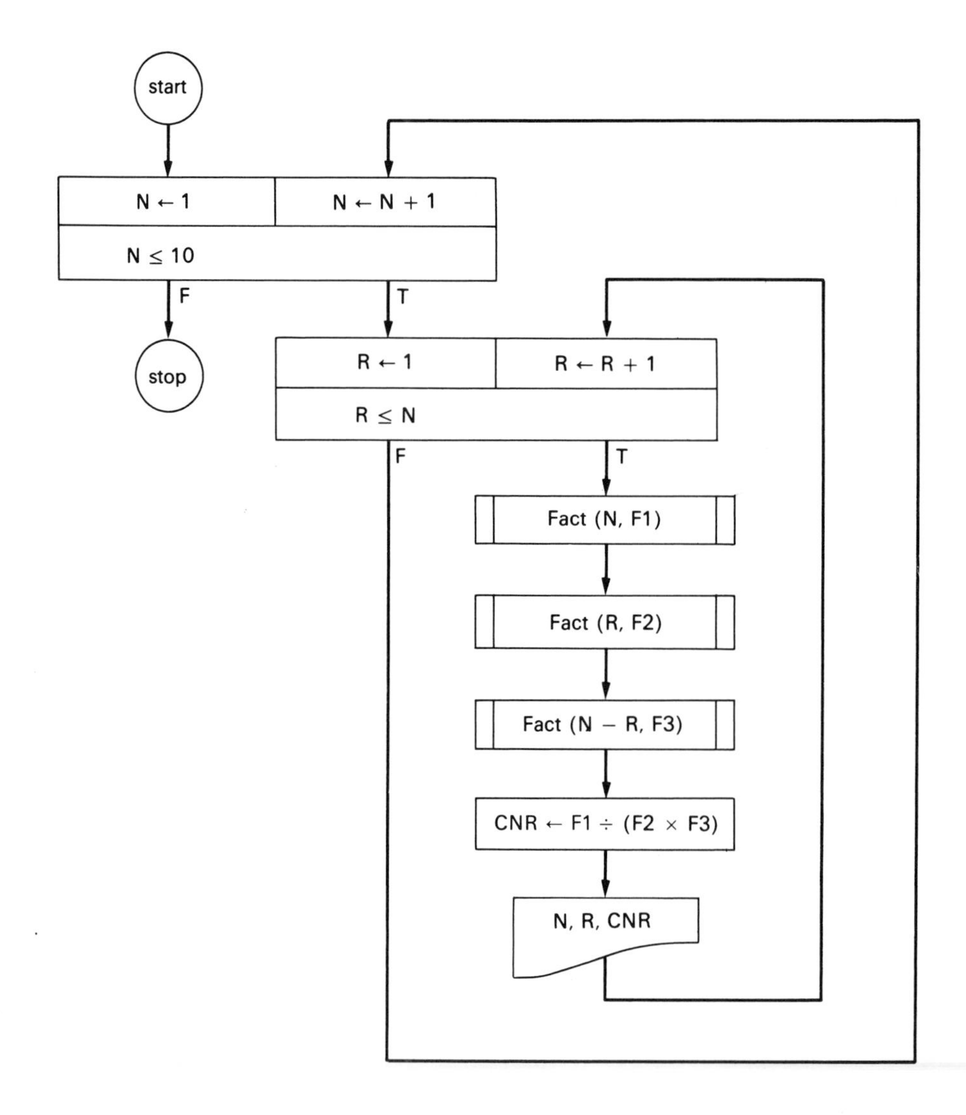

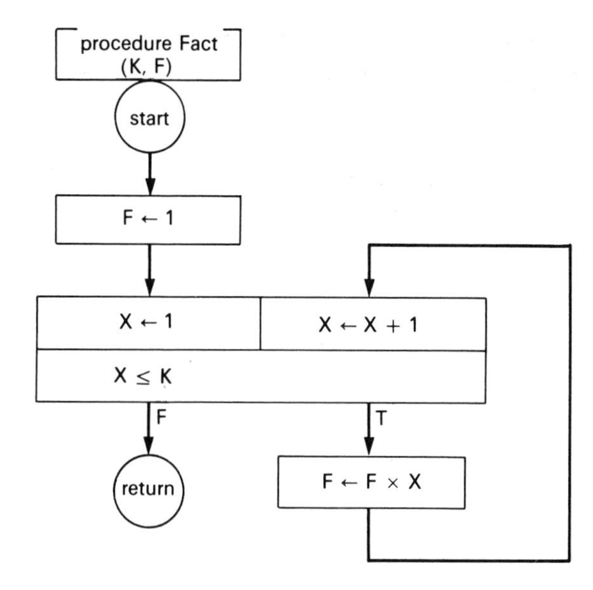

In preparation for each call to Fact, pointers to be used for the parameters K and F are set. In the first two calls, these pointers refer to the cells named N and F1, or R and F2. Less obvious is the meaning of the third call, in which one of the arguments is the *expression* N − R. The compiler interprets this as an instruction to compute the value of the expression, store it in a cell inaccessible to the programmer, and assign the address of this cell as a value of the pointer. Constants used as expressions are handled in the same way, the value being stored in an inaccessible cell referred to by a pointer (except in Fortran). For variables, however, the pointer refers to the cell named by the variable and not to an inaccessible cell to which the value is moved. This permits variables to be used as arguments indicating where the procedure is to store the *results* of a calculation, as well as to indicate where the procedure should obtain its data.

One difference should be noted between the use of global variables and the use of pointers. The pointers are set *before* the call is executed, even though some of the arguments are the names of variables whose values are generated inside the procedure. Using global variables, on the other hand, we may need to move such values *after* the procedure is called. (Compare this version of the Pascal's triangle program with the version in Section 7.1.)

Functions There is another way of returning a value from a procedure to the main program. Certain procedures, called *functions*, associate a value with the function name itself. As we have seen, some functions are provided by the computer system, such as functions for computing absolute value, integer part, or square root. A function reference can be included in an arithmetic expression. Thus, the use of functions, where appropriate, tends to make the program neater than the use of procedures with the call instruction.

When we write a function, we must indicate the expression whose value is to be returned (associated with the name of the function in the main program). We write this expression in the return instruction. For example, we can make the factorial procedure into a function, because its purpose is to generate a single value:

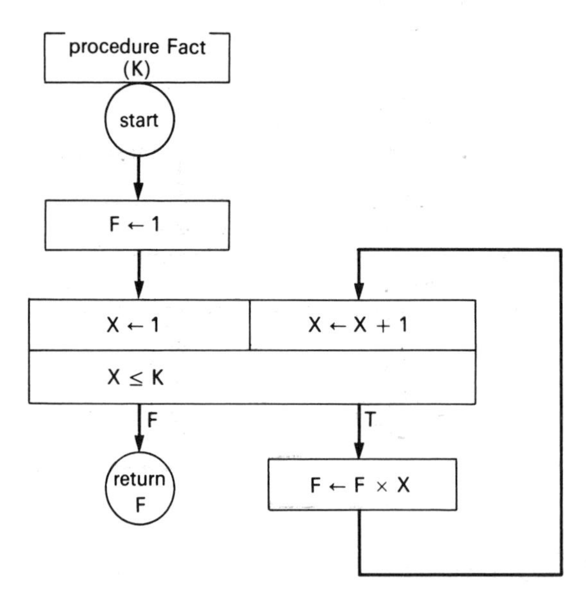

The expression F in the return instruction designates the value to be returned. Notice how short the Pascal's triangle program becomes when the call instructions are replaced by function references:

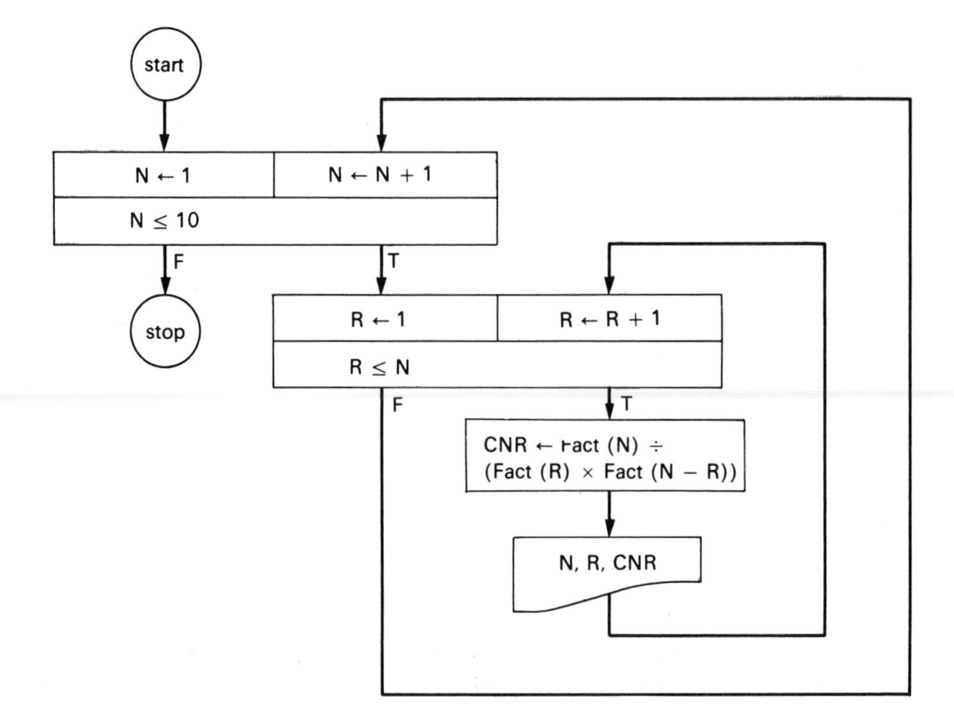

Labels as Parameters The compiler, when it generates the instructions of the machine-oriented object program, allocates a cell to each machine instruction. Thus, there is a machine address corresponding to each label and each variable in the human-oriented source program. Labels are included in the symbol table along with variables and other names. Therefore, the compiler can create a pointer whose value is the address of an instruction, corresponding to a label.

Thus, we can have procedures with labels as arguments. For example, a factorial procedure might detect an error when its argument is negative. In the main program, we might want to print a message if this occurred. Labeling the print instruction, we give the label to the procedure as an argument:

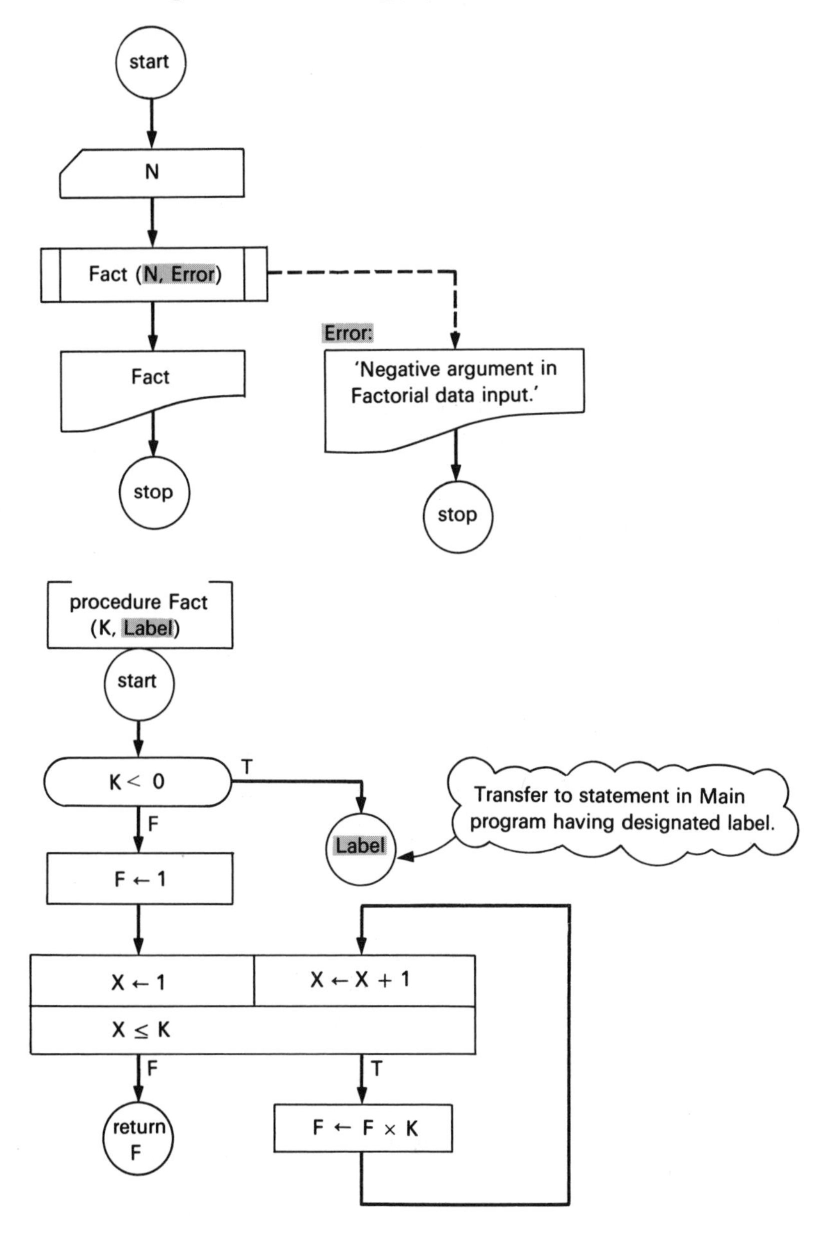

In another important case, we have a procedure that calls another procedure. We may want the first procedure to call different procedures when we use it at different points in the main program. For example, suppose we want to evaluate several different functions at the points (0.00, 0.01, 0.02, . . . , 0.99, 1.00) between 0 and 1 in steps of 0.01, and we want to find out how many of the 101 function values, for each function, are negative. We can write a procedure to find and count the 101 function values; this procedure will in turn call the procedure corresponding to the function being evaluated.

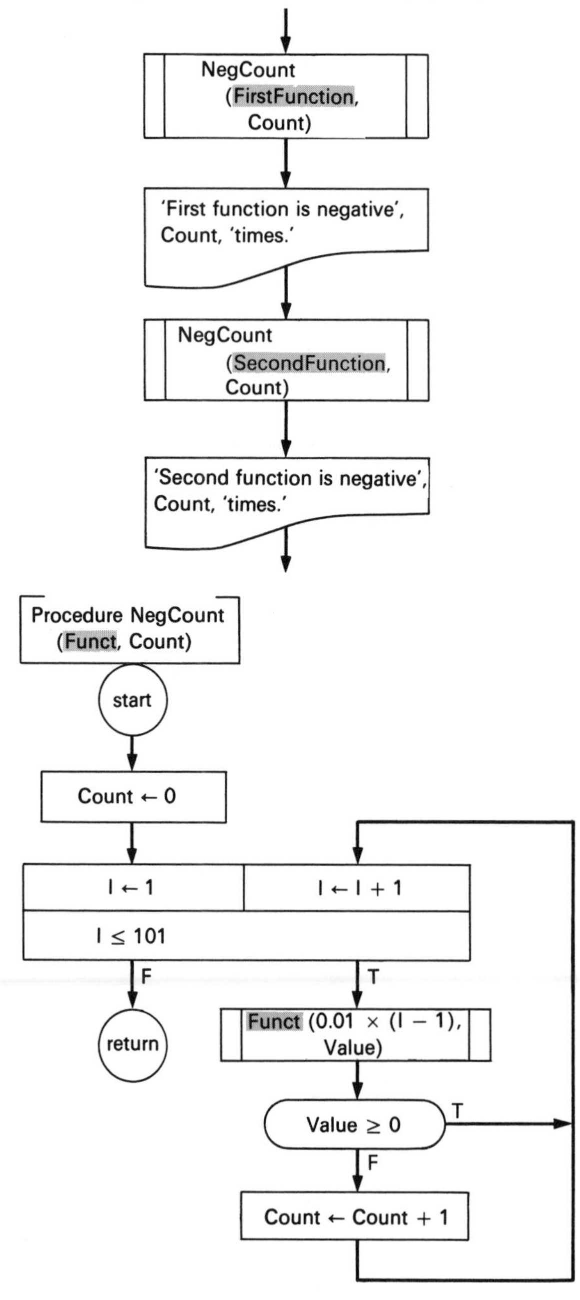

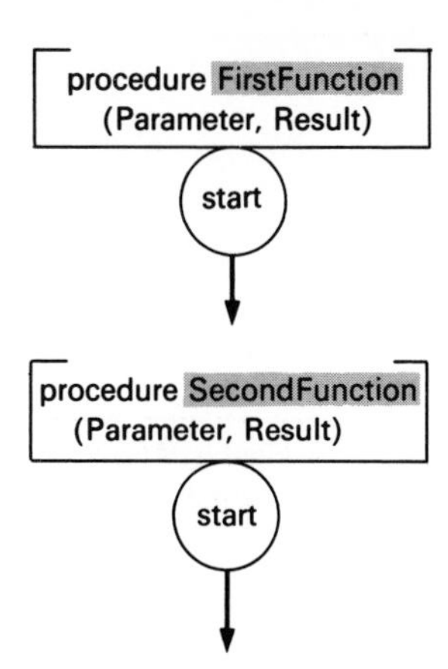
procedure FirstFunction
(Parameter, Result)
start
procedure SecondFunction
(Parameter, Result)
start

Flowchart

Lists and variables as arguments:

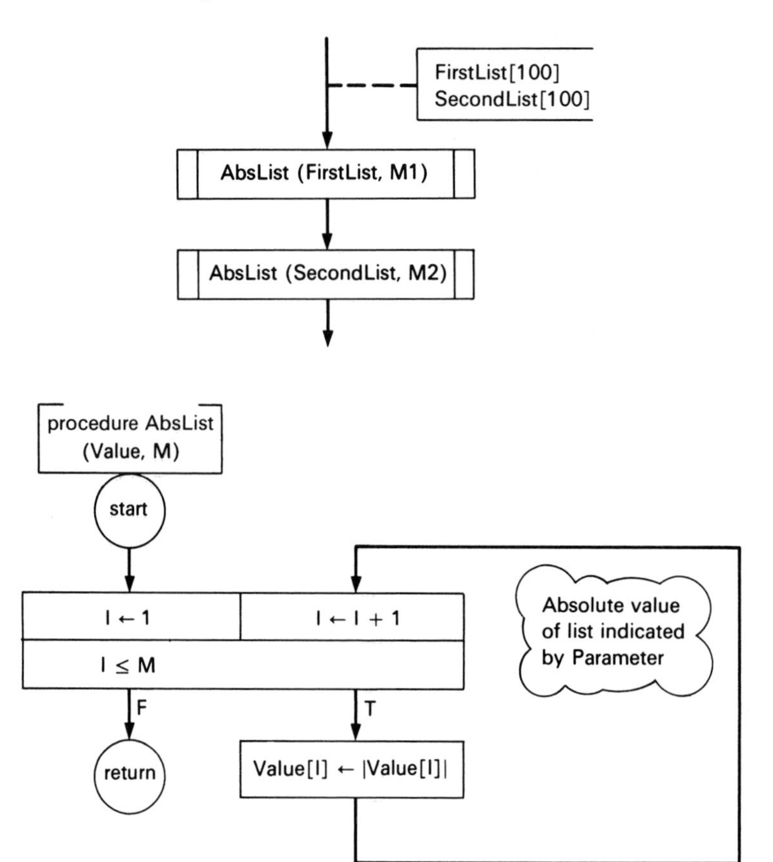

Variables and expressions as arguments:

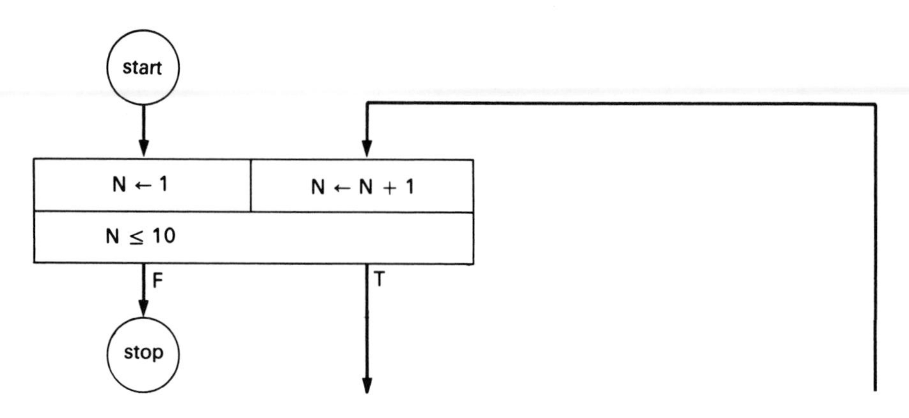

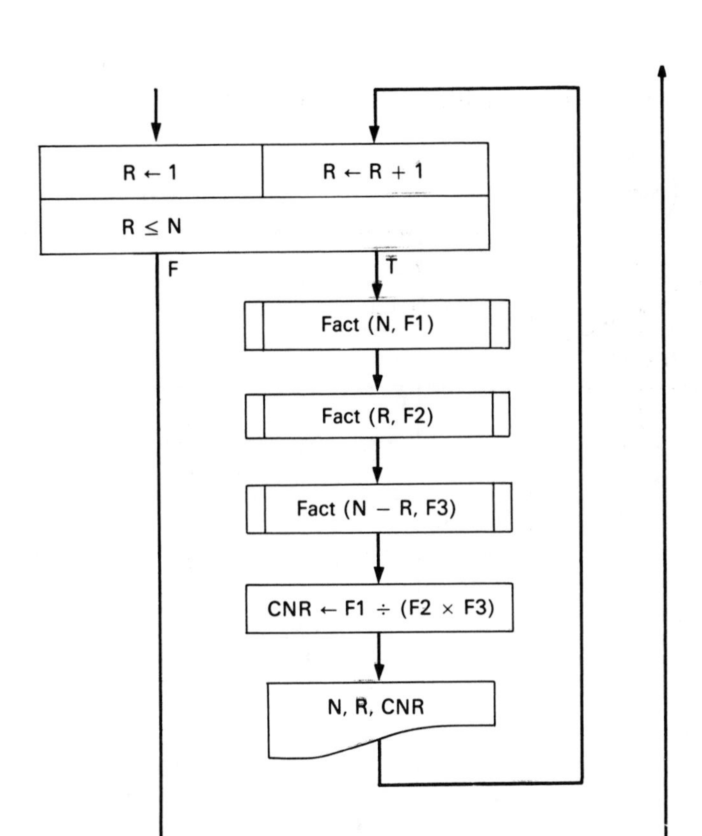
R ← 1
R ← R + 1
R ≤ N
F
T
Fact (N, F1)
Fact (R, F2)
Fact (N − R, F3)
CNR ← F1 ÷ (F2 × F3)
N, R, CNR

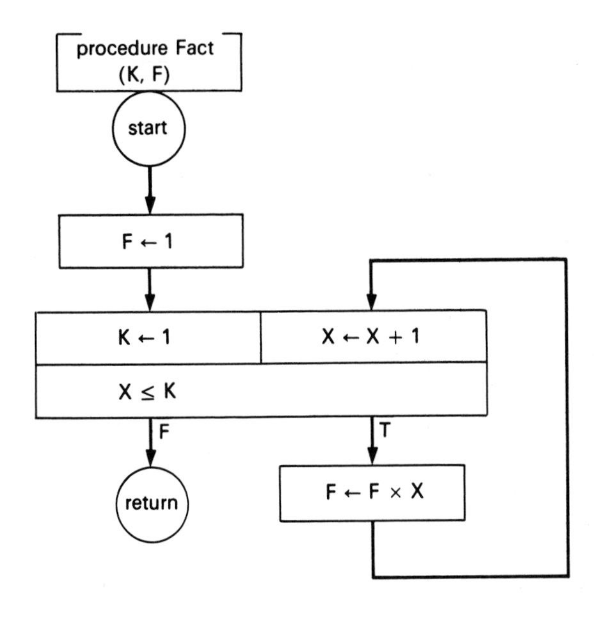
procedure Fact
(K, F)
start
F ← 1
K ← 1
X ← X + 1
X ≤ K
F
T
return
F ← F × X

Pascal's triangle program with factorial function:

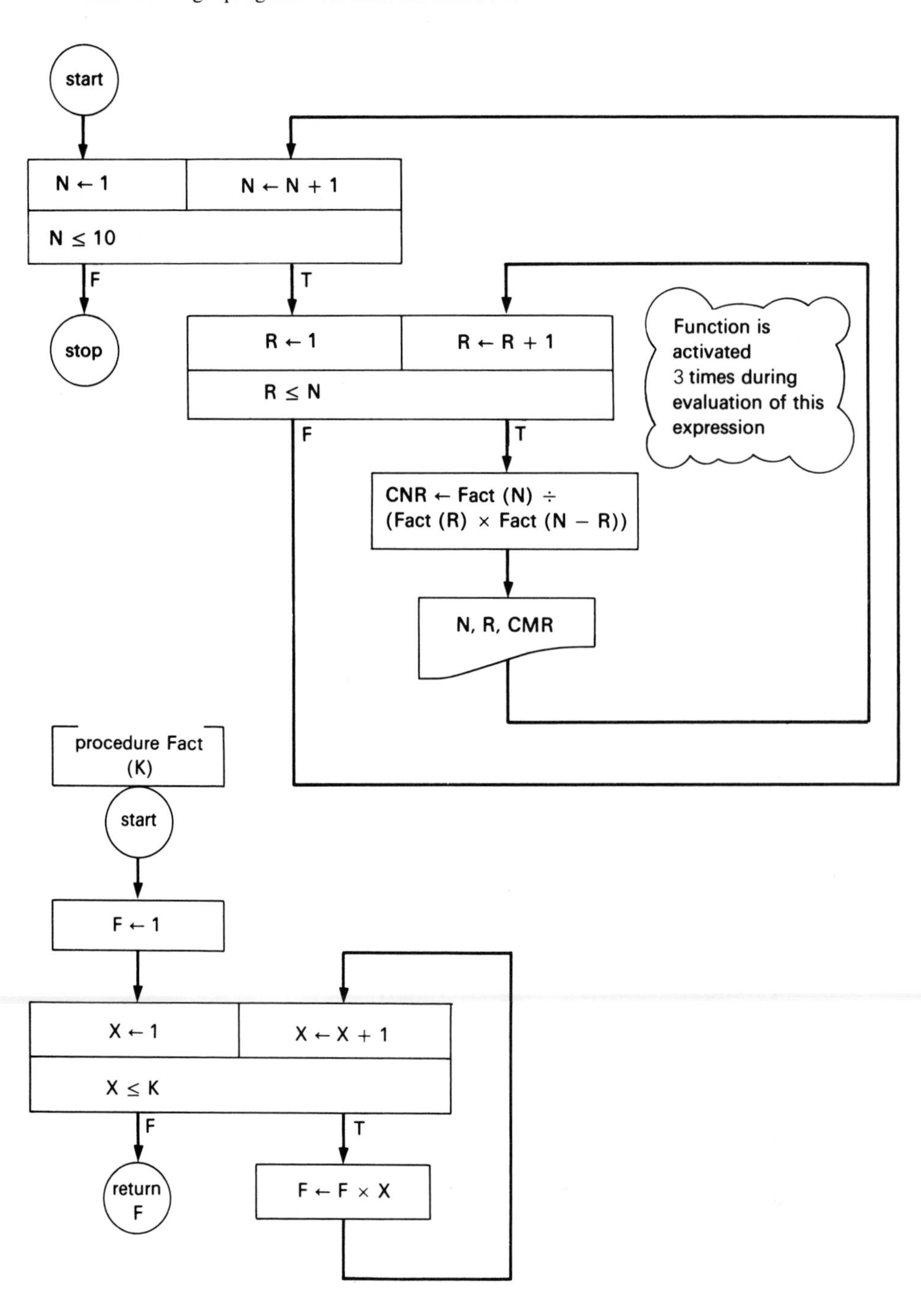

Basic Arguments in Basic can be used only with *functions*. A function must have a name consisting of three letters, the first two of which must be FN. The group of instructions[1] defining the function must begin with a DEF instruction and end with FNEND, and must include an assignment instruction with the function name on the left side.

Function:

```
110 FOR N = 1 TO 10
120    FOR R = 1 TO 10
125    REM: FUNCTION ACTIVATED 3 TIMES
127    REM: TO EVALUATE THIS EXPRESSION
130    LET C=FNF(N)/(FNF(R)*FNF(N-R))
140       PRINT N, R, C
150       NEXT R
160    NEXT N

200 DEF FNF (K)
210    LET F = 1
220    FOR X = 1 TO K
240       NEXT X
250    LET FNF = F
260 FNEND

900 END
```

[1] Many computers restrict the use of functions in Basic to those that can be defined by a single expression, in the form

DEF *name* (*parameter list*) = *expression*

where the expression may involve any or all of the listed parameters.

Fortran

Lists and variables as arguments:

```
      DIMENSION FIRST(100), SECOND(100)
      . . .
      CALL ABSL (FIRST, M1)
      CALL ABSL (SECOND, M2)
      . . .
      END

      SUBROUTINE ABSL (VALUE, M)
C     ABSOLUTE VALUE OF LIST
C     INDICATED BY PARAMETER
      DIMENSION VALUE(1)
      DO 19 I = 1, M
        VALUE(I) = ABS (VALUE(I))
19      CONTINUE
      RETURN
      END
```

Variables and expressions as arguments:

```
      DO 119 N = 1, 10
        DO 118 IR = 1, N
          CALL FACT (N, F1)
          CALL FACT (IR, F2)
          CALL FACT (N - IR, F3)
          CNR = F1 / (F2 * F3)
          WRITE (6, 87) N, IR, CNR
87        FORMAT (2 I 10, F 15.3)
118       CONTINUE
119     CONTINUE
      STOP
      END

      SUBROUTINE FACT (K, F)
      F = 1.0
      DO 219 IX = 1, K
        F = F * FLOAT (IX)
219     CONTINUE
      RETURN
      END
```

Function:

```
      DO 119 N = 1, 10
        DO 118 IR = 1, N
C         FUNCTION IS ACTIVATED THREE
C         TIMES DURING EVALUATION OF
C         THIS EXPRESSION
          CNR = FACT (N) / (FACT (IR) *
     1      FACT (N - IR))
          WRITE (6, 87) N, IR, CNR
```

```
87        FORMAT (2 I 10, F 15.3)
118       CONTINUE
119     CONTINUE
      STOP
      END

      FUNCTION FACT (K)
      F = 1.0
      DO 219 IX = 1, K
        F = F * FLOAT (IX)
219     CONTINUE
      FACT = F
      RETURN
      END
```

PL/1

Lists and variables as arguments:

```
LARGE: procedure options (main);
  declare (FIRSTLIST(1:100),
     SECONDLIST (1:100), M1, M2)
     float;
  . . .
  call ABSLIST (FIRSTLIST, M1);
  call ABSLIST (SECONDLIST, M2);
  . . .
ABSLIST: procedure (VALUE, M);
  /* Absolute value of list indicated by
     parameter */
  declare VALUE (1:*) float;
  declare M fixed;
  do I = 1 to M;
     VALUE(I) = abs (VALUE(I));
     end;
  return;
  end ABSLIST;
```

Variables and expressions as arguments:

```
PASCAL: procedure options (main);
  declare (N, R, F1, F2, F3, CNR) fixed;
  do N = 1 to 10;
     do R = 1 to N;
        call FACT (N, F1);
        call FACT (R, F2);
        call FACT (N - R, F3);
        CNR = F1 / (F2 * F3);
        put skip list (N, R, CNR);
        end;
     end;
FACT: procedure (K, F);
  declare (K, F, X) fixed;
  F = 1;
  do X = 1 to K;
     F = F * X;
     end;
  return;
  end FACT;
  end PASCAL;
```

Function:

```
PASCAL: procedure options (main);
  declare (N, R, CNR, FACT returns) fixed;
  do N = 1 to 10;
     do R = 1 to N;
        /* Function is activated 3 times
           during evaluation of this
           expression. */
```

```
        CNR = FACT (N) / (FACT (R) *
          FACT (N - R));
        put skip list (N, R, CNR);
        end;
      end;
FACT: procedure (K) returns fixed;
   declare (K, X, F) fixed;
   F = 1;
   do X = 1 to K;
      F = F * X;
      end;
   return (F);
   end FACT;
   end PASCAL;
```

Words to Remember

parameter argument function

Exercises

1. Rewrite as functions the following procedures presented in Section 7.1. Use parameters to eliminate all global variables.
   (a) the Euclidean algorithm (return GCD).
   (b) absolute value of a complex number (see exercise 1b, Section 7.1).
   (c) cube root.
2. Write a program that will read a table of distances and elevations, such as Table 6.1a, arranged in order of increasing distances. Include also a function which, when given this table and a distance value as arguments, will search for the nearest table entry and perform linear interpolation (as described in Section 6.1). The function should return, as its value, the elevation corresponding to the given distance value.
3. Write a function to count the number of negative items in a list. Write a main program to read the items of a list, use the function, and print the number of negative items read.
4. Write a function to count the number of *blanks* in a string. If the string is carefully prepared, this is the same as the number of words. Write a main program to read two strings, and print the string having the largest number of words (that is, the largest number of blanks).
5. Write a function to evaluate a polynomial, given the degree, the list of coefficients, and a value of the variable.
6. (Requires features not available in Basic.) Write a procedure to find an approximate root of a function by the "method of bisection." The procedure must be given two argument values, A[1] and A[2], such that the values of the function at A[1] and A[2] are opposite in sign. (Check for this, and assume that an error has occurred if this requirement is not met.)

   Evaluate the function at the midpoint, 0.5 × (A[1] + A[2]). Replace either A[1] or A[2] by this midpoint value, making the choice in such a way that the values of the function at the midpoint, and at the end that is not replaced, are still opposite in sign. Repeat this operation, until the distance between A[1] and A[2] is less than a given tolerance.

   The arguments to the procedure are
   (a) the name of the function whose root is to be found;
   (b) a list of length two, consisting of A[1] and A[2];
   (c) a variable, whose value is the tolerance to determine when to stop.

   When the distance has been reduced below the required tolerance, the procedure will return with the new "interval of uncertainty" in the list A.
7. (Requires features not available in Basic.) Given a function, as well as the left and right endpoints of an interval on which the function is defined, and the number of subdivisions desired, the *trapezoidal* approximation to the *integral* of the function may be calculated as follows:

Let x_0 be the left endpoint, and let x_n be the right endpoint, where n is the given number of subdivisions. For i between 0 and n, let

$$x_i = x_0 + i \cdot (x_n - x_0)/n$$

The approximate integral of the function is then

$$Trap = \tfrac{1}{2}f(x_0) + f(x_1) + f(x_2) + f(x_3) + \cdots + f(x_{n-2}) + f(x_{n-1}) + \tfrac{1}{2}f(x_n)$$

where f is the given function.

Write a procedure whose arguments are the function name, the endpoints of the interval, and the number of subdivisions. The procedure should compute the trapezoidal approximation to the integral of the given function over the designated interval.

8. This section includes a Sort procedure that uses explicit pointers based on allocated storage. Simplify this procedure, taking advantage of the automatic parameter–argument correspondence feature.

7.3 *Program Organization*

In Section 7.1, it was pointed out that program structure may be introduced for any of several reasons:

(a) to segment a large, complicated program into comprehensible chunks;
(b) to minimize unwanted effects due to changes in a part of the program;
(c) to isolate a program segment in order to locate a mistake;
(d) to make it easier to reuse a group of instructions in a different program;
(e) to avoid rewriting a group of instructions that is used more than once in a program.

So far, we have emphasized the first and last of these reasons. We now consider how a program can best be organized to isolate segments, and to make it easier to reuse a procedure in another program.

Local Variables A procedure may involve both global variables and pointers (which are simply a special category of global variables). Besides these, however, there are usually a number of other variables in a procedure—for instance, X in Fact; I, J, K, and T in Sort; J in CharC; R in QbRt; I, J, and R in Euclid. These are *local* variables—they are supposed to have no meaning outside the procedure. However, the existence of these extra variables can be an annoyance to the programmer. If the program is long, he may inadvertently use the same variable elsewhere in the program. It will then be treated as a global variable, and will refer to the same cell whether it is used in the procedure or elsewhere. When the procedure is executed, the contents of this cell may change, with unexpected results.

Isolating the procedure will not be effective unless we can "localize" the variables that need not be globally known, thus minimizing undesirable side effects.

An extreme approach would be to outlaw global variables, and use pointers (via parameters lists, for example) exclusively. The only global variables would be pointers to arguments, which are inaccessible to the programmer. However, even though the use of parameters may slightly simplify the writing of a program, the use of pointers introduces some inefficiency. It would be nice to permit both local and global variables in a program, and let the programmer decide which to use.

The problem is solved in most programming languages by *declaring* local variables in the procedure. For example,

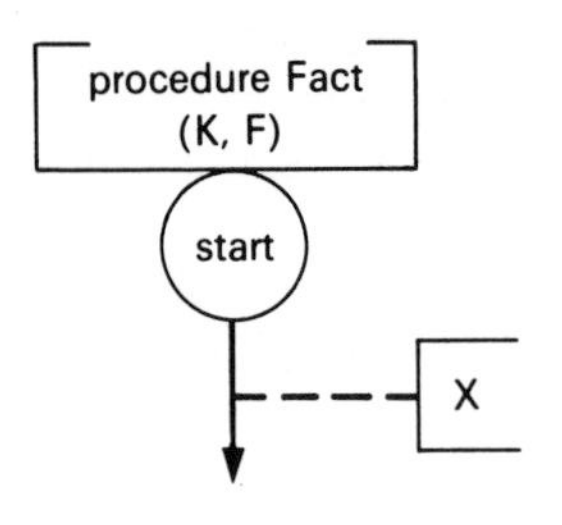

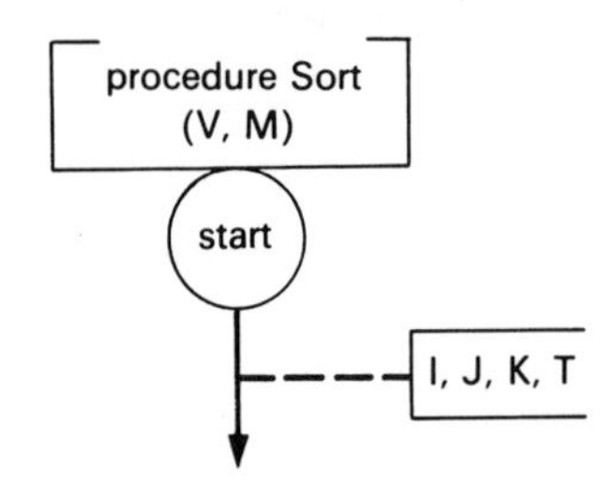

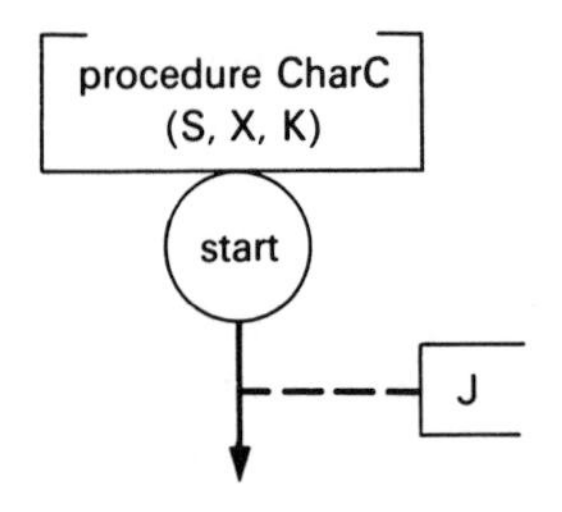

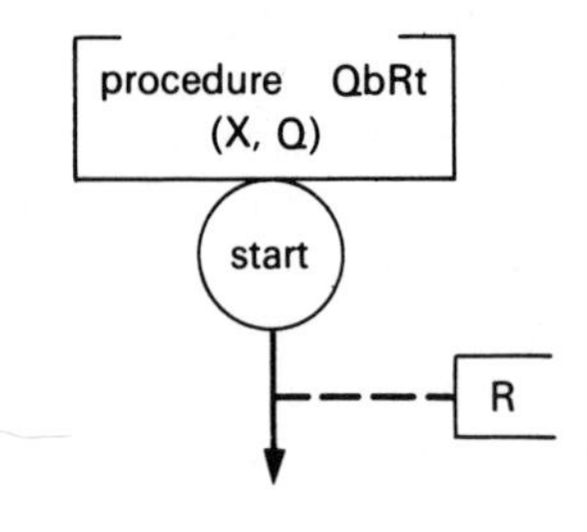

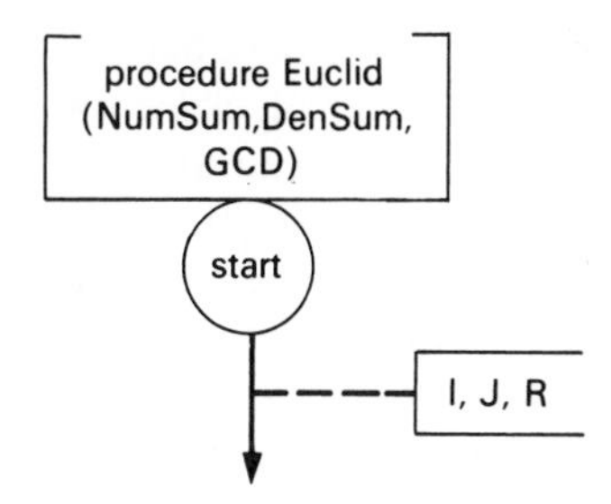

In Basic, however, all variables are global; in Fortran all variables except parameters and COMMON variables (see Section 7.1) are local.

Local Variables in Blocks A block is a group of instructions that is to be treated as a unit for some purpose. A procedure is a special kind of block that is activated by a call instruction and deactivated when a return instruction inside the procedure is executed; accordingly, a procedure is usually written in a place apart from the normal "flow" of instructions. A procedure must have a label and may have parameters and local variables. The principal difference between procedures and other blocks is that the latter are written "in line" and executed when reached in the normal instruction sequence. Thus, a block that is not a procedure does not have a return instruction.

Any block, however, may contain declarations for local variables, including lists, strings, and rectangular arrays. The advantage, for the programmer, is that he can use a name in a block (to refer to a local cell or group of cells) which is also used outside the block to refer to a different cell.

Blocks may contain other blocks inside them; this means in particular that a block may contain procedures or a procedure may contain blocks. Thus the terms "global" and "local" are relative. A more useful concept is the *scope* of a name. A global variable has the whole program as its scope. If an inner block includes declarations, the scope of names declared there is limited to the inner block. The scope of a name is a certain portion of a program, characterized by the fact that the name refers to the *same cell*, throughout. Declaring the same name in an inner block associates a *different cell* with those references to the name that occur within the inner block; hence, the inner block is not included in the scope of the first version of the name.

In many modern programming languages, new storage is allocated *at the time of entry* to a block (including a procedure) for all the names appearing in declarations at the beginning of the block. This dynamic storage allocation makes it possible to provide for *recursive* procedure calls. That is, a procedure may call itself, because new storage is allocated for the local variables each time the procedure is called. Especially for recursive procedure calls, the concept of *scope* of names attains paramount importance.

External Procedures In the program structures we have been considering so far, the instructions form a single unit that may contain procedures and other blocks inside it. Names not declared in any of the internal blocks are, by default, global to the main program. The compiler must have access to the entire program when it is generating machine code for the inner blocks, because the interpretation of names, expressions, instructions such as read and allocate, and so forth, may depend on declarations and other information relating to the main program.

However, sometimes we want to create a procedure in a completely independent environment from the main program. We may want to produce a separate portion of object program for a procedure, so that it can be used with various main programs at different times without having to be recompiled. Such an *external procedure* would have much the same relation to the main program as the standard functions (square root, absolute value, and integer part), which can be used in a program although the instructions defining them are not included in the program unit.

Combining segments of object programs that have been generated separately, at different times, involves one extra task. The actual cells to be occupied by the instructions (and the data) of each segment cannot be predicted when the object program instructions are created. Therefore "relative" addresses are allocated—each instruction, variable, and other unit of data used in a segment is given an address *relative* to the beginning of the segment. The object program, in this unfinished form with relative addresses, is written out on a tape or otherwise made available for later use.

Just before execution, the segments that are needed must be gathered together, and the actual "absolute" cells to be used for each segment must be determined. Then the object program instructions are converted to finished form, all the relative addresses being modified to produce the absolute addresses needed during execution. This work of gathering the segments and modifying them is done by a *loader* program. The loader first loads the main program, detecting references to procedures that are not included within the main program. Then it scans the "user library" of program segments that have been generated previously by the user. The loader will presumably find in the user library all the procedures referred to in the main program that have not already been found, except for those standard procedures provided by the system. It then scans the "system library" and loads these standard functions. All the segments found on the user library and on the system library must be modified because they are in relative address form. After the loader completes all this work, execution can finally begin.

Words to Remember

local variable	external procedure	library
scope	loader	

Appendix A: Glossary

Absolute value (2.8)* The numerical value obtained by ignoring the sign of a number. A standard function for performing this operation is available in most programming languages.

Access (1.2) Finding a desired mark on an information storage medium. If the storage device must be physically moved, this may decrease the efficiency of the computation process.

Address (2.9) The internal name of a cell used in machine-oriented instructions. It consists of a number giving the consecutive position of the cell among all the cells in the computer. (Compare *Byte addressing.*)

Algorithm (3.1) A network of paths, leading from each initial state (of some system having a finite number of states) to some terminal state. A path is specified by designating exactly one transition from any state that it reaches. Thus paths may join, but must not fork nor loop.

Allocate (6.3) To make available to the program an area of storage, from a pool of available space. The allocate instruction includes the name of a structure (thus determining the amount of storage to be allocated) and of a pointer (whose value is set, to permit the area to be referenced).

* The number in parentheses refers to the section in the text where the concept is first used or defined.

Alphabetic information (1.1) Information (marks) in the form of letters, words, etc., used primarily in non-numerical applications of computers.

Argument (2.8, 7.2) An expression or name, included in an instruction that activates a procedure or function, to specify a particular cell or group of cells to be operated on.

Arithmetic operator (2.5) One of the four symbols $+$, $-$, $\times$, and $\div$ used in an expression to specify an operation of ordinary arithmetic. For typographical reasons, many languages use $*$ to designate multiplication and / for division.

Arithmetic unit (2.5) The portion of the internal computer hardware where arithmetical manipulations are performed.

ASCII (5.1) American Standard Code for Information Interchange—a widely used standard code, with 96 characters including upper- and lower-case letters, each character being coded in 7 bits (or 8 bits including parity).

Assertion (3.4) Part of a conditional instruction. If the assertion is true, the contingent instruction will be executed; if the assertion is false, the contingent instruction will be ignored.

Assign (2.2) To give a value to a variable, by storing a number in the cell referred to by the variable.

Assignment instruction (2.3) An instruction that includes a variable and an expression whose value is to be computed and assigned to the variable.

Automatic return (7.1) A feature whereby information is provided to a procedure or function when it is activated, so that on completion of its task it can cause the sequence of instructions to resume from the point where it was activated. This information is processed automatically, without any extra effort by the programmer.

Basic (2.1) Often written BASIC; a programming language that is relatively simple and easy to learn. (See Appendix B.)

Binary numeral (4.3) A string of 1 and 0 digits, representing a number in the binary system. Each digit position has twice the value of the next position to its right (analogous to a decimal numeral, where each digit position has ten times the value of the next position to its right).

Bit (4.1) A fundamental unit of information; a position at which an elementary mark can be either present or absent.

Branching (3.4) Conditionally changing the sequence of operations in a computation by executing a conditional instruction.

Byte (5.2) A space inside the computer that is just big enough to hold one character. Thus, a byte is smaller than a cell.

Byte addressing (5.2) A property of some computing machines, including IBM 360 and 370, characterized by the fact that each byte has its own internal name and can therefore be referenced directly by machine-oriented instructions.

Call (7.1) An instruction that activates a procedure. A call may include arguments corresponding to the parameters in the procedure. On completion, the procedure will return to the instruction that follows the call.

Cell (2.2) A space inside the computer that is just big enough to hold one number.

Character (5.1) Any of a specific repertory of symbols used in non-numeric computing, including letters, digits, punctuation, and mathematical signs. On most computers, the character repertory is either the ASCII or the EBCDIC character set.

Coded information (4.2) A way of representing any of a predetermined set of patterns by a code. The code is usually chosen for economy of transmission and storage and has little or no relation to the pictorial form of the pattern it represents. (Compare *Pictorial information.*)

Column (6.2) In a matrix or rectangular array, a set of elements all of which have the same second subscript, while the first subscript takes on all values in sequence. (Compare *Row.*)

Comparison (5.3) Determination of the desired ranking of a pair of items during a sort operation. The number of comparisons is often used as a measure of the efficiency of a sorting algorithm.

Compiler (2.9) A sophisticated program that converts the instructions of a human-oriented (source) program to machine-oriented (object) form. The compiler subdivides expressions into elementary steps, and it changes symbolic cell names to numerical addresses. A system that compiles and executes each source-language instruction individually is an *interpreter.*

Compression (4.2) A process of systematically replacing certain bit patterns by other patterns, selected so as to reduce redundancy and thus minimize the total number of bits to be stored or transmitted.

Computation (1.2) Manipulation, storage, and transmission of marks. Derivatives of the word "compute" are still used for historical reasons, although most of the steps in a computation are usually nonarithmetical.

Computer (1.2) A machine for the automatic manipulation of information.

Condition (3.4) Any characteristic of the state of a computer that can be detected and used to control the sequence of execution of instructions in a program.

Constant (2.2) The name of a cell whose value never changes. The value is set before the instruction containing the constant is executed. The value is determined from the appearance of the constant as it is written in the program.

Constant string (5.2) A string constant, written by enclosing a string of characters between single quote marks.

Contents (2.2) In a cell, the number (or pattern of bits) stored there.

Contingent instruction (3.5) The final part of a conditional instruction. If the assertion is true, the contingent instruction will be executed; if the assertion is false, the contingent instruction will be ignored.

Control instruction (3.4) Any instruction (such as if, go to, stop, call, or return) that can affect the sequence of execution of the instructions of a program.

Count control (3.6) Determining the number of times to execute a loop, by using a counter variable whose value increases by a fixed increment (usually *one*) for each execution.

Data (2.4) Values to be used in a calculation that are not specified as constants when the program is written but are read into the computer during execution of the program.

Datum (6.3) In a file, that portion of the structure used to store information other than pointers or keys.

Dimension declaration (5.3) A declaration of the maximum length of a list, or of the maximum size and arrangement of a rectangular array or matrix.

Directory (6.4) A list of pointers organized in sequence according to some *key*, to provide access to items in a file.

Doubly linked list (6.3) A flexible data structure, in which each item includes a forward pointer (to the following item) and a backward pointer (to the preceding item).

Dynamic allocation (6.3) A system for making storage available to the program during execution. This increases flexibility, as the amount allocated for each of several different purposes may change as the calculation progresses.

EBCDIC (5.1) Extended Binary-Coded Decimal Interchange Code—an eight-bit code, including upper- and lower-case letters (used on IBM 360 and 370 series computers).

Efficiency (4.2) A measure of reduction in the cost of storing and transmitting information, achieved in some manner, such as through code compression (use of fewer bits to represent patterns that occur more frequently).

Error detection (4.2) Recognition of errors that have been introduced in a bit pattern during storage or transmission. It is usually made possible by such means as incorporating extra redundant parity bits.

Euclidean algorithm (3.3) An algorithm attributed to Euclid (c. 300 B.C.), for finding the greatest common divisor of a pair of integers.

Execute (2.1) To carry out the instructions of a program.

Exponent (4.3) Part of the scale factor of a floating-point numeral, indicating how many places the decimal (or binary) point should be moved to the right. (If the exponent is negative, the point should be moved to the left.)

Expression (2.5) A group of symbols representing a numerical value. (See also *String expression.*)

External procedure (7.3) A procedure that is compiled independently and separately from the other parts of the program in which it will be used.

Factor (2.10) A portion of an expression that is connected to the rest of the expression by a multiplication or division operator.

Fetch (2.3) To retrieve or obtain the numerical value from a cell where it has been stored, so that it can be put to further use in a calculation.

Fibonacci sequence (3.3) A sequence of numbers, wherein each number is the sum of the two previous numbers. The sequence was investigated by Leonard of Pisa, surnamed Fibonacci, about 1200 A.D.

File (6.3) An ordered set of items. Usually each item is rather extensive, and the items are often stored on disc or tape devices rather than in the internal portion of the computer.

Finite state automaton (3.3) A machine having a finite number of states. A computer is such a machine.

Floating binary (4.3) A system for representing numbers by numerals consisting of a significant part, which is a binary numeral having a value between one-half and one, and a scale factor indicating the power of two by which the significant part should be multiplied.

Floating decimal (4.3) A system for representing numbers by numerals consisting of a significant part, which is a decimal numeral having a value between one-tenth and one, and a scale factor indicating the power of ten by which the significant part should be multiplied.

Flowchart (2.1) A graphical representation of the sequence of steps of a calculation.

Fortran (2.1) Often written FORTRAN; a widely used programming language. (See Appendix B.)

Function (2.8, 7.2) A procedure (provided by the system or defined by the programmer) that returns a *value*.

Global list name (7.1) A name associated with the same group of cells throughout all parts of the program, including a procedure and the part of the program where the procedure is activated. It is often more efficient to use pointers than to use global list names.

Global variable (7.1) A name associated with the same cell throughout all parts of the program, including a procedure and the part of the program where the procedure is activated.

go to *instruction* (3.4) An instruction that unconditionally interrupts the normal sequence of execution of the steps of a program.

Graph (6.2) A set of nodes, some of which may be connected. A graph may be represented in a computer by its "connection matrix."

Greatest common divisor (3.3) For a pair of integers, the largest integer that divides exactly into each of them. It may be calculated by the Euclidean algorithm.

Huffman code (4.2) A code for representing a string with the fewest possible bits. The code can be determined for any set of characters whose relative frequency is known.

Increment (3.7) The amount by which the index variable value is increased at each execution of a count-controlled loop.

Index (3.7) The counter or variable used to control a count-controlled loop.

Indirect sorting (6.1) A sorting method wherein the data items are not moved, but pointers referring to the items are manipulated.

Information (1.1) The marks (letters, digits, symbols, etc.) manipulated by a computer.

Initial state (3.1) Any of the states of a system that can legally be used as the starting point for a path of an algorithm.

Initialize (2.6, 3.6) To give an initial value to a variable, before executing a program or a loop.

Input list (2.4) List of variables whose values are to be provided as data, in the "list" form of a read instruction.

Insert (6.3) To place a new item between two existing items in a file. If the file is organized as a doubly linked list, this is easily accomplished by rearranging pointers.

Instruction (2.1) The portion of a program that prescribes a single operation.

Integer part (2.8) The largest integer that does not exceed the given number. In Fortran, truncation of negative numbers produces an integer whose *absolute value* is the integer part of the *absolute value* of the given number.

Integer quotient (2.8) The integer part of a quotient. (See also *Remainder*.)

Intermediate goal (3.2) In problem solving, a means for focusing attention on a portion of the problem in hopes that this will contribute to the overall solution.

Interpolate (6.1) To estimate a value intermediate between two table entries.

Interpreter (2.9) A compiler system wherein each source-language instruction is translated to machine-language form just before it is executed.

Item (6.3) An entry in a file.

Iteration instruction (3.7) A special instruction, available in some form in most programming languages, that provides a shorthand for writing count-controlled loops. This simplifies the control instructions required and reduces mistakes.

Join (5.1) To connect a pair of strings together, thus forming a longer string.

Key (6.1) The portion of an item that is used as the basis for comparison in sorting, searching, ranking, etc.

Label (3.4) A name that refers to an instruction.

Length of string (5.1) The number of characters in a string. Stored as the value of an auxiliary variable associated with the string, it is modified when characters are added to or deleted from the string.

Library (7.3) A set of procedures in partially compiled form, that have been generated previously. These can include procedures developed previously by the user, or standard procedures and functions provided as part of the computer system.

Linkage (7.1) The process of communicating between a procedure and the program that activates it; also, the information involved in this process.

List (5.3) A group of items referred to by a single name along with a subscript indicating a particular item in the group. (See also *Input list, Output list, Doubly linked list.*)

Loader (7.3) A program provided by the computer system that combines a main program with external procedures and library procedures that have been previously prepared in the form of partially compiled ("relocatable") object programs.

Local variable (7.3) A name used in a procedure to refer to a different cell than the one referred to when the same name is used in another part of the program.

Loop (3.4) A group of instructions to be executed more than once.

Machine-oriented language (2.9) A program representation that is in suitable form to be interpreted by the limited devices that handle information inside a computer.

Mark (1.2) A detectable change in the physical universe that can be used to store or transmit information.

Matrix (6.2) A rectangular array, represented in most programming languages as a set of doubly subscripted variables.

Morse code (4.2) A code used in telegraphy. It is an early example of efficiency through the use of shorter patterns for more frequent letters.

Negation (2.5) Changing the sign of a numerical value. It is indicated by a minus sign to the left of a subexpression in contrast to subtraction, indicated by a minus sign between two subexpressions.

Nested iteration (3.7) A loop (often count-controlled) in which the block of instructions to be repeated includes one or more entire inner loops.

Noise (4.2) Errors that interfere with the use (especially transmission) of information.

Normalize (4.3) To convert a floating-point numeral to a standard form, in which the significant part has no leading zeros (that is, the leftmost digit is nonzero).

Null string (5.1) A string whose length is zero. A null string contains no characters, although it has a name and its length has the definite value zero.

Null transition (3.1) In a finite-state machine, a transition that accomplishes nothing—that returns directly to the same state. (Thus, a machine stops when a null transition is executed repeatedly.) The terminal state, for each path in an algorithm, must be a state from which the null transition is permitted.

Numerical information (1.1) Information in the form of numbers, used primarily for scientific, engineering, statistical, and commercial applications of computers.

Object program (2.9) Program in machine-oriented form. This term is used especially when the program is preserved in this form for later use.

Off-line (2.9) Operation of a computer where the user is not present during the time his program is being executed and hence cannot interact with the computation.

On-line (2.9) Operation of a computer where the user is available for interaction as his program is being executed.

Operand (2.9) Either of the two numbers operated upon during execution of a single machine instruction that prescribes an arithmetic operation.

Operation (2.1) A portion of a computation, corresponding to the execution of a single instruction.

Operation code (2.9) A numeric code, in the machine-oriented version of a program, that indicates which of the machine's repertory of operations is to be performed.

Output list (2.4) A list of expressions whose values are to be printed in order to display the results of a calculation.

Packed string (5.2) A string stored as compactly as possible in the computer. Except in machines that have byte addressing, the individual characters in a packed string cannot be stored or fetched directly, since the contents of a cell consists of several characters.

Parameter (7.2) A name included at the beginning of a procedure or function. When the same name appears in the procedure body, the occurrence of the name as a parameter signals to the compiler that an indirect reference is to be made, via a pointer, to the corresponding *argument.*

Parity (4.2) An extra redundant bit, attached to a bit pattern so that the number of "on" or 1 bits will be *odd.*

Parser (2.9) The portion of a compiler that subdivides an expression into its elementary components.

Path (3.1) A sequence of states in a finite-state machine, leading from an initial state to a terminal state. For each state reached along the path, exactly one transition to a subsequent state must be designated.

Pattern (3.2) A repetitive group of states. Recognition of patterns is an important part of the process of solving problems or of creating algorithms.

Pictorial information (1.1) Information displayed in the form of graphical representations, drawings of objects, etc. Used in many ways in special computer applications.

PL/1 (2.1) Pronounced "P L one"; a very complete language, containing many features besides those covered in this text (used primarily on IBM 360 and 370 computers). Most programmers master only a portion of the language. (See Appendix B.)

Pointer (5.4) A variable used primarily to indicate the position of another variable within a list or in allocated storage.

Prescribe (2.1) To designate or specify, in an imperative way. An instruction prescribes a particular operation; a program prescribes an entire calcuation.

print *instruction* (2.4) An instruction by means of which the results of a calculation can be displayed outside the computer for human use.

Printer (2.4) The device on which results of a calculation can be displayed outside the computer for human use.

Procedure (7.1) A program segment or block that is not executed in normal sequence but is activated by a *call* or by a function reference.

Program (2.1, 3.3) Any sequence of instructions that is written in accordance with the rules of some human-oriented or machine-oriented programming language.

Programming language (2.1) A set of rules that determine how a program should be written. The compiler embodies these rules and operates according to them, to prepare the program for execution.

Quantity of information (4.1) A measure of a message or pattern, to determine how much time will be required to transmit it (over a given channel) or how much space will be required to store it (in a given set of devices).

Queue (2.4) The set of numbers waiting to be read by the reader.

Rational number (4.4) Any number that can be expressed exactly as the ratio of two integers.

read *instruction* (2.4) An instruction by means of which data prepared by a human can be entered into a computer for use during the group; a matrix.

Redundancy (4.2) The use of extra information, beyond that absolutely necessary use during a calculation.

Rectangular array (6.2) A group of items referred to by a single name along with a pair of subscripts indicating a particular item in the group; a matrix.

Redundancy (4.2) The use of extra information, beyond that absolutely necessary to achieve the principal intended purpose, in order to serve some secondary purpose (such as detection of errors or easier recognition).

Reference variable (5.4) A pointer.

Register (2.5) A storage area inside the computer, analogous to a cell. Registers are faster and more accessible to the arithmetic unit, but more expensive, than cells. Registers cannot be directly referred to by a programmer using a human-oriented language.

Relation (3.4) An assertion as to the relative size of the values of two expressions. Used as part of a conditional instruction.

Release (6.3) To make available an area of storage that has been previously allocated, when it is no longer needed. This permits it to be allocated for some other purpose at a later time.

Remainder (2.8) In integer division, the portion of a quantity that is left, when as many multiples as possible of a divisor have been removed.

Results (2.4) Values produced during the course of a calculation that are to be displayed for use by humans.

Retrieve (2.2) To fetch, or obtain, a value from a cell or other storage area.

Return (7.1) 1. To leave a procedure and resume the computation sequence at the point from which the procedure was activated. 2. The value associated with the name of a function procedure, on completion.

Rounding (2.8) Finding the integer nearest a given numerical value. Usually accomplished by adding one-half unit to the given value and then finding the integer part.

Row (6.2) In a matrix or rectangular array, a set of elements all of which have the same *first* subscript, while the second subscript takes on all values in sequence. (Compare *Column.*)

Scale factor (4.3) The portion of a floating-point numeral that tells how to modify the indicated significant part to obtain the actual value being represented.

Scope (7.3) That portion of a program within which a given name is associated with a fixed cell. The scope of a global variable is the entire program; the scope of a local variable is the procedure or block in which it is declared.

Sentinel (3.6) A data value that is recognizable as being outside the expected range of values for the variable to which it is assigned; included deliberately with other data values in the reader to signal that no further data follows.

Shift (5.4) To modify the index of a count-controlled loop, so as to begin with a subscript value other than one.

Significant part (4.3) The portion of a floating point numeral that indicates a value to be multiplied by the scale factor. (See also *Normalize.*)

Simple expression (2.10) An expression containing no arithmetic operators. A variable, constant, or function expression.

Sort (5.3) To arrange the items of a list or file in (increasing) order, on the basis of some key.

Source program (2.9) A program in human-oriented form that is ready to be processed by a compiler.

Square root (2.8) The value which, when multiplied by itself, will produce a designated number. A standard function for performing this operation is available in most programming languages.

State (3.1) Any of the possible conditions of a finite-state machine.

Store (1.2) To make a more or less permanent physical change that can be detected at a later time. (Compare *Transmit.*)

String (5.1) A group of characters, taken in a certain sequence, and used for the manipulation of letters and words of text, in non-numerical computer applications.

String comparison (5.1) Determining whether one string precedes another in sequence, according to some designated rule of alphabetization. Where only upper-case letters and simple punctuation are involved, little difficulty arises. With more complicated symbols, however, a more sophisticated program may be needed.

String constant (5.1) A constant string. It is written by enclosing the desired sequence of characters between single quote marks.

String expression (5.1) An expression whose value is a string. It may be a string name, a string constant, or a function which operates on strings.

Structure (6.3) A designated arrangement of variables, lists, strings, etc., that may be used to form dynamic arrangements of information, such as doubly linked lists.

Subscript (5.2, 5.3) An expression whose value indicates a particular one of the items of a list.

Subscripted variable (5.3) One of the items of a list. It is designated by the list name along with the value of the subscript.

Substring (5.2) Any subset consisting of a group of consecutive characters from a given string. It may be a null string or the entire given string, as extreme examples.

Symbol table (2.9) A table used by a compiler, to associate the machine-oriented address with the symbolic name (variable or constant) of a cell in the source program.

Symbolic information (1.1) Information in the form of symbols, such as mathematical or musical signs, used in non-numerical computing.

Table (6.1) A group of two or more correlated lists. It is normally used by applying the same subscript value to several of the lists, to refer to related information.

Term (2.10) A portion of an expression that is connected to the rest of the expression by means of an addition, subtraction, or negation operator.

Terminal state (3.1) A state of a finite-state machine in which the machine can be allowed to remain indefinitely. Each path of an algorithm must lead to some terminal state.

Tolerance (4.4) An acceptable deviation from the ideal, that will be permitted if the ideal value cannot be attained (for example, because of the limitations of the number representation of a computer).

Transition (3.1) A change from one state of a finite-state machine to the next, in accordance with the rules of operation of the machine.

Transmit (1.2) To make a physical change that can be detected at a different place. (Compare *Store.*)

Type declaration (4.3) A specification, usually at the beginning of a program or program segment, that the value stored in a cell is coded in one of several different numerical forms (such as floating binary point form, etc.), so that the arithmetic unit can be instructed to interpret the value properly.

Undefined (2.2) Referring to the contents of a cell, or the value of a variable, before it has been specified in some manner during the execution of a program.

Unpacked string (5.2) A string stored inside the computer in such a way that each individual character can be stored or fetched directly. Except in machines that have byte addressing, each cell will contain only one character.

Upper limit (3.7) The value against which the index of a count-controlled loop is compared to determine when the iteration has been completed.

Value (2.2) The particular number associated with a variable at a particular time.

Variable (2.2) The name of a cell whose value may change during a computation. The value is undefined until a number has been stored in the cell.

Appendix B: List of Recommended Basic, Fortran, and PL/1 Texts

Basic

Kemeny, J. G., and T. E. Kurtz, *Basic Programming*, second ed. New York: Wiley, 1971.

Gately, W. Y., and G. G. Bitter, *Basic for Beginners*. New York: McGraw-Hill, 1970.

Coan, J. S., *Basic Basic: An introduction to computer programming in Basic language*. New York: Hayden, 1970.

Spencer, D., *A Guide to Basic Programming: A Time-Sharing Language*. Reading, Mass.: Addison-Wesley, 1970.

Hatfield, L. L., and D. C. Johnson, *Computer Assisted Mathematics Program: A First Course*. Glenview, Ill.: Scott, Foresman, 1968.

Walther, J. W., and D. C. Johnson, *Computer Assisted Mathematics Program: A Second Course*. Glenview, Ill.: Scott, Foresman, 1969.

Fortran

* Meissner, L. P., *Rudiments of Fortran*. Reading, Mass.: Addison-Wesley, 1971.

Sturgul, J. R., and M. J. Merchant, *Applied Fortran Programming*. Belmont, Calif.: Wadsworth, 1973.

† Murrill, P. W., and C. L. Smith, *An Introduction to Fortran IV Programming: A General Approach*. New York: International Textbook, 1970.

†‡ Kennedy, M., and M. B. Solomon, *Ten Statement Fortran Plus Fortran IV*. Englewood Cliffs, N.J.: Prentice-Hall, 1970.

‡ Peterson, W. W., and J. L. Holz, *Fortran IV and the IBM 360*. New York: McGraw-Hill, 1971.

† Organick, E. I., and L. P. Meissner, *Fortran IV Featuring Standard Fortran plus Watfor–Watfiv*. Reading, Mass.: Addison-Wesley, 1974.

McCracken, D. D., *A Guide to Fortran IV Programming*, second ed. New York: Wiley, 1972.

†‡ Cress, P., P. Dirksen, and J. W. Graham, *Fortran IV with Watfor and Watfiv*. Englewood Cliffs, N.J.: Prentice-Hall, 1970.

Kallin, S., *Introduction to Fortran*. Auerbach, 1972.

PL/1

Bates, F., and M. L. Douglas, *Programming Language One*. Englewood Cliffs, N. J.: Prentice-Hall, 1967.

Fike, C. T., *PL/1 for Scientific Programmers*. Englewood Cliffs, N.J.: Prentice-Hall, 1970.

* Anger, A. L., *Computer Science: The PL/1 Language*. New York: Wiley, 1972.

Bohl, M., and A. Walter, *Introduction to PL/1 Programming and PL/C*. Chicago: Science Research Associates, 1973.

Kennedy, M., and M. B. Solomon, *Eight Statement PL/C (PL/Zero) Plus PL/One*. Englewood Cliffs, N.J.: Prentice-Hall, 1972.

* Covers only a portion of the language.
† Includes Watfiv.
‡ Recommended for IBM-360 and IBM-370 users only.

Index